RENEWABLE ENERGY: RESEARCH, DEVELOPMENT AND POLICIES

RENEWABLE ENERGY

TRADE AND INVESTMENT IN ESSENTIAL SERVICES

RENEWABLE ENERGY: RESEARCH, DEVELOPMENT AND POLICIES

Additional books in this series can be found on Nova's website under the Series tab.

Additional E-books in this series can be found on Nova's website under the E-book tab.

RENEWABLE ENERGY: RESEARCH, DEVELOPMENT AND POLICIES

RENEWABLE ENERGY

TRADE AND INVESTMENT IN ESSENTIAL SERVICES

EBONY L. DUNBAR
EDITOR

nova publishers
New York

For permission to use material from this book please contact us:
Telephone 631-231-7269; Fax 631-231-8175
Web Site: http://www.novapublishers.com

Library of Congress Cataloging-in-Publication Data

ISBN: 978-1-62948-941-4

Published by Nova Science Publishers, Inc. † New York

CONTENTS

PREFACE

Global investment in renewable energy has grown sharply in the past five years as countries strive to meet growing energy demands, reduce carbon dioxide emissions, and strengthen energy security. Given that a broad group of services are indispensable to the development and functioning of renewable energy projects, the rapid expansion in renewable energy investment and installed capacity worldwide implies a similarly vibrant global market for renewable energy services. This book offers estimates of the U.S. and global markets for trade and investment in services essential to energy production in the solar, wind, small hydropower, and geothermal sectors, as well as discusses trade barriers affecting these services. The book also describes federal programs that provide grants, loans, loan guarantees, and other direct or indirect incentives for energy efficiency, energy conservation, and renewable energy.

Chapter 1 – *Renewable Energy and Related Services: Recent Developments* offers estimates of the U.S. and global markets for trade and investment in services essential to energy production in the solar, wind, small hydropower, and geothermal sectors, as well as discusses trade barriers affecting these services. The services span a range of industries, including consulting, engineering, construction, and equipment maintenance and repair. Global demand for such services has grown rapidly in the past five years as more and more countries strive to meet rising energy needs, reduce carbon output, and strengthen energy security by developing renewable energy. Global capacity in the field more than doubled to 653 gigawatts between 2007 and 2012, while global investment stood at a record $244 billion in 2012, up 71 percent during the period. Europe, the United States, and Asia, particularly China, are consistently among the largest markets for renewable energy services. Trade in renewable energy services occurs chiefly through foreign direct investment, in which a firm sets up a commercial presence abroad. Although the United States is a leading supplier and consumer of renewable energy services, evidence suggests it is likely a net importer, given the large presence of foreign affiliates providing these services in the U.S. market. Nonetheless, U.S. providers export substantial amounts of renewable energy services, primarily to Canada, while Mexico, other Latin American countries, and other large emerging markets present opportunities for U.S. service providers. Local-content requirements are the most significant trade barrier in this field. Although largely applied to renewable energy equipment, these requirements often act as de facto barriers to services exports because many renewable energy equipment manufacturers also provide services in support of their products. Restrictions on investment and on temporarily moving employees into foreign markets also hinder exports of renewable energy services. Some regional and bilateral trade negotiations are now working to liberalize the market by loosening these requirements.

Chapter 2 - Energy is crucial to the operation of a modern industrial and services economy. Recently, there have been growing concerns about the availability and cost of energy and about environmental impacts of fossil energy use. Those concerns have rekindled interest in energy efficiency, energy conservation, and the development and commercialization of renewable energy technologies. Many of the existing energy efficiency and renewable energy programs have authorizations tracing back to the 1970s. Many of the programs have been reauthorized and redesigned repeatedly to meet changing economic factors. The programs apply broadly to sectors ranging from industry to academia, and from state and local governments to rural communities. Since 2005, Congress has enacted several major energy laws: the Energy Policy Act of 2005 (EPACT 2005; P.L. 109-58); the Energy Independence and Security Act of 2007 (EISA; P.L. 110- 140); the Energy Improvement and Extension Act (EIEA), enacted as Division B of the Emergency Economic Stabilization Act (EESA; P.L. 110-343); and the American Reinvestment and Recovery Act (ARRA; P.L. 111-5). Each of those laws established, expanded, or modified energy efficiency and renewable energy research, development, demonstration, and deployment (RDD&D) programs. The Department of Energy (DOE) operates the greatest number of efficiency and renewable energy incentive programs. The Department of the Treasury and the Department of Agriculture (USDA) operate several programs. A few programs can also be found among the Departments of Interior (DOI), Labor (DOL), Housing and Urban Development (HUD), Veterans Affairs (VA), and the Small Business Administration (SBA).

This report describes federal programs that provide grants, loans, loan guarantees, and other direct or indirect incentives for energy efficiency, energy conservation, and renewable energy. For each program, the report provides the administering agency, authorizing statute(s), annual funding, and the program expiration date. The appendixes provide summary information in a tabular format and also list recently expired programs.

In: Renewable Energy
Editor: Ebony L. Dunbar

ISBN: 978-1-62948-941-4

Chapter 1

RENEWABLE ENERGY AND RELATED SERVICES: RECENT DEVELOPMENTS*

United States International Trade Commission

ABSTRACT

Renewable Energy and Related Services: Recent Developments offers estimates of the U.S. and global markets for trade and investment in services essential to energy production in the solar, wind, small hydropower, and geothermal sectors, as well as discusses trade barriers affecting these services. The services span a range of industries, including consulting, engineering, construction, and equipment maintenance and repair.

Global demand for such services has grown rapidly in the past five years as more and more countries strive to meet rising energy needs, reduce carbon output, and strengthen energy security by developing renewable energy. Global capacity in the field more than doubled to 653 gigawatts between 2007 and 2012, while global investment stood at a record $244 billion in 2012, up 71 percent during the period. Europe, the United States, and Asia, particularly China, are consistently among the largest markets for renewable energy services. Trade in renewable energy services occurs chiefly through foreign direct investment, in which a firm sets up a commercial presence abroad. Although the United States is a leading supplier and consumer of renewable energy services, evidence suggests it is likely a net importer, given the large presence of foreign affiliates providing these services in the U.S. market. Nonetheless, U.S. providers export substantial amounts of renewable energy services, primarily to Canada, while Mexico, other Latin American countries, and other large emerging markets present opportunities for U.S. service providers. Local-content requirements are the most significant trade barrier in this field. Although largely applied to renewable energy equipment, these requirements often act as de facto barriers to services exports because many renewable energy equipment manufacturers also provide services in support of their products. Restrictions on investment and on temporarily moving employees into foreign markets also hinder exports of renewable energy services. Some regional and bilateral trade negotiations are now working to liberalize the market by loosening these requirements.

---

* This is an edited, reformatted and augmented version of United States International Trade Commission, Investigation No. 332-534, USITC Publication 4421, dated August 2013.

Acroynms

AC	alternating current
APEC	Asian Pacific Economic Forum
ARRA	American Recovery and Reinvestment Act of 2009 (Public Law 111-5)
a-Si	amorphous silicon
BNDES	Banco Nacional de Desenvolvimento Econômico e Social
BOS	balance of systems
BSW-Solar	German Solar Industry Association
BTU	British thermal unit
CdTe	cadmium telluride
CIGS	Copper indium gallium (di)selenide
CSP	concentrated solar power
CPC	Central Production Classification
CPV	concentrating photovoltaics
CSI	California Solar Initiative
CREIA	Chinese Renewable Energy Industries Association
c-Si	crystalline silicon
DSIRE	Database of State Incentives for Renewables and Efficiency
DC	direct current
EGSA	Environmental Goods and Services Agreement
EIA	Energy Information Administration
ENR	Engineering News-Record
EPC	engineering, procurement, and construction
EPIA	European Photovoltaic Industry Association
EPRI	Electric Power Research Institute
EU	European Union
FHFA	Federal Housing Finance Agency
FIT	feed-in tariff
GATT	General Agreement on Tariffs and Trade
GATS	General Agreement on Trade in Services
GE	General Electric
ICT	information and communications technology
IEA	International Energy Agency
IOU	investor-owned utility
IPPs	independent power producers
IRS	Internal Revenue Service
ISP	independent service provider
ITC	investment tax credit
JPEA	Japan Photovoltaic Energy Association
KEMCO	Korea Energy Management Corporation
LBNL	Lawrence Berkeley National Laboratory
METI	Ministry of Economy, Trade, and Industry (Japan)
NREL	National Renewable Energy Lab

OEM	original equipment manufacturer
O&M	operations and maintenance
PACE	Property-Assessed Clean Energy
PPA	power purchase agreement
PTC	production tax credit
PV	photovoltaic
PVPS	Photovoltaic Power Systems Programme
RESOP	Renewable Energy Standard Offer Program
ROE	return on equity
RPS	renewable portfolio standards
SOE	state-owned enterprise
TRIMS Agreement	Agreement on Trade-Related Investment Measures
WTO	World Trade Organization

GLOSSARY

Balance of system (BOS): Refers to all components of a renewable energy power station other than the electricity-generating apparatus, such as a wind turbine or solar panel. These can include supporting infrastructure, including land, as well as wiring, switches, support racks, inverters, and batteries.

Biofuels: Liquid or gaseous fuel for transport produced from biomass.

Biomass energy: Energy derived from any plant-derived organic matter available on a renewable basis, including dedicated energy crops and trees, agricultural food and feed crops, agricultural crop wastes and residues, wood wastes and residues, acquatic plants, animal wastes, municipal wastes, and other waste materials.

BTU: British thermal unit, a common measure of energy consumption that refers to the quantity of heat required to raise the temperature of one pound of liquid water by one degree Fahrenheit at the temperature at which water has its greatest density (approximately 39 degrees Fahrenheit).

Crystalline silicon (c-Si) module: This module uses c-Si as the photosensitive material. They were the first modules commercialized and account for most global production.

Feed-in tariffs: Government incentives that offer a set price for renewable energy to investors.

Independent power producer (IPP): An IPP is an entity that primarily produces electricity for sale on the wholesale market. It is not a utility, does not own electricity transmission, and does not have a designated service area.

Investor-owned utility (IOU): An IOU is a for-profit utility.

Net metering: Allows owners of distributed energy systems (e.g., residential and commercial PV and wind systems) to receive credit for excess electricity that is fed into the grid. These credits can be used to offset utility charges for grid power used at other times. In some cases, system owners may receive a payment at the end of the year for any unused credits.

Power purchase agreement (PPA): A long-term agreement to purchase electricity. In the wholesale market, this is generally between a utility and an independent power producer. There are also PPAs between renewable energy firms and on-site users.

PV cells: A cell converts sunlight into electricity and is the basic element of a module.

PV module: A module (also commonly referred to as a panel) is made up of interconnected cells encapsulated between a backing material and a clear plastic or glass front.

Publicly owned utility: A nonprofit, state, or local government utility. Retail power market: The market for the sale of electricity to consumers.

Thin-film module: This module uses a thin layer of a raw material (most commonly a-Si, CdTe, or CIGS) as the photosensitive material and is a newer (2nd generation) PV technology.

Turnkey project: A turnkey project is one in which one party, usually a developer, manages all stages from design through construction and commissioning and then sells the project to the owner/operator.

Total energy consumption: For the purposes of this report, refers to the consumption of primary energy sources (e.g., fossil fuels like crude petroleum or natural gas) before they are transformed into other forms of energy (i.e., electricity).

Watt: A unit of electrical power equaling the amount of power produced from the expense of one joule of energy in one second. Wattage is expressed as follows:

1,000 watts (W) =	1 kilowatt (kW)
1,000 kilowatts (kW)=	1 megawatt (MW)
1,000 megawatts (MW) =	1 gigawatt (GW)
1,000 gigawatts (GW) =	1 terawatt (TW)

Watt-hour: A measure of electricity consumption. One watt-hour (Wh) is equal to the steady expense of one watt of power over one hour. Electricity consumption is expressed as follows:

1,000 watts-hours (Wh) =	1 kilowatt-hour (kWh)
1,000 kilowatts-hours (kWh) =	1 megawatt-hour (MWh)
1,000 megawatts-hours (MWh) =	1 gigawatt-hour (GWh)
1,000 gigawatts-hours (GWh) =	1 terawatt-hour (TWh)

Wholesale power market: The market for the sale of electricity from electricity generators to utilities/entities that resell the electricity on the retail market.

EXECUTIVE SUMMARY

Overview

Global investment in renewable energy has grown sharply in the past five years as countries strive to meet growing energy demands, reduce carbon dioxide emissions, and strengthen energy security. Given that a broad group of services are indispensable to the development and functioning of renewable energy projects, the rapid expansion in renewable energy investment and installed capacity worldwide implies a similarly vibrant global market for renewable energy services.

In a letter dated July 30, 2012, the U.S. Trade Representative (USTR) requested that the Commission provide a report on renewable energy and related services that among other things defines types of renewable energy and related services, identifies leading suppliers, and generally describes the relationship of renewable energy services to the development of renewable energy projects worldwide. As requested, this report estimates the size of the U.S. and global markets, discusses trends, and identifies barriers to U.S. trade and investment in renewable energy services. The USTR also asked that the report focus on services incidental to the development, generation, and distribution of renewable energy, with particular emphasis on wind energy (onshore and offshore) and solar energy, and other technologies that the Commission's research shows to be of significance. The USTR defined such services to include scientific and technical consulting, services incidental to energy distribution, professional services, construction and engineering services, management consulting and related services, and maintenance and repair of equipment, among others.

This report arrives at several major conclusions. First, the U.S. and global markets for renewable energy services are growing rapidly, particularly for solar and wind energy. The United States, the European Union (EU), Canada, and China are the leading markets for wind and solar energy services, but many emerging markets in Asia and Latin America are also seeing growth in spending on installations and services. Services related to other renewable energy sources (mainly small hydropower and geothermal) are also on the rise, though to a lesser degree and in more widely distributed areas.

Second, though data on trade in renewable energy services are not available, it is likely that large U.S. and EU services firms lead in global exports of renewable energy services. Similarly, it is likely that the United States and the EU account for much of the global importation of such services, though to a lesser degree, as many countries in the early stages of developing renewable energy markets lack domestic providers and must import the services.

Third, while there are few barriers that specifically target trade in renewable energy services, broad barriers to investment and to the movement of natural people appear to affect firms' ability to operate in certain markets. Local-content requirements in some markets may discourage services imports as well. While most such rules are intended to foster domestic industry and discourage imports of energy equipment, they may have the secondary effect of limiting services imports, as manufacturers frequently provide such services such as engineering, design, installation, and operations and maintenance along with the sale of their goods.

Key Findings and Observations

Renewable energy is growing rapidly, particularly wind and solar.

Global renewable energy capacity is growing quickly; it is estimated to have more than doubled to 653 gigawatts (GW) between 2007 and 2012 (figure ES.1).[1] Solar energy was the fastest-growing renewable energy, fueled by growth in solar installations in Europe (primarily in Germany, Italy, and Spain), China, and the United States. These five countries collectively accounted for nearly two-thirds (421 GW) of total global renewable energy capacity in 2012.

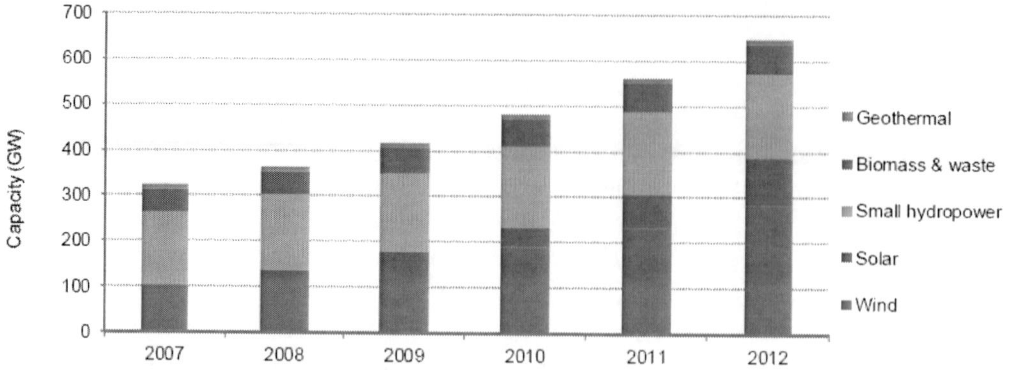

Source: Bloomberg New Energy Finance database.

Notes: Small hydropower refers to hydropower sources with generating capacities equal to or less than 50 MW. While included in the overall global renewable energy capacity figures, the total capacity data for marine energy installations are too small relative to the other technologies represented here to appear in the figure.

Figure ES.1. Global renewable energy capacity by source, 2007–12.

Total U.S. renewable energy capacity nearly doubled during 2007–12 to 110 GW (figure ES.2). Further, over the same period annual installations of renewable energy capacity more than tripled, rising to 17 GW in 2012. These installations accounted for the largest source of capacity growth among all energy sources. Wind energy accounted for over half of U.S. renewable energy capacity in 2012, up from about 30 percent in 2007. In fact, new U.S. wind energy capacity topped 13 GW in 2012, or almost 80 percent of total new renewable energy installations that year.

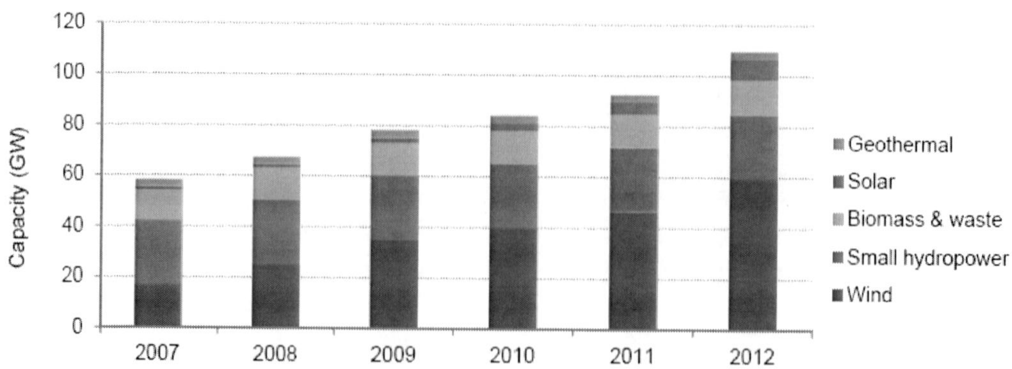

Source: Bloomerg New Energy Finance database.

Note: Small hydropower refers to hydropower sources with generating capacities equal to or less than 50 MW.

Figure ES.2. U.S. renewable energy capacity by source, 2007–12.

The United States and the European Union are likely the largest suppliers and consumers of renewable energy services.

The United States and the European Union collectively account for roughly 60 percent of both global capacity and investment in renewable energy, and the two economies are likely the largest suppliers and consumers of renewable energy services as well. Countries that have demonstrated an interest in increasing their use of renewable energy resources yet lack the domestic capacity to design, implement and maintain renewable energy projects, present an opportunity for U.S. exporters of renewable energy services. Even in countries that have the domestic expertise to provide renewable energy services, project developers may choose to import services from an international supplier based on cost, quality, or other needs. In this case, U.S. exporters of renewable energy services may also find opportunities in more mature markets.

Market Size for Renewable Energy Services

The United States is consistently among the largest markets for renewable energy services.

Comprehensive published data on the value of the global and U.S. markets for renewable energy services are unavailable. For this study, the U.S. International Trade Commission (the Commission or USITC) estimated market size on a sector-by-sector basis by obtaining industry estimates of the share each service contributes to total project cost and then multiplying them by installed capacity for a given market. The findings indicate that Europe, the United States, and Asia, particularly China, are consistently among the largest markets for renewable energy services.

Solar Photovoltaic Energy Services

The global market for solar photovoltaic (PV) services has grown rapidly in response to the explosive rise in solar PV installations worldwide; installed capacity grew more than 10-fold from 2007 to 2012. The value of global solar PV services associated with installations was estimated to be $34 billion in 2011, or 36 percent of the broader market for solar PV installations (including equipment and services). The largest markets for solar PV services in 2011 were Italy ($9.8 billion), Germany ($5.1 billion), the United States ($3.1 billion), and Japan ($3.0 billion).

The major global services industries associated with solar energy—large-scale project development; engineering, procurement, and construction (EPC); and operations and maintenance (O&M)—are highly fragmented, but tend to be dominated by multinational firms based in the three largest markets, Europe, China, and the United States. The residential and commercial PV services markets tend to be dominated by domestic firms and are most vibrant in the United States, the European Union, and Japan.

Wind Energy Services

The global market for wind energy services has grown in concert with the sharp increase in globally installed wind energy capacity, which more than quintupled from 2007 to 2012. In 2011, the value of services associated with wind installations was estimated at nearly $23 billion, or roughly 32 percent of the broader global market for wind power installations (including equipment and services). The largest markets in 2011 for wind services were China ($9.0–$11.8 billion), the United States ($5.5–$7.1 billion), Germany, ($2.3–$3.5 billion), and Canada ($1.2 billion).

The value of global O&M services to the wind energy sector also continues to rise steadily. The global market for wind O&M services was estimated at $6.2–$7.2 billion in 2011, with Europe accounting for about half of that market. Germany, Spain, and the United Kingdom collectively accounted for 85 percent of wind O&M services in Europe, reflecting their large installed base. Original equipment manufacturers that produce wind turbines are thought to account for the vast majority of O&M services provided globally, with smaller roles played by independent service providers and wind farm operators.

Hydropower Energy Services

The global market for services associated with all hydropower installations in 2010 was estimated at $72 billion. The global market for services related to small-capacity hydropower—less than 50 megawatts (MW)—is substantially smaller, and is estimated to account for less than 5 percent ($2.3 billion) of the broader market for hydropower installations of all sizes. The largest markets for services related to small hydropower projects were China, Brazil, Japan, and India. The value of services related to a particular small hydro power project varies significantly based on the characteristics of the site, and on whether the project represents new development or the addition of new capacity to an existing dam.

Geothermal Energy Services

U.S. and global growth in geothermal energy services has been slow in recent years, but is expected to accelerate in the near term as countries add to installed capacity or develop new geothermal resources. The value of the global geothermal EPC services market was estimated at $315 million in 2010, and the O&M services market was estimated at $2.5 billion. In the United States, the value of the geothermal EPC services market in 2010 was estimated at $34 million, while the O&M services market was estimated at $594 million. Indonesia, the Philippines, Kenya, Rwanda, and Ethiopia all represent significant potential markets for new geothermal capacity, and thus for geothermal services.

Factors Affecting Supply and Demand

Government incentives have played a key role in the global development of renewable energy.

Renewable energy has historically cost more to generate than energy from conventional sources like fossil fuels (e.g., coal, petroleum, and natural gas). As a result, renewable energy has largely been uncompetitive with the alternatives absent policy mechanisms that lower its generating costs and encourage (or require) its use. These policy mechanisms include fiscal incentives, such as tax credits to offset the cost of generating renewable energy, and regulatory policies that mandate the use of renewable energy and set its price. Often, a combination of these mechanisms are used to develop renewable energy resources and technologies and ensure their deployment. Over 125 countries have explicit policy targets in place to promote renewable energy.

Econometric analysis indicates that feed-in tariffs have a direct, positive impact on installed capacity.

The Commission examined the relationship between feed-in tariffs[2] and installed wind capacity in 54 countries during 2006–10, and found that having a feed-in tariff policy correlated with an additional 1,856 MW of installed wind capacity.

Growth in renewable energy installations and advances in technology boost demand for renewable energy services.

Demand for renewable energy underpins overall demand for renewable energy services. Specific factors affecting services demand include the growth in installations of renewable technologies, the size and age of existing renewable capacity, and improvements or advances in renewable technologies.

Falling equipment prices have driven up the demand for services as new capacity is developed.

Increased competition among renewable technology manufacturers, greater economies of scale in manufacturing, and advances in technology and manufacturing processes have driven down the prices paid by project developers for many renewable technologies, such as wind turbines and solar panels. Lower prices for these technologies, in turn, have led to growth in project development and renewable energy installations, and thus in renewable energy services.

Trade and Investment

Trade

Trade in renewable energy services typically occurs when a services firm establishes a commercial presence abroad, as well as when a firm temporarily moves an individual or individuals to provide services in a client's territory. While data on U.S. and global trade in renewable energy services are unavailable, certain broad trends can be discerned.

U.S. firms dominate the United States' domestic PV services market, but they are also actively exporting.

In the United States, domestic firms dominate the installation of PV systems at all levels. In 2012, U.S.-based companies performed almost all U.S. installations of residential systems, nearly 90 percent of U.S. nonresidential installations, and more than half of U.S. installations for utility-scale projects. Nevertheless, foreign participation is rising as the scale of installation increases.

Existing U.S. project developers are increasingly exporting project development services. Some are expanding in Canada, while many that are already active abroad are looking to expand beyond their traditional foreign markets. Some U.S. project developers also provide EPC services in foreign markets for other project developers, as well as O&M services after project completion. Nonetheless, U.S. firms' foreign activities cover only a small share of the global market; for example, their share of the project development market outside of the United States was likely about 5 percent in 2012.

The United States is a net importer of wind energy services, though U.S. exports likely grew during the 2007–12 period.

The United States is a net importer of wind energy services, as many of the largest wind services firms operating in the country are subsidiaries of foreign multinationals.

Approximately one-third of newly installed U.S. wind capacity was developed by U.S. affiliates of foreign parents in 2012.

U.S. wind energy service firms have a limited global presence, and as a result, U.S. exports likely account for a small share of global exports of wind services. Canada is reportedly the largest export market for U.S. wind services, with Mexico and other countries in Latin America becoming increasingly attractive markets for U.S. firms.

Data on imports and exports of wind energy technologies indicate that European firms dominate the wind equipment sector, and likely the service sector as well, in most major markets.

The largest global exporters of wind energy equipment are also likely to be the largest suppliers of wind energy services; indeed, service provision is often written into contracts for equipment sales. Germany, Denmark, and Spain collectively accounted for over 75 percent of equipment exports in 2012 (the United States accounted for 6 percent), and the three countries' wind services providers are active in many markets around the world. An exception is China, whose market is dominated by Chinese equipment and service providers.

The largest global importers of renewable energy technologies are also likely to be the largest consumers of renewable energy services. For instance, the United States is the largest importer of wind-powered generating sets, reflecting the growth in U.S. installed wind capacity and implying corresponding growth in wind services supplied by domestic and foreign firms alike.

Investment

Global investment in renewable energy projects rose by 71 percent from 2007 to 2012, with Europe, China, and the United States attracting the majority of funds.

Data on global and U.S. investment specific to renewable energy services are unavailable. However, global investment in renewable energy overall stood at a record $244 billion in 2012, up 71 percent compared with 2007, although down 12 percent from 2011 levels (figures ES.3 and ES.4).

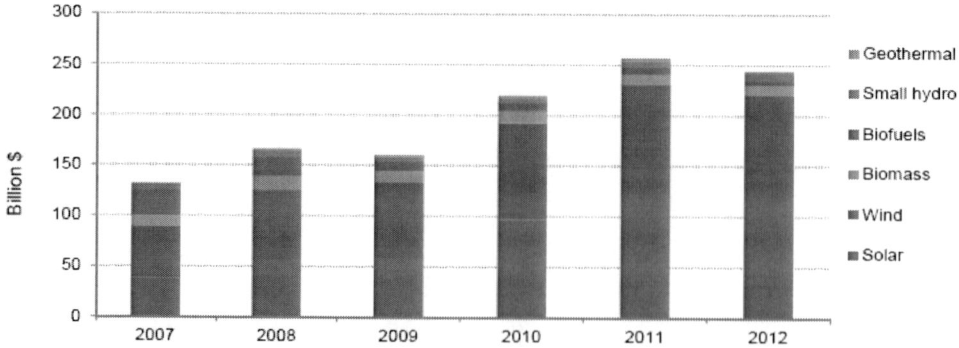

Source: Bloomberg New Energy Finance database.

Note: While included in the overall global renewable energy investment figures, the total investment data for marine energy installations are too small relative to the other technologies represented here to appear in the figure.

Figure ES.3. Global renewable energy investment by sector, 2007–12.

These figures broadly reflect investment associated with the development of renewable energy projects and the installation of renewable energy generating capacity. Because renewable energy services are needed to develop these projects, investment in renewable energy overall serves as a useful proxy for trends and investment in renewable energy services.

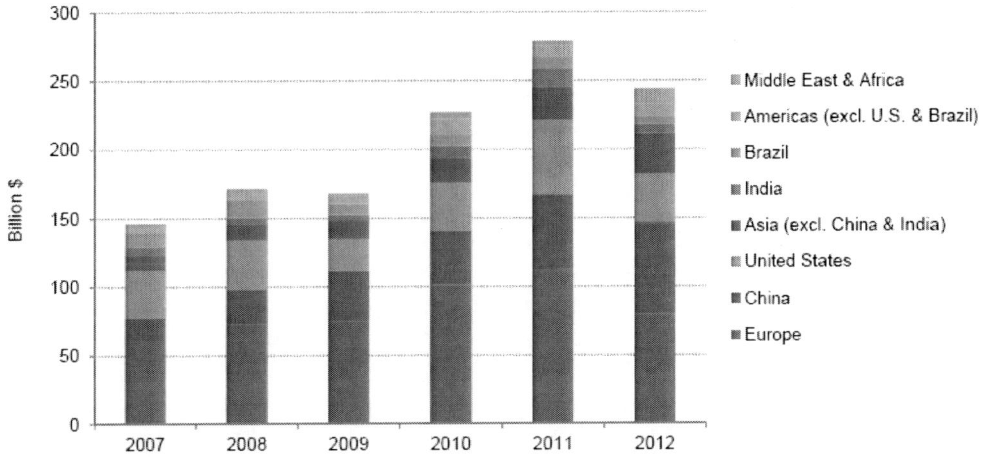

Source: REN-21, Renewables 2013, 2013, 62-63.

Figure ES.4. Global renewable energy investment by country and region, 2007–12.

Barriers to Trade

U.S. firms cite local-content requirements as the most significant trade barrier in the provision of renewable energy services.

The local-content requirements found in many markets typically mandate that a certain share of renewable energy equipment and/or services be locally sourced, often driving up prices and making it cost-prohibitive for foreign providers to enter the market. Even a local-content requirement that applies only to equipment may act as a de facto barrier to services provision, as many wind and solar energy equipment manufacturers provide services in tandem with the sale of their goods.

Restrictions on investment and on the movement of service providers also hinder exports of renewable energy services, though the problem is being addressed as part of some regional and bilateral trade negotiations.

Barriers to trade in services, including horizontal (non-sector-specific) restrictions on a firm's ability to set up a commercial presence in a foreign country or temporarily move service providers there, continue to hamper efforts by providers of renewable energy services to penetrate foreign markets. Whereas recent multilateral efforts to address trade barriers in services overall have gained some traction, multilateral efforts to specifically address barriers to trade in renewable energy services have largely stalled. By contrast, regional and bilateral efforts to specifically address barriers affecting renewable energy services have met with more success, and include commitments to address trade in environmental goods and services within the Asia-Pacific Economic Cooperation (APEC) forum and the proposed Trans-Pacific Partnership (TPP) Agreement.

SECTION 1. INTRODUCTION

Purpose and Background

Global investment in renewable energy has grown rapidly in the past five years as countries strive to meet growing energy demands, reduce carbon dioxide emissions, and strengthen energy security. As a result, not only has demand for renewable energy equipment and related materials risen, so too has demand for the services required to develop, construct, and operate renewable energy facilities. These services, which span a range of industries including consulting, engineering, construction, and maintenance and repair of equipment, are broadly referred to in this report as "renewable energy services." At present, there is very little literature on renewable energy services, and data on market size or trade activity are equally limited. However, services are becoming increasingly important in the renewable energy value chain as equipment manufacturers look to expand their offerings beyond goods[1] and as independent service providers look to expand beyond their home markets.

The United States and Europe are home to many of the world's largest renewable energy service providers, since that is where much of the growth in renewable energy has been concentrated in recent years. However, economic conditions and other factors have slowed growth in these two markets, while renewable energy investments in Asia, Latin America, and the Middle East are rising. As a result, U.S. and European firms are seeking to increase their presence in those markets, with mixed results. While the road to establishing business relationships and accessing the market is fairly straightforward in some economies, trade and investment barriers inhibit full participation in others.

On July 30, 2012, the U.S. International Trade Commission (Commission or USITC), received a request from the United States Trade Representative (USTR) asking that the Commission conduct an investigation and provide a report that examines renewable energy services.[2] In the request letter, the USTR noted that the Commission had provided a report on trade in this sector to the USTR in 2005 and asked that the Commission prepare a new report that would address recent developments in the renewable energy services sector.

As requested by USTR, this report defines the types of renewable energy and related services, lists leading suppliers, and generally describes the relationship of renewable energy services to the development of renewable energy projects worldwide. Further, the report estimates the size of the U.S. and global markets for certain renewable energy services, identifies key export and import markets for such services, and examines factors affecting supply and demand. The report focuses on U.S. and global renewable energy services trade during 2007–11, highlighting recent trends in investment in renewable energy projects and firms, identifying barriers to U.S. trade and investment in renewable energy services, and examining recent efforts to liberalize trade in leading markets for such services. Finally, the Commission uses econometric analysis to examine the role of clean energy incentive programs in encouraging investment in and creating markets for wind energy goods and services.

Scope

This report focuses on services that are essential to the development, generation, and distribution of renewable energy. These services include scientific and technical consulting, professional services, construction and engineering services, management consulting and related services, maintenance and repair of equipment, and other services required throughout the life cycle of a renewable energy project. As requested, the principal focus of this report is on wind and solar energy services. However, the USTR also directed the Commission to include discussions of other renewable technologies that the Commission's research showed to be significant. Based on a number of factors, including growth potential, this report therefore includes discussions of services related to small hydroelectric and geothermal energy.[3]

Services, including renewable energy services, are traded through two principal channels. The first channel, cross-border trade, entails sending people, information, or money across national borders. [4] The second channel, affiliate transactions, entails selling services through affiliated firms set up or acquired by multinational companies in foreign markets. Such affiliates are funded through foreign direct investment. The Commission estimates that the majority of renewable energy services are traded through affiliate transactions.[5]

Classification of Renewable Energy Services

Most countries that have signed the General Agreement on Trade in Services (GATS) use the Services Sectoral Classification List (W/120) drawn up by the World Trade Organization (WTO) to organize and define the scope of specific commitments they have made in their national schedules. However, energy services, including renewable energy services, do not appear separately on this list. As a way to apply the GATS agreement to energy services, as well as other services not separately listed on the W/120, WTO members have devised a "checklist" approach to making GATS commitments. Under this approach, members create a list of services identified in the W/120 that they consider relevant to the sector in question and that they agree to represent the scope of that sector for scheduling purposes.[6] In its 2003 GATS offer, the United States proposed such a checklist as a way for GATS members to make commitments in energy services. The services on this checklist are listed in table 1.1. Commitments on energy services apply to renewable energy services, as long as they are not specifically exempted from a country's commitments.[7]

Aside from specific market access and national treatment commitments, there are several general obligations that apply to virtually all service sectors. These obligations can help promote trade in services, even when the services are not listed in a country's schedule of specific commitments (table 1.2). Among the GATS framework principles that apply to nearly all services sectors are most-favored-nation treatment, contained in Article II, and transparency, contained in Article III. In addition, where commitments have been scheduled, the framework contains disciplines on domestic regulation in Article VI and limits on the actions of monopolies and exclusive suppliers in Article VIII.[8]

Table 1.1. Checklist of energy-related services included in the U.S. GATS offer, 2003

Central product classification code	Description
5115, 883	Services incidental to mining
8675	Certain related scientific and technical consulting services
887	Services incidental to energy distribution
861, 862, 863, 8672, 8673, 9312, 93191, 932	Certain professional services, including engineering and integrated engineering services
6111, 6113, 6121, 621, 622, 631, 632	Distribution services, including commission agents, wholesale trade, and retail trade services that apply to fuels, related products, and brokerage of electricity
633, 8861-8866	Maintenance and repair of equipment, except transport-related equipment
865	Management consulting and related services
511-518	Construction and related engineering services
7131	Pipeline transportation of fuels
7422	Storage and warehouse services, particularly bulk storage services of liquids and gases
8676	Technical testing and analysis services

Source: WTO, "Council for Trade in Services—Special Session—Communication from the United States—Initial Offer," TN/S/O/USA, September 4, 2003.

Approach and Data Sources

The focus of this report is on identifying and analyzing trends in renewable energy services markets. Data on specific renewable energy services are not generally available. The Commission therefore developed estimates of market size for these services by obtaining industry estimates of the share of a project cost that each service constitutes and then applying the estimates to installed capacity for a given market. Furthermore, the Commission undertook econometric analysis to examine the relationship between feed-in tariffs (a type of incentive supporting renewable energy) and installed wind energy capacity to determine whether one has a direct bearing on the other.

In collecting information for this report, the Commission conducted primary research and consulted a wide range of secondary sources to collect both qualitative and quantitative information. Analysts conducted in-person and telephone interviews with representatives of renewable energy service providers, domestic and foreign governments, industry and trade associations, educational institutions, nongovernmental groups, and international organizations. Commission staff undertook fieldwork in several U.S. cities, as well as in Brazil, Chile, China, France, Germany, Japan, Indonesia, and Spain. Interviews were conducted with large multinational firms as well as small and medium-sized enterprises. Secondary information sources included industry journals and websites, U.S. and foreign government publications, and publications by international organizations such as the World Bank, the Organisation for Economic Co-operation and Development (OECD), and the WTO.

Table 1.2. General obligations of the GATS

Obligation	Description
Nondiscrimination	Article II provides for most-favored-nation treatment (MFN). Under MFN, WTO members commit that they will treat services and service suppliers of other WTO member countries no less favorably than they treat like services and service suppliers of any other country in the world. Members must adhere to MFN principles except in those areas in which they have listed exemptions.
Transparency	GATS transparency obligations are listed in Article III, which requires members to: • promptly publish relevant measures that will be applied generally; • notify the WTO of significant changes in laws, regulations, or administrative guidelines with significant bearing on services trade; • set up enquiry points that other WTO members can use to submit relevant questions; and • respond promptly to requests for information from other WTO members.
Domestic regulation[a]	GATS domestic regulation obligations, as contained in Article VI, require WTO members to: • avoid using regulatory powers so as to create barriers to trade in services; • ensure that measures that will be applied generally are administered in a reasonable, objective, and impartial way; and • ensure that licensing and qualification requirements or technical standards for sectors in which members have made specific commitments on market access or national treatment, (1) are based on objective and transparent criteria, (2) are not more burdensome than necessary, and (3) in the case of licensing procedures, are not in themselves a restriction on the supply of the service.
Monopolies and exclusive suppliers[a]	Article VIII of the GATS states that WTO members should ensure that if a monopoly supplier competes in supplying a service outside the scope of its monopoly rights, it does not abuse its monopoly position in a way that limits market access or national treatment.

Source: World Trade Organization, General Agreement on Trade in Services.

[a] Note: Articles VI and VIII apply only to industries for which countries have made specific commitments.

Data on renewable energy capacity and generation were incorporated from a number of sources, including Bloomberg New Energy Finance, the International Energy Agency (IEA), the International Renewable Energy Agency (IRENA), and the Energy Information Administration (EIA) of the U.S. Department of Energy. Since data on trade in renewable energy services are not generally collected by governments or available from industry, the discussion of such trade is largely anecdotal, giving estimates of trade activity where possible.

Organization

This section presents the background and scope of the report. Section 2 provides an overview of the global market for renewable energy services and examines the factors affecting supply and demand for such services. Sections 3 through 5 present more detailed information and analysis of the markets for solar, wind, and small hydropower and geothermal energy services, respectively. These sections generally begin with an overview of the relevant services and provide information on the U.S. and global markets for such services, trade and investment, barriers to trade, and profiles of individual country or regional markets. Section 6 identifies and describes policy tools implemented by nations to cultivate renewable energy markets, coupled with econometric analysis intended to assess the relationship between such policies and markets. Appendix A includes the original and amended request letters from USTR. Appendix B contains the Commission's *Federal Register* notices for this investigation. Appendix C summarizes the information and views contained in written submissions filed with the Commission by interested parties.

SECTION 2. OVERVIEW OF THE MARKETS FOR RENEWABLE ENERGY AND RENEWABLE ENERGY SERVICES

Introduction

This section provides a broad overview of the markets for renewable energy and renewable energy services. It gives information on and describes the size and growth of renewable energy markets, as well as investment in renewable energy worldwide. It also describes trends in policies supporting the development of renewable energy and in factors affecting the supply of and demand for renewable energy and renewable energy services. Very few data are available on either global or U.S. trade in renewable energy services, so it is difficult to ascertain how large this trade is or its growth rate. However, given that renewable energy services are indispensable to developing and maintaining renewable energy projects, the rapid growth in renewable energy investment and installed capacity worldwide implies a similarly vibrant global market for renewable energy services.

Government policies have played a key role in the development of renewable energy in many economies, including the European Union (EU) and the United States, where both manufacturers and service providers for these technologies are well established. The EU and United States collectively account for roughly 60 percent of global capacity and investment in renewable energy.[1] As a result, the two economies are likely the largest suppliers and consumers of renewable energy services. Countries that have experienced recent growth in installed renewable energy capacity, yet lack the domestic capacity to design and implement renewable energy projects, present an opportunity for U.S. exporters of renewable energy services. Even in countries that have the domestic capacity and expertise to provide renewable energy services, project developers may choose to import services from an international supplier based on cost, quality, or other needs. In this case, opportunities for U.S. exporters of renewable energy services may also exist in more mature markets.

At the same time, forced localization measures, such as local-content requirements, can act as a barrier to trade in both technologies and services related to renewable energy. Relevant barriers to trade in services include restrictions on commercial presence and on the temporary movement of service providers; both hamper efforts by providers of renewable energy services to penetrate foreign markets. While recent multilateral efforts to address trade in services more broadly have gained some traction, similar efforts to specifically address barriers to trade in renewable energy services have largely stalled. By contrast, regional and bilateral efforts to reduce barriers to trade in renewable energy services have met with more success. These include commitments to address trade in environmental goods and services within the Asia-Pacific Economic Cooperation (APEC) forum and the proposed Trans-Pacific Partnership (TPP).

Overview of Global and U.S. Renewable Energy Markets

Global Market

Renewable energy sources, including traditional biomass, geothermal energy, hydropower, and solar and wind power (box 2.1),[2] accounted for 19 percent of global energy consumption in 2011.[3] Traditional biomass, such as wood used for heating, accounted for a little over half of energy consumption from renewable sources in 2011, or 9.3 percent of global energy consumption (figure 2.1).[4] Excluding traditional biomass, which is outside the scope of this report, renewable energy sources accounted for about 9.7 percent of global energy consumption in 2011, a slight uptick compared with previous years.[5] Hydropower is the largest single source of renewable energy other than biomass, accounting for 3.7 percent of global energy consumption. All other renewable energy sources, including wind, solar, and geothermal sources, together accounted for about 6 percent of global energy consumption.

Renewable energy sources can substitute for fossil fuels for electricity generation, heating and cooling, and transportation (e.g., biofuels for gasoline).[6] Electricity is the fastest-growing segment of the global energy market, and renewables are the fastest-growing source of electricity, although from a relatively small base.[7] In 2011, global renewable energy-generating capacity exceeded 1,360 gigawatts (GW), or more than 25 percent of total global electricity-generating capacity (5,360 GW in 2011).[8] Hydropower is the largest source of electricity generation from renewable energy sources, accounting for about 15 percent of global electricity production.[9] In comparison, fossil fuels account for about two-thirds of global electricity generation, while nuclear fuels account for about 8 percent.

Global renewable energy capacity is growing quickly: it is estimated to have more than doubled to 653 GW between 2007 and 2012 (figure 2.2).[10] Solar energy was the fastest-growing renewable energy, fueled by growth in solar installations in Europe (primarily in Germany, Italy, and Spain), China, and the United States, as noted in section 3. These five countries collectively accounted for nearly two-thirds (421 GW) of total global renewable energy capacity in 2012.[11] Wind-generated power capacity surpassed small-capacity hydropower in 2010 to become the second-largest renewable energy source globally. The top countries with renewable power-generating capacity by sector are shown in table 2.1. Renewable energy capacity has grown substantially in other markets in recent years—although, again, capacity there has grown from a smaller base. Eastern Europe, Latin America, the Middle East, and North Africa are among the fastest-growing markets for renewable energy, driven in large part by growth in solar and wind-energy installations.

BOX 2.1. Renewable energy is derived from a variety of sources

Renewable energy is derived from a variety of sources, and is used to produce electricity, heat, and fuels. Renewable energy principally includes bioenergy, geothermal energy, hydropower, ocean energy, solar energy, and wind energy.[a]

Bioenergy is renewable energy derived from biological sources (i.e., biomass) to be used for heat, electricity, or transport fuel.[b] Traditional biomass includes primarily wood used for heating. Heat produced from the combustion of other forms of biomass in a boiler can be used to generate electricity through a steam turbine. Through various processes, biomass can also be transformed into biogases that are used to produce electricity as well. Electricity produced from bioenergy accounted for about 1.2 percent of global electricity production in 2009. Biofuels (e.g., ethanol) derived from biomass feedstocks like sugarcane, corn, or wheat accounted for about 2 percent of all transport fuels in 2009.

Geothermal energy is derived from heat from the earth either to be used directly as heat or to generate electricity. Geothermal sources include hot water or steam reservoirs deep in the earth, which are accessed by drilling; geothermal reservoirs; and shallower ground closer to the earth's surface.[c] In 2009, the United States accounted for almost 30 percent of global geothermal power capacity, and for about 15 percent of global electricity generated from geothermal sources. Iceland, El Salvador, Kenya, the Philippines, and Costa Rica supply a significant share of their total electricity demand from geothermal sources.

Hydropower is derived from the energy of flowing water. Turbines placed along flowing rivers or at dams convert the water's kinetic energy (energy based on motion) to mechanical, then electrical, energy. Hydropower accounted for about 16 percent of global electricity production in 2009. The largest global producers of hydropower include China, Brazil, Canada, the United States, and Russia.

Ocean energy is derived primarily from the potential and kinetic energy of tides, currents, and waves. Tidal and wave power have been in development since the 1970s, although technologies to harness ocean energy are still not widely deployed. In 2011, the 254 megawatt (MW) Silhwa Lake tidal power plant in the Republic of Korea (Korea) overtook the Rance tidal plant in France to become the world's largest tidal power station. Together, these two tidal power plants account for nearly all global tidal energy capacity (527 MW).[d]

Solar energy is derived from sunlight that is converted into heat and electricity. Solar photovoltaic (PV) systems convert solar energy into electricity. Concentrating solar power (CSP) plants use mirrors or lenses to concentrate sunlight and create temperatures to drive turbines or engines to produce electricity. Solar heating technologies collect solar energy to produce heat.[e]

Wind energy is derived from the energy of air flow. Wind turbines convert the wind's kinetic energy passing through a rotating turbine's blades into electricity. Offshore wind turbines are located in coastal regions, where wind resources are often better than those onshore. Offshore projects face more challenges than their onshore counterparts, mainly due to infrastructure challenges, weather, distance from users, and water depth, among others.

[a] This section is based mainly on IEA, *Renewable Energy: Markets and Prospects by Technology*, 2011, various pages.

[b] USDA, ERS website, http://www.ers.usda.gov/topics/farm-economy/bioenergy.aspx (accessed January 30, 2013).

[c] NREL, "Geothermal Energy Basics," http://www.nrel.gov/learning/re geothermal.html (accessed January 30, 2013).

[d] REN-21, *Renewables 2012*, 2012, 46. [e] SEIA, "Solar Heating & Cooling" (accessed January 30, 2013).

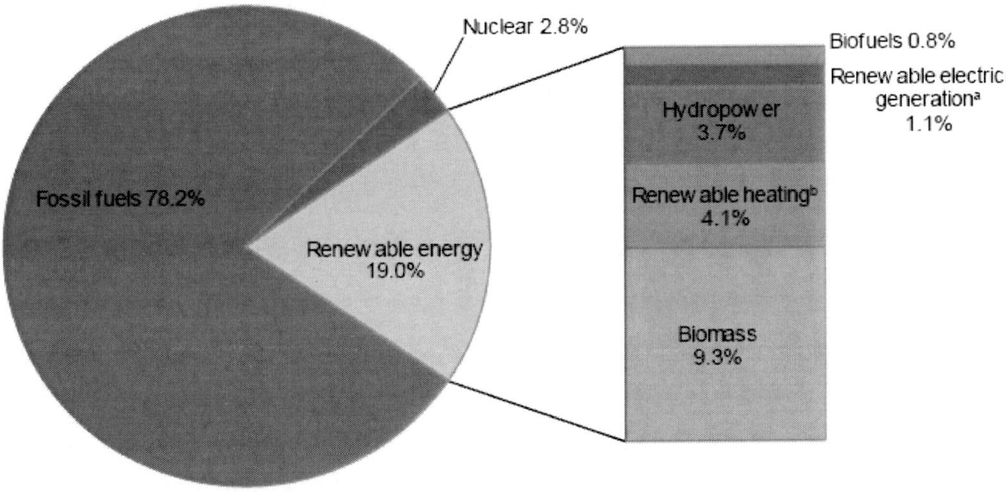

Total = 531 quadrillion BTUs

Sources: REN-21, Renewables 2013, 2013, 21; EIA.

[a] Includes biomass, geothermal, solar, and wind resources.

[b] Includes biomass, geothermal, and solar resources.

Figure 2.1. Global energy consumption by source, 2011.

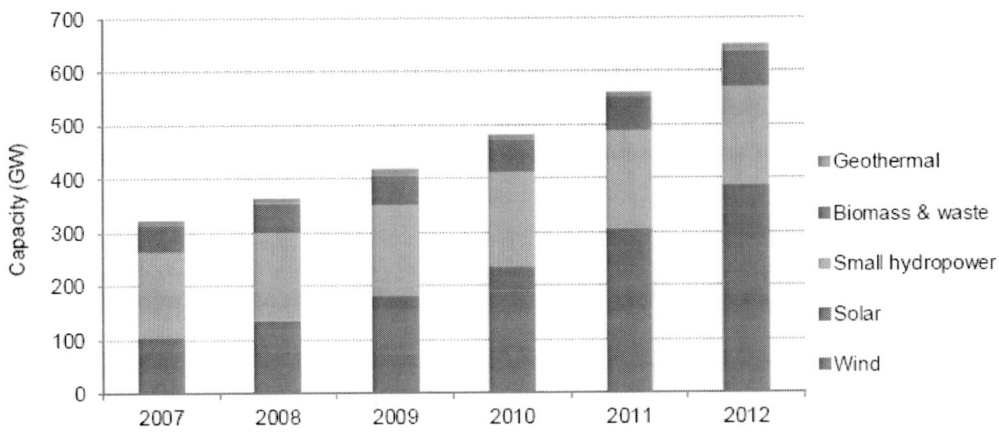

Source: Bloomberg New Energy Finance database.

Notes: Small hydropower refers to hydropower sources with generating capacities equal to or less than 50 MW. While included in the overall global renewable energy capacity figures, the total capacity data for marine energy installations are too small relative to the other technologies represented here to appear in the figure.

Figure 2.2. Global renewable energy capacity by source, 2007–12.

Table 2.1. Countries with the largest installed renewable power generating capacity, by sector, 2012

Renewable energy sector	Country	Installed capacity (MW)	Share of total (%)
Wind[a]	China	63,474	26.8
	United States	46,459	19.6
	Germany	28,608	12.1
	Spain	21,779	9.2
	India	15,732	6.6
	Subtotal	176,052	74.2
	All other	61,176	25.8
	World total	237,228	100.0
Small hydropower[a]	China	62,123	35.0
	United States	25,291	14.3
	Japan	13,234	7.5
	Switzerland	3,700	2.1
	India	3,400	1.9
	Subtotal	107,748	60.7
	All other	69,966	39.3
	World total	177,414	100.0
Solar photovoltaic (PV)	Germany	31,369	31.0
	Italy	16,283	16.1
	United States	7,527	7.4
	Japan	7,002	6.9
	Spain	4,464	4.4
	Subtotal	66,645	65.9
	All other	34,556	34.1
	World total	101,201	100.0
Biomass and waste[a]	United States	12,994	20.9
	Brazil	8,770	14.1
	Germany	5,380	8.7
	China	4,887	7.9
	Japan	3,362	5.4
	Subtotal	35,393	57.0
	All other	26,674	43.0
	World total	62,067	100.0
Geothermal	United States	3,113	27.3
	Philippines	1,880	16.5
	Indonesia	1,354	11.9
	Mexico	1,008	8.8
	Italy	882	7.7
	Subtotal	8,237	72.1
	All other	3,187	27.9
	World total	11,424	100.0

Source: Bloomberg New Energy Finance database (accessed February 11, 2013). a2012 data are unavailable. Figures are from 2011.

Note: Excludes renewable power generating capacity from concentrated solar power (CSP) technologies, which use sunlight to heat a fluid to generate electricity in large power plants. In 2011, the latest year for which data are available, global CSP capacity amounted to 1.6 GW. Spain accounted for 60 percent (953 GW) of CSP capacity, while the United States accounted for 34 percent (542 GW) that year.

U.S. Market

Fossil fuels have historically supplied the majority of the energy consumed in the United States. In 2012, U.S. consumption of renewable energy totaled 9 quadrillion British thermal units (BTUs),[12] or about 9 percent of total U.S. energy consumption (figure 2.3), up from 6.5 percent in 2007.[13] In comparison, petroleum, natural gas, and coal still collectively accounted for over four-fifths of U.S. energy consumption in 2012. In recent years, consumption of renewable energy, particularly from wind, solar, and geothermal sources, has grown rapidly, although from a small base. By contrast, domestic consumption of petroleum and coal has somewhat declined in recent years.[14]

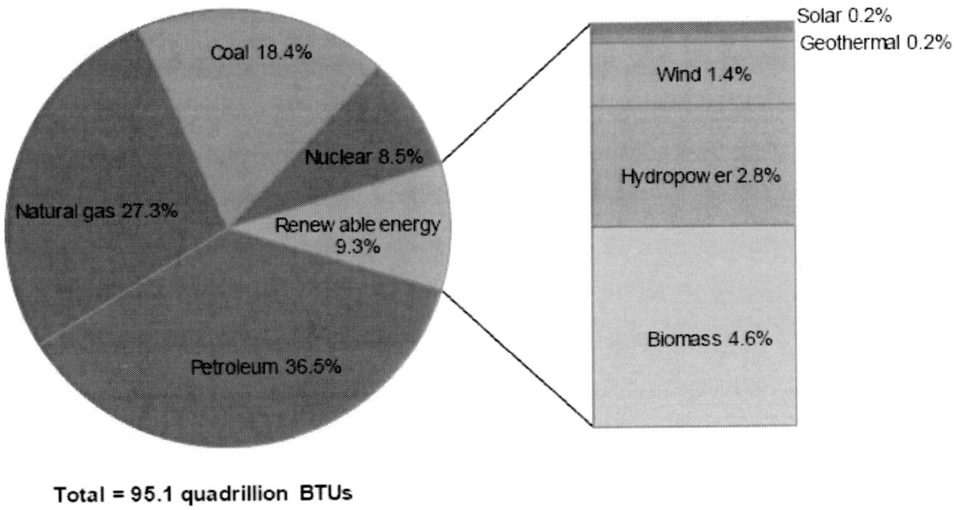

Source: EIA, *Monthly Energy Review*, March 2013, table 1.3.

Figure 2.3. U.S. energy consumption by source, 2012.

In the United States, over half of renewable energy is used to produce electricity.[15] Between 2007 and 2012, the share of net electricity generation from renewable energy sources grew from 8.5 percent to almost 13 percent, driven primarily by growth in wind energy. However, almost two-thirds of net electricity generation from renewable sources was supplied by hydropower in 2012.[16] In contrast, fossil fuels, including coal, petroleum, and natural gas, accounted for 68 percent of U.S. net electricity generation in 2012, down from 72 percent in 2007.[17]

U.S. renewable energy capacity (other than large hydropower) is growing, and between 2007 and 2012, nearly doubled to 110 GW (figure 2.4). Indeed, annual installations of renewable energy capacity more than tripled to 17 GW during the same period (figure 2.5), accounting for the largest source of capacity growth among energy sources.[18] Wind energy accounted for over half of renewable energy capacity in 2012, up from about 30 percent in 2007. In fact, new wind energy capacity topped 13 GW in 2012, or almost 80 percent of total new installations that year, driven in part by the looming expiration of the wind production tax credit, which is described in sections 4 and 6.

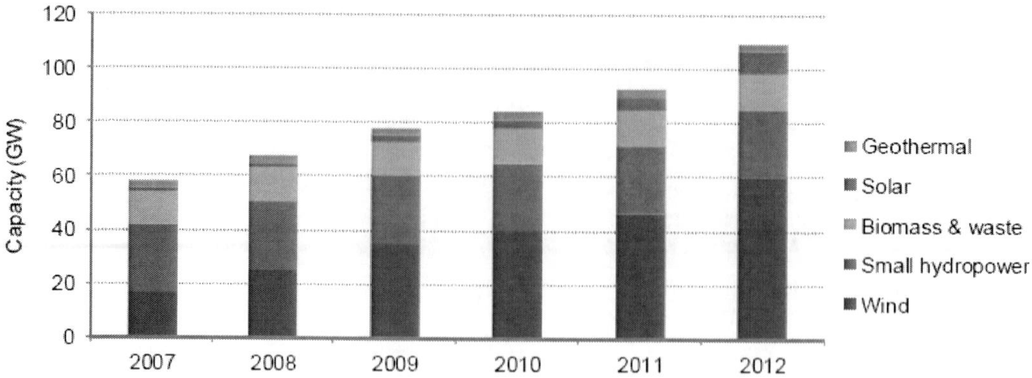

Source: Bloomerg New Energy Finance database.

Note: Small hydropower refers to hydropower sources with generating capacities equal to or less than 50 MW.

Figure 2.4. U.S. renewable energy capacity by source, 2007–12.

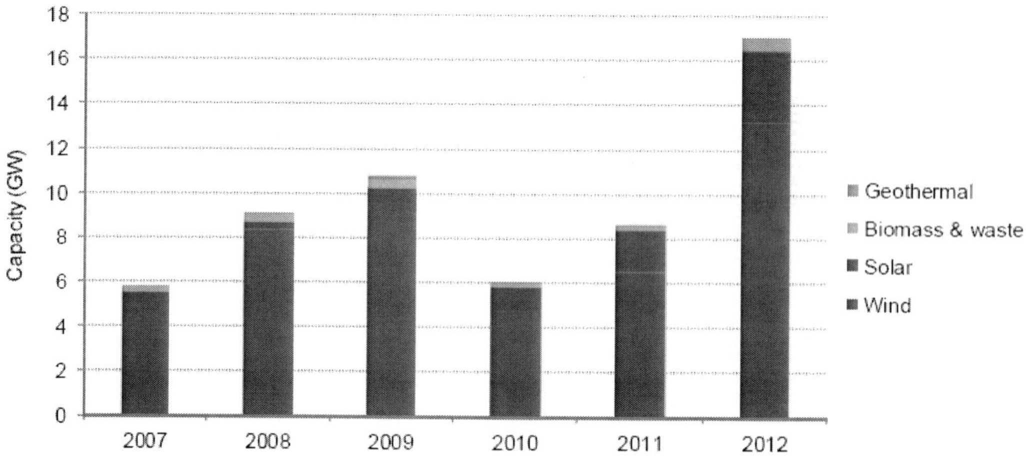

Source: Bloomberg New Energy Finance database.

Figure 2.5. U.S. annual installations of renewable capacity, 2007–12.

Factors Affecting Supply and Demand for Renewable Energy

Policies Supporting Renewable Energy

Government policy has played a key role in the global development of renewable energy. Growing concerns about energy security and the potential negative environmental and economic effects of increased greenhouse gas emissions (and, more broadly, climate change) have spurred efforts in many countries to promote and harness renewable energy sources.[19] Indeed, over 125 countries have explicit policy targets in place to promote renewable energy in one form or another, including targets to increase renewable energy as a share of energy supply and to increase installed capacity of specific technologies, among others.[20]

Renewable energy has historically cost more to generate than energy from conventional sources like fossil fuels (e.g., coal, petroleum, and natural gas). As a result, renewable energy has largely been uncompetitive with cheaper alternatives absent policy mechanisms that lower its generating costs and encourage (or mandate) its use. These mechanisms include fiscal incentives, such as tax credits to offset the cost of generating renewable energy, and regulatory policies that mandate the use of renewable energy and set its price. Often, governments implement a combination of these policies and measures to fuel the development and deployment of renewable energy resources and technologies.

Fiscal Incentives

Fiscal incentives are designed to lower the cost of generating renewable energy to make it more competitive with fossil fuels. The number of countries offering fiscal incentives to encourage the development of renewable energy has grown in recent years. Over 100 countries offer some form of fiscal incentive, including grants and rebates, production or investment tax credits, and energy-production payments.[21]

In the United States, a wide array of fiscal incentives are available to lower the cost of renewable energy project development, including various tax credits, cash grants in lieu of credit, and loan guarantees (table 2.2). Tax credits are the most common form of fiscal incentive. At the federal level, renewable energy production tax credits (PTCs) encourage renewable energy production by providing a tax credit of up to 2.3¢ per kilowatt-hour (kWh) of energy produced from qualifying renewable sources. Similarly, business energy investment tax credits (ITCs) encourage investment in renewable energy projects by reducing tax liabilities by up to 30 percent of the cost of qualified renewable technologies, including biomass, geothermal, solar, and wind, among others.[22] Both types of tax credits have been modified and extended at various times (most recently in 2012), which has affected both the timing and scope of the development of renewable energy projects, particularly for wind projects.[23]

In 2008, the financial crisis tightened credit markets and significantly reduced financing options for renewable energy project developers. In response to the crisis, the American Recovery and Reinvestment Act of 2009 (ARRA, or the Stimulus Bill)[24] created several temporary mechanisms to address the shortfall in financing and to stimulate investment in renewable energy production and manufacturing of renewable technologies. Chief among these were a program offering cash grants in lieu of ITCs for qualified renewable energy technologies, loan guarantees for renewable energy projects, and a manufacturing tax credit to encourage investment in facilities manufacturing clean energy products (see table 2.2).

Regulatory Policies

Regulatory policies are a principal driver behind the creation of renewable energy markets worldwide. Two of the most widespread regulatory policies are renewable portfolio standards (RPSs, also known as renewable energy standards) and feed-in tariffs (FITs). RPSs primarily affect demand for renewable energy by mandating its use, while FITs primarily affect the supply of renewable energy by paying a renewable energy generator a known rate of return. Similar to fiscal incentives, the number of countries implementing RPSs and FITs has grown in recent years to encourage the development of renewable energy markets.

An RPS (also known as a renewable energy standard) is a regulatory mandate that requires entities that supply electricity, such as utility companies, to generate or buy a portion of their retail electricity sales from renewable sources such as wind, solar, and biomass.[25] Because it effectively creates demand for renewable energy by mandating its use, an RPS can also be seen as promoting increased energy production from renewable sources.[26] Twenty-two countries and 54 jurisdictions at the state, provincial, or regional level maintain RPSs.[27] In the United States, RPSs are set and administered at the state level—29 states and the District of Columbia have RPS policies in place.[28]

Table 2.2. Selected U.S. fiscal incentives to promote renewable energy deployment

Type of incentive	Description	Renewable sector	Expiration date
Production tax credit (PTC)	10-year, production-based credit equal to 2.3¢/kWh	Biomass (closed-loop), geothermal, solar, wind	Project must be under construction by end of 2013.
	10-year, production-based credit equal to 1.1¢/kWh	Biomass (open-loop), hydropower, marine, landfill gas, trash combustion	Project must be under construction by end of 2013.
Investment tax credit (ITC)	Credit equal to 30 percent of eligible capital expenditures	Fuel cells, solar, small-capacity wind (projects of 100kw or less)	Project must be commissioned by end of 2016 for 30 percent ITC (10 percent ITC after 2016, without expiration).
		Biomass, geothermal hydropower, marine/tidal, wind	End of 2013.
	Credit equal to 10 percent of eligible capital expenditures	Geothermal	No expiration.
		Combined heat and power (CHP) systems	End of 2016.
Cash grant program (Treasury 1603 program)	Cash grant equal to up to 30 percent of eligible capital expenditures in lieu of the ITC	Same as those that qualify for the ITC	Project must be under construction by end of 2011 and completed by end of 2016.
Loan guarantee program (DOE 1705 loan program)	Authorized $16 billion in loan guarantees, mostly for wind and solar generation projects	Biomass, geothermal, solar, wind	Must have begun construction on project before September 30, 2011.
Manufacturing tax credit (MTC)[a]	Allocated $2.3 billion in investment tax credits up to 30 percent of investment in manufacturing facilities of clean energy products	Batteries, biomass, fuel cells, solar, wind	Project must have been commissioned before February 17, 2013.

Source: Bloomberg, *Sustainable Energy in America 2013*, 2013, 21–23; DOE, Loan Programs Office, https://lpo.energy.gov/?page id=45 (accessed February 5, 2013); North Carolina Solar Center, DSIRE database, "Federal Incentives/Policies for Renewables and Efficiency," January 3, 2013; IRS, "Fact Sheet: $2.3 Billion in New Clean Energy Manufacturing Tax Credits," January 8, 2010; IRS, "Advanced Energy Credit for Manufacturers (IRC 48C)," n.d. (accessed February 5, 2013); EIA, "Biomass for Electricity Generation," n.d. (accessed February 6, 2013).

Note: "Closed-loop" refers to biomass materials that are grown exclusively to produce energy. "Open-loop" refers to biomass materials considered either as waste or a byproduct that is used to produce energy.

[a]Qualified Advanced Energy Project Credit, enacted by the American Recovery and Reinvestment Act of 2009 (February 17, 2009), as section 48C of the Internal Revenue Code (IRS).

A feed-in tariff (FIT) offers a guarantee of payments to renewable energy developers for the electricity they produce.[29] Payments are based on a certain price per kilowatt-hour (kWh) at which electricity is purchased, typically as part of a long-term agreement set over a period of 15–20 years.[30] Because FITs generally guarantee payments at a known rate of return, they promote the supply of renewable energy. Seventy-one countries and 28 jurisdictions at the state or provincial level have implemented FITs.[31] In the United States, FITs are set and administered at the state level—five states have FITs in place (California, Hawaii, Oregon, Rhode Island, and Vermont).[32]

Changes to these policy mechanisms can affect the growth in the supply of renewable energy. For instance, in 2012, Spain suspended its FITs for new renewable energy installations in response to the country's economic crisis.[33] As a result, the growth in installations in Spain is expected to decline rapidly, thereby limiting the growth in the supply of renewable energy.[34]

Other Factors Affecting Demand and Supply for Renewable Energy

As noted previously, demand for renewable energy is almost entirely policy driven because renewable energy has largely not been cost-competitive with fossil fuel-generated electricity.[35] As a result, other factors affecting the development of renewable energy are principally supply driven, and include declining costs of renewable energy technologies, natural resource endowment, adequate electrical grid connections, and relative prices of alternative fuels.

Declining costs for renewable energy technologies can affect the supply for renewable energy. Advances in technology, greater economies of scale, and increased competition have contributed to declining prices for renewable technologies paid by project developers, leading to an increase in the supply of renewable energy projects. In response to declining costs, several countries have modified their renewable energy policies to decrease the level of support, primarily through reductions in FITs.[36]

Natural resource endowments can also affect the supply of renewable energy. Viable natural resources are needed to generate electricity, such as adequate and consistent wind resources or sunlight. In addition, adequate electricity grid connection is needed to connect renewable energy generating sources to consumption centers. For instance, inadequate grid connection can hamper efforts to develop solar and wind energy projects in many regions throughout the world, including in the United States, as described in sections 3 and 4. Another factor affecting the supply of renewable energy is the relative prices of alternative fuels, such as petroleum or natural gas. For instance, the recent boom in U.S. natural gas production has resulted in record low prices for natural gas, making electricity generated from renewable sources less competitive absent policy supports.

Overview of Renewable Energy Services

Renewable energy services broadly comprise a range of services associated with the generation, transmission, distribution, and sale of electricity and/or heat produced from renewable energy sources.[37] These also include services related to the planning, design, construction, and operation and maintenance of renewable energy installations, among others.[38] Figure 2.6 provides examples of different types of renewable energy services associated with renewable energy production. Although some of these services may overlap, services that are typically offered in the early stages of project development include planning, design, and engineering services (so-called "front-end" services). Project management and financial services are typically offered during the entire development of the project. Operations and maintenance (O&M) services—so-called "back-end" services—are important once the project is up and running.

Firms may offer one or more of these services. For instance, some firms may provide consulting services that focus on the planning stages of a renewable energy project, including geological analysis, site and resource evaluation, and environmental permitting. Other firms may provide a broader array of service offerings, such as project management services (including design services) and engineering, procurement, and construction (EPC) services. O&M services may be offered by original equipment manufacturers (OEMs) that produce the technologies that generate renewable energy, such as wind turbines or solar modules; may be done in-house; or may be carried out by unaffiliated independent service providers (ISPs).

Consumers of renewable energy services are principally the developers and operators of larger-scale renewable energy projects, such as wind or solar farms that generate electricity for sale to public or private electric utilities. Other consumers of renewable energy services may include private landowners who install smaller-scale renewable energy systems (e.g., rooftop solar installations) and industrial consumers of electricity, or other entities or individuals not connected to an electricity grid.[39]

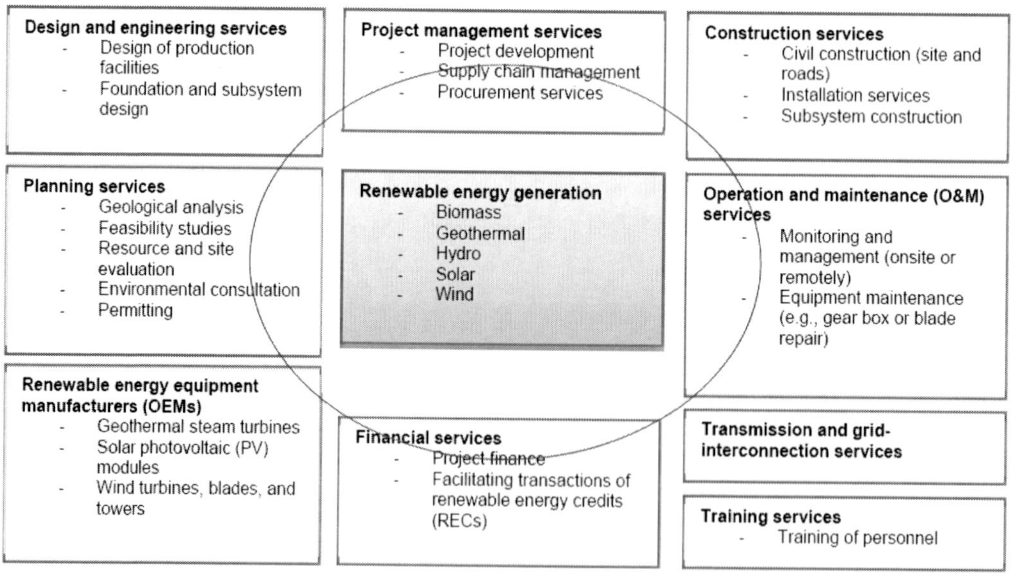

Sources: Adapted from Steenblik and Geloso Grosso, "Trade in Services Related to Climate Change," 2011, 7–11; USITC, *Renewable Energy Services*, 2005, 3-7; Quanta Power Generation company website http://www.quantarenewable.com/ (accessed February 19, 2013); Mastec company website, http://www.mastec.com/en/about/ (accessed February 19, 2013); Monkelbaan, "Sustainable Energy Services in a SETA," 2013.

Figure 2.6. Renewable energy projects encompass a broad range of goods and services.

Factors Affecting Demand and Supply for Renewable Energy Services

Demand

Demand for renewable energy underpins overall demand for renewable energy services. As noted previously, government policies that lower generating costs (through fiscal incentives like tax credits), mandate its use (through RPSs), and set its prices (through FITs) are principal drivers of the demand for and supply of renewable energy. Specific factors affecting the demand for renewable energy services, in turn, include the growth in installations of renewable technologies, the size and age of existing renewable capacity, and improvements or advances in renewable technologies.

Growth in new installations of renewable technologies is a primary driver of demand for front-end renewable energy services. These include balance of systems (BOS) services,[40] such as design and engineering, construction, and installation, and preconstruction services, such as environmental consultation and permitting services. Capacity growth, in turn, is affected by many different factors, as outlined in the previous section.

In addition, increased competition among renewable technology manufacturers, greater economies of scale in manufacturing, and advances in technology and manufacturing have driven down the prices paid by project developers for many renewable technologies, such as wind turbines and solar panels.[41] Lower prices for these technologies, in turn, have led to growth in project development services and renewable energy installations.

The growth and age of the installed capacity of renewable technologies are primary drivers of demand for back-end renewable energy services, particularly operation and maintenance (O&M) services. For instance, wind turbines are typically covered under a warranty provided by the turbine's manufacturer, during which time maintenance and repair services are included under the terms of the contract. As warranty periods expire and turbines age, demand for out-of-warranty O&M services typically increases. Indeed, O&M can account for up to 35 percent of the annual cost of power generation at the end of a wind turbine's life, reflecting greater maintenance and upkeep, compared with 10 percent for a new turbine.[42]

Improvements or advances in renewable technologies may weaken demand for certain O&M services. For instance, wind turbines incorporating newer direct-drive turbine technologies do not require gearboxes to generate electricity, and therefore eliminate the need for gearbox maintenance, a type of O&M service.[43] In addition, because direct-drive technologies use fewer parts, there are relatively lower maintenance requirements overall.[44] In another example, wind-farm operators that monitor wind farm conditions remotely can improve the ordering of spare parts, work scheduling, and planning activities, thereby reducing the frequency and improving the efficiency of onsite maintenance.[45] Finally, technologies that measure and control the tension in the bolted joints of turbines can prevent joint failures that arise from insufficient bolt tension in turbine installations, and can reduce associated maintenance costs by up to 50 percent.[46] Despite these trends, the substantial increase in global installed wind capacity continues to drive growth in the O&M market.

Supply

Factors affecting the supply of renewable energy services include the availability and technical expertise of skilled workers, access to infrastructure, geography, and weather conditions. For instance, the rapid growth in renewable energy installations, particularly in wind energy, has reportedly outpaced the supply of skilled workers who perform certain renewable energy services, such as EPC or O&M services. [47] In addition, qualified technicians are needed to service various technologies, including hydraulics, mechanics, and information technology, often in challenging environments. The lack of qualified technicians to undertake O&M services in wind parks has been cited as one of the biggest challenges facing wind energy service providers globally.[48] O&M service providers, including OEMs, ISPs, and wind-farm operators, reportedly intend to triple their workforce in the coming years in response to continued growth in wind installations, underscoring the need for skilled labor.[49]

Geography, access to infrastructure, and weather conditions can affect the supply of O&M services. In the case of offshore wind turbines, the availability of marine vessels and port services affects the ability of service providers to travel to offshore locations to provide O&M services. [50] Long distances to remote locations, particularly offshore, as well as harsh weather conditions, can also limit firms' ability to provide such services by impeding physical access to such locations.

Estimates of the Size of the Global and U.S. Market for Renewable Energy Services

Size of the Global Market

Comprehensive published data on the value of the global market for all renewable energy services in all renewable energy sectors are unavailable.[51] However, given that renewable energy services are indispensable to the development and maintenance of renewable energy projects, the rapid growth in renewable energy investment and installed capacity worldwide indicates a similarly vibrant global market for renewable energy services.

Solar PV

The global market for solar PV services has grown rapidly in response to growth in solar PV installations worldwide. The value of global solar PV services associated with installations was estimated to be $34 billion in 2011, or 37 percent of the broader market for solar PV installations (including equipment and services), as described in section 3. The largest markets for solar PV services in 2011 were Italy ($9.8 billion), Germany ($5.1 billion), the United States ($3.1 billion), and Japan ($3.0 billion).[52] According to one published source, the global market for solar PV O&M services is expected to grow annually by 43 percent to reach $18.4 billion by 2017.[53] Europe is expected to be the largest market for solar PV O&M services, due to continued growth in solar PV installations in the coming years.[54]

Wind

The global market for wind services has grown as global wind installations have increased. In 2011, the value of services associated with wind installations was estimated at nearly $23 billion, or roughly 32 percent of the broader global market for wind power installations (including equipment and services), as noted in section 4. The largest markets for wind services are China ($9.0–$11.8 billion), the United States ($5.5– $7.1 billion), Germany ($2.3–$3.5 billion), and Canada ($1.2 billion).[55]

Similarly, the value of global O&M services also continues to rise steadily, as the volume of global wind-energy capacity more than doubled between 2007 and 2012. Although estimates vary, the global market for wind O&M services was estimated at \$6.2– \$7.2 billion in 2011, with Europe estimated to account for about half of that market.[56] OEMs that produce wind turbines are thought to provide almost two-thirds of wind O&M services in the European market. Other O&M service suppliers include ISPs and wind-farm operators.[57]

Hydropower and Geothermal

The global market for services associated with all hydropower installations was estimated at nearly \$72 billion, as outlined in section 5.[58] The global market for services associated with installations of small-capacity hydropower (i.e., less than or equal to 50 MW) is substantially smaller, and is estimated to account for less than 5 percent (\$2.3 billion) of the broader market for hydropower installations of all sizes. Finally, the global market for geothermal services was estimated at \$2.8 billion in 2010, with O&M services accounting for the bulk of those services (\$2.5 billion), as noted in section 5.

Size of the U.S. Market

Similar to the global market, comprehensive published data on the total value of the U.S. market for services in all renewable energy sectors are unavailable. However, the United States is consistently among the largest markets for renewable energy services.

Solar PV

The U.S. market for solar PV services grew rapidly between 2007 and 2012, driven by declining prices in solar PV installations and by state and federal government policies. The value of services associated with solar PV installations is estimated to be \$3.1 billion in 2011, or 37 percent of the broader U.S. market for solar PV installations (including equipment and services), as described in section 3. By comparison, the U.S. market for solar PV O&M services was likely around \$100 million in 2011. Although the value of O&M services was comparatively small—solar PV panels require considerably less maintenance than wind turbines—the growing installed solar PV base offers increasing opportunities for O&M service providers.

Wind

Similarly, the U.S. market for wind-energy services grew rapidly between 2007 and 2012, as wind-energy capacity increased by nearly 30 percent annually. The value of services associated with wind installations is estimated at \$7.4 billion in 2012, or 32 percent of the broader U.S. market for wind power installations (including equipment and services).[59] The U.S. market for O&M services in the wind-energy sector was substantially smaller; it was estimated to be \$1.7 billion in 2012, up from \$467 million in 2007.

Hydropower and Geothermal

The U.S. market for hydropower and geothermal-energy services is considerably smaller than that for wind and solar services.[60] The value of services associated with the construction and installation of all hydropower projects (i.e., including those with capacities greater than 50 MW) in the United States was estimated at \$322 million in 2010, although the value of O&M services is likely to be considerably larger. The value of services associated with the construction and installation of geothermal services was estimated at $34 million in 2010, while O&M services to support existing installed capacity were likely around $590 million.

Global and U.S. Investment in Renewable Energy Services

Global Investment

Data on global and U.S. investment specific to renewable energy services are unavailable. However, global investment in renewable energy in general stood at a record $244 billion in 2012, up 71 percent compared with 2007, although down 12 percent compared with 2011 levels (figure 2.7). These figures broadly reflect investment associated with the development of renewable energy projects and installation of renewable energy generating capacity. Because renewable energy services are needed to develop these projects, investment in renewable energy more broadly serves as a useful proxy for trends and investment in renewable energy services.[61]

The solar sector accounted for the majority of global investment in renewable energy in 2012 ($143 billion, or 58 percent), having eclipsed wind power in 2010 to take the top spot. Global investment in solar power grew the fastest among all renewable energy sectors, having nearly quadrupled between 2007 and 2012 (although having declined in 2012 by 9 percent compared with 2011). Factors contributing to the growth in solar investment include falling prices for solar PV modules; robust demand for rooftop solar PV installations in Europe, particularly in Germany and Italy; growth in small-scale solar PV installations in other countries, including China and the United Kingdom; and the financing of large-scale solar thermal electricity generation projects in Spain and the United States.[62]

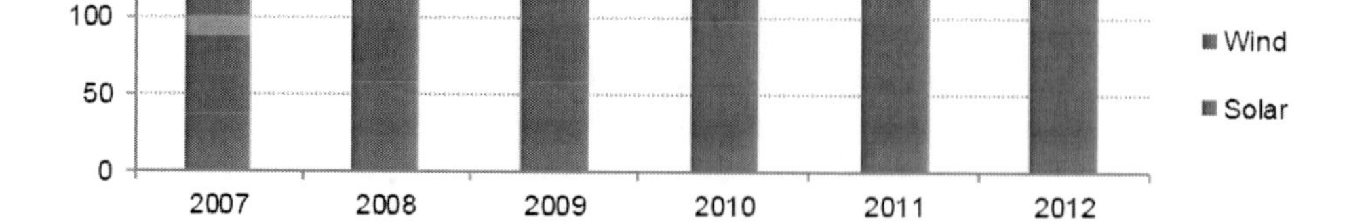

Source: Bloomberg New Energy Finance database.

Note: While included in the overall global renewable energy investmentfigures, the total investmentdata for marine energy installations are too small relative to the other technologies represented here to appear in the figure.

Figure 2.7. Global renewable energy investment by sector, 2007–12.

Despite increasing by almost two-thirds between 2007 and 2011, global investment in wind power tapered off and declined to $78 billion in 2012, 13 percent lower than in 2011. Factors contributing to the recent decline include overcapacity in global turbine manufacturing, resulting in lower turbine prices and reduced output; uncertainty in Europe over policy measures to support renewable energy; and slower growth in wind installations in China due, in part, to inadequate grid connections.[63]

Global investment in renewable energy grew in most regions of the world between 2007 and 2012 (figure 2.8), due in large part to government policies supporting renewable energy as outlined in the previous section.[64] Although Europe is the largest regional market for renewable energy investment ($79.9 billion in 2012), China surpassed the United States in 2009 to become the largest single-country market for such investment ($66.6 billion in China compared to $36 billion in the United States in 2012). Indeed, China's investment in renewable energy more than quadrupled between 2007 and 2012, and grew by 22 percent between 2011 and 2012, reflecting continued growth in installed capacity to generate renewable energy. Despite a lack of adequate grid connections, China added nearly 20 GW of wind-energy capacity in 2011, up from 17 GW in 2010. Moreover, 2.2 GW of solar PV capacity was commissioned in 2011, and solar thermal electricity generation continues to grow as well.[65] Although not as large, investment in renewable energy in India increased by almost two-thirds between 2010 and 2011, due in part to a national program to develop 20 GW of solar power capacity by 2022.[66]

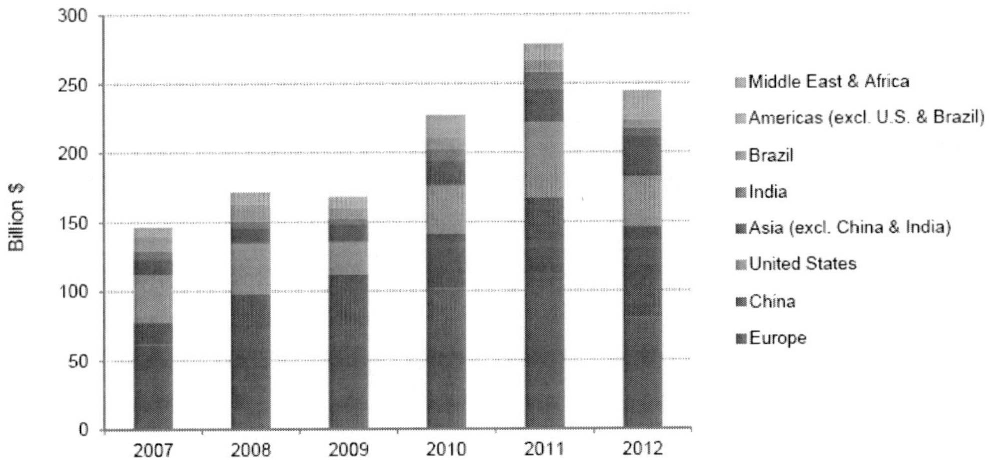

Source: REN-21, Renewables 2013, 2013, 62-63.

Figure 2.8. Global renewable energy investment by country and region, 2007–12.

U.S. Investment

Although the United States lagged behind China, U.S. domestic investment in renewable energy grew by almost 60 percent between 2007 and 2011 to $54.8 billion; investment, however, fell by almost 35 percent between 2011 and 2012 to $36 billion. The growth in U.S. investment in 2011 alone was driven, in part, by the looming expiration of federal renewable energy incentive programs. For instance, the federal loan guarantee program, which covered $16.1 billion in project debt, expired in September 2011. Renewable energy developers sought to finance projects in time to take advantage of these programs before their expiration in 2011.[67] Uncertainty surrounding the future of the incentive programs dampened renewable energy investment in 2012.

Global and U.S. Trade in Renewable Energy Services

Renewable energy services are typically supplied by establishing a commercial presence and by temporarily moving service providers into a client's territory. In addition, improvements in information and communications technologies (ICTs) have made the provision of some renewable energy services, such as remote monitoring, possible through cross-border supply.[68] Other examples in which ICTs have boosted the cross-border supply of renewable energy services include the transmission of architectural and engineering specifications, design plans, environmental assessments, and computer modeling simulations.[69]

To some degree, the supply of certain renewable energy services broadly mirrors the movement of certain renewable energy technologies, as the services and technologies often complement each other.[70] Often, manufacturers that produce and export renewable energy technologies, such as wind turbines, also provide associated services in the market to which the product was exported. Some manufacturers of wind turbines may provide both front-end and back-end services, including financing of turbine sales; project development and installation services; and training, monitoring, and repair services. For instance, in 2013 Vestas, a U.S. producer of wind turbines, received an order to produce and export over 150 wind turbines for the 299 MW Blackspring Ridge Wind Project in Alberta, Canada. The firm will also provide maintenance services on the turbines as part of a 20-year service and maintenance agreement included in the order.[71]

Other manufacturers of renewable energy technologies may provide design, construction, and installation services as part of their overall product offering.[72] One source surveyed the top firms that design and build wind and solar energy projects. According to this source, global exports of design and construction services in the wind sector totaled $829 million in 2012, while global exports of design and construction services in the solar sector totaled $1.2 billion.[73]

Although data on global exports of different renewable energy technologies are largely unavailable due to issues with the way exports are classified,[74] exports of certain renewable energy equipment like wind-powered generating sets have grown in recent years.[75] These exports are likely representative of broader trade trends, reflecting the potential growth in trade in wind energy services. Between 2007 and 2012, global exports of wind-powered generating sets increased by 72 percent to $6.8 billion.[76] Germany, Denmark, and Spain collectively accounted for over 75 percent of these exports in 2012 (the United States accounted for 6 percent). The largest global exporters of renewable energy technologies are also likely to be the largest suppliers of renewable energy services. Indeed, Germany, Denmark, and Spain have well-developed and sophisticated providers of wind energy services, as noted in section 4. Likewise, the largest global importers of renewable energy technologies are also likely to be the largest consumers of renewable energy services. For instance, the United States is the largest importer of wind-powered generating sets, reflecting the growth in U.S. installed wind capacity and wind services supplied by domestic and foreign firms alike.

Trade in renewable energy services also depends on the extent to which countries where renewable energy projects are located have firms with the capacity to provide all the necessary goods and services.[77] A number of countries, particularly developing countries, reportedly lack the manufacturing base to produce renewable energy technologies and have little to no domestic capacity to design renewable energy projects, including wind, solar, or geothermal energy projects. [78] As a result, both the renewable energy technologies and many of the renewable energy services must be imported.[79]

Countries that have experienced growth in installed capacity of renewable energy, yet lack the domestic capacity to design renewable energy projects, present an opportunity for U.S. exporters of renewable energy services. For instance, some firms offering project development services in the solar PV sector are targeting emerging solar PV markets like Chile and South Africa, as noted in section 3. Even in developed countries that have the domestic capacity and expertise to provide renewable energy services, project developers may choose to import services from an international supplier based on cost, quality, or other considerations.[80] In this case, opportunities for U.S. exporters of renewable energy services exist in mature markets as well.

Barriers to Trade and Investment in Renewable Energy Services

As noted in section 1, renewable energy services comprise a broad range of services that largely fall within the following three Central Product Classification (CPC) groups: services incidental to energy distribution; other professional, technical, and business services; and construction services. Measures that act as barriers to trade and investment in renewable energy services tend to be horizontal in nature; that is, they affect a broad range of services, such as laws that pertain to a country's general investment regime, or to its immigration policy.

Horizontal Measures Affecting Trade and Investment in Renewable Energy Services

As noted above, renewable energy services are typically supplied by setting up a commercial presence and by temporarily moving service providers into a client's territory. Barriers that limit either activity are thus the most likely to impede firms' ability to supply renewable energy services in foreign markets.

Certain barriers to renewable energy services can be identified by World Trade Organization (WTO) members' schedules of country-specific General Agreement on Trade in Services (GATS) commitments that are classified within the CPC groups noted above.[81] Among the countries that trade these services the most,[82] investment restrictions are the most common barriers to the establishment of a commercial presence for renewable energy service providers in foreign markets. These include restrictions on legal form, foreign equity limits, and economic needs tests. [83]

Measures that hinder the temporary movement of foreign workers can act as a barrier to trade in renewable energy services. Common restrictions include limits on entry and duration of stay, quotas, and labor market tests. Other impediments include qualification and licensing restrictions, including nationality or residency requirements.[84] Burdensome requirements for visa applications, work permits, and residence permits can hinder a company's ability to rapidly deploy technically skilled workers needed to perform maintenance or repair services on renewable energy equipment in foreign countries. For instance, visa restrictions in Brazil reportedly limit the ability to bring in skilled wind technicians from the United States in the absence of a reciprocal visa agreement between the two countries.[85] According to General Electric (GE), a manufacturer of renewable energy technologies and a service provider, it may not be feasible to hire fully trained, domestic workers that are needed in each country in which GE operates, underscoring the need to allow the movement of persons to perform services in a timely manner.[86]

Specific Measures Affecting Trade and Investment in Renewable Energy Services

Forced localization measures, such as local-content requirements, can act as a barrier to trade in both technologies and services for renewable energy.[87] For example, local-content requirements may require renewable energy projects like wind or solar farms to include a minimum share of goods or services of domestic origin to qualify for an FIT. Some of these requirements have been challenged at the WTO as being inconsistent with WTO obligations, as described in sections 3 and 4. Local-content requirements are most often directed at renewable technologies, as a part of efforts to spur domestic manufacturing and job creation, rather than specifically targeting services.[88] However, many project developers that provide renewable energy technologies also supply renewable energy services. Project developers that do not produce goods locally in markets that require local content can therefore be shut out from providing services in those markets as well.

Local-content requirements can limit market access for service providers in other ways. For instance, they may require workers to be nationals of a particular country, require specific procedures to take place within a country, or require firms to have local partners or majority local ownership.[89] Other impediments may include cumbersome project approval processes, country-specific standards, or certifications that do not recognize foreign providers of renewable energy services.[90]

Efforts to Liberalize Trade and Investment in Renewable Energy Services

Many countries have lowered barriers to trade and investment in environmental goods and services (including renewable energy services) to encourage the use and deployment of environmental technologies in response to environmental and energy concerns and with the goal of mitigating climate change. Multilateral efforts in recent years to specifically address barriers to trade in renewable energy goods and services have yet to produce binding commitments in this area. However, multilateral efforts to address trade in services more broadly have gained some traction. At the same time, regional and bilateral efforts to address trade and investment in renewable energy services—and environmental goods and services more broadly—have advanced at a faster pace.

No agreements specific to renewable energy services have been reached in recent years. In 2009, the United States proposed establishing an Environmental Goods and Services Agreement (EGSA), a plurilateral initiative within the WTO but outside of the Doha round, to reduce barriers to trade in certain "climate-friendly" technologies.[91] However, the EGSA never materialized, largely due to definitional concerns among several WTO member countries, as well as differing views on product coverage. Member countries continue to discuss classification issues related to renewable energy services.[92] In June 2013, the United States announced its intention to re-launch negotiations to establish a plurilateral initiative to reduce barriers to trade in environmental goods, including clean energy technologies such as solar, wind, hydro, and geothermal power.[93]

There has been recent progress in other multilateral efforts to address liberalization of trade in services more broadly. For instance, the United States and 20 of its trading partners plan to address barriers to services trade by initiating negotiations on a new trade agreement on services (the so-called Trade in Services Agreement) in 2013. In addition, in June 2013 the United States announced its intention to work in the Trade in Services Agreement negotiations towards lowering barriers to trade in environmental services, including renewable energy services.[94]

Regional and bilateral efforts to liberalize trade in renewable energy goods and services are also underway. For instance, in 2012 members of the Asia-Pacific Economic Cooperation (APEC), of which the United States is a member, agreed to reduce applied tariff rates to 5 percent or less by the end of 2015 on 53 separate environmental goods, including wind turbines and blades, generator sets, and solar panels. [95] In 2013, the United States stated its intention to continue to work with other APEC members to further facilitate trade in environmental goods and services, including renewable energy services.[96] Some WTO members are said to view the APEC agreement on trade in environmental goods and services as a platform to revive efforts at the WTO to address barriers to trade in environmental goods and services.[97] However, other WTO members, including Argentina, Bolivia, India, and Venezuela, reportedly are reluctant to endorse such an approach.[98]

The United States and 10 other Asia-Pacific countries are also currently negotiating a Trans-Pacific Partnership (TPP) agreement, and efforts to develop text on trade-related environmental issues have included discussions on reducing barriers to trade in environmental goods and services.[99] Other bilateral efforts to reduce barriers to trade and investment in environmental goods and services have broadly centered on cooperative efforts to address threats to the environment, encourage the deployment of environmental technologies to mitigate climate change, and promote environmental services (see table 2.3).

Table 2.3. Selected bilateral efforts to reduce barriers to trade and investment in environmental goods and services

Agreement	Description
U.S.-Bahrain FTA	Parties state their intention of "[p]romoting the growth of the environmental technology business sector."[a]
U.S.-CAFTA FTA	"The Parties recognize that strengthening their cooperative relationship on environmental matters can enhance environmental protection in their territories and may encourage increased trade and investment in environmental goods and services" (Article 17.9).
U.S.-Colombia FTA	"The parties are committed to expanding their cooperative relationship on environmental matters, recognizing it will help them achieve their shared environmental goals and objectives, including the development and improvement of environmental protection, practices, and technologies" (Article 18.10).
U.S.-Jordan FTA	Parties state their interest in pursuing "[s]ponsorship of trade missions to encourage the use of environmental technologies."[b]
U.S.-Korea FTA	"The Parties are committed to expanding their cooperative relationship in bilateral, regional, and multilateral forums on environmental matters, recognizing that such cooperation will help them achieve their shared goals and objectives, including the development and improvement of environmental protection, practices, and technologies" (Article 20.8).

Table 2.3. (Continued)

Agreement	Description
U.S.-Morocco FTA	"The Parties recognize that strengthening their cooperative relationship on environmental matters can encourage increased bilateral trade and investment in environmental goods and services" (Article 17.3).
U.S.-Oman FTA	Parties state their intention of "[s]trengthening capacity to establish a Cleaner Production Center and promoting the growth of the environmental business technology sector".[c]
U.S.-Panama TPA	"Parties recognize that strengthening their cooperative relationship on environmental matters can enhance environmental protection in their territories and may encourage increased trade and investment in environmental goods and services" (Article 17.10).
U.S.-Peru TPA	"Parties are committed to expanding their cooperative relationship on environmental matters, recognizing it will help them achieve their shared environmental goals and objectives, including the development and improvement of environmental protection, practices, and technologies" (Article 18.10).
EU-Chile FTA	"The EC and Chile will undertake to consolidate economic relations in key sectors such as hydroelectricity, oil and gas, renewable energy, energy-saving technology, and rural electrification" (Article 22).
EU-Colombia/Peru FTA	Parties state their intention to consider actions to mitigate climate change by "[f]acilitating the removal of trade and investment barriers to access to, innovation, development, and deployment of goods, services and technologies that can contribute to mitigation or adaptation, taking into account the circumstances of developing countries" (Article 275).
EU-Korea FTA	"The Parties shall strive to facilitate and promote trade and foreign direct investment in environmental goods and services, including environmental technologies, sustainable energy, energy efficient products and services and ecolabelled goods, including through addressing related nontariff barriers" (Article 13.6).
India-Japan Agreement	"Each Party shall endeavor to [...] encourage trade and dissemination of environmental goods and services" (Article 8).
Japan-Switzerland Agreement	Parties state their intention to "[e]ncourage trade and dissemination of environmental products and environmental-related services in order to facilitate access to technologies and products that support the environmental protection and development goals, such as [...] sustainable promotion of renewable energy and climate-change related goals (Article 9).

Sources: WTO, "Energy Services," 2010, 21; Monkelbaan, "Sustainable Energy Services in a SETA," 2013, 46; USTR website, http://www.ustr.gov (accessed March 22, 2013).

[a] Memorandum of Understanding (MOU) on Environmental Cooperation between the Government of the United States and the Government of the Kingdom of Bahrain.

[b] U.S.-Jordan Joint Statement on Environmental Technical Cooperation.

[c] MOU on Environmental Cooperation between the Government of the United States and the Government of the Sultanate of Oman.

References

Asia-Pacific Economic Cooperation (APEC). "Annex C—APEC List of Environmental Goods." *Leaders' Declaration,* September 8–9, 2012. http://www.apec.org/Meeting-Papers/LeadersDeclarations/2012/2012_aelm/2012_aelm_annexC.aspx (accessed February 7, 2013).

Bartholl, Carsten, and Sven Oleownik. “Wind Services: New Growth Opportunities.” Presentation given at Cleantech Forum Europe, Munich, Germany, April 18, 2012.

Bloomberg New Energy Finance (BNEF) database (subscription only, accessed various dates).

———. *Market Sizing*, November 2012.

Brightsource Energy. “Ivanpah and the DOE Loan Guarantee Program.” Press release, June 5, 2012. http://www.brightsourceenergy.com/ivanpah-and-the-doe-loan-guarantee-program (accessed February 4, 2013).

Broad, Will. “Post-Warranty Wind O&M Costs Proven to Increase by 25%.” *Wind Energy Update*, January 18, 2013. http://www.renewableenergyworld.com/rea/partner/first-conferences/news/article/2013/01/post-warranty-wind-o-m-costs-proven-to-increase-by-25?cmpid=WNL-Wednesday-January23-2013.

Clover, Ian. “Spain Suspends Feed-In-Tariffs, Receives EU Criticism.” *InfraEurope*, February 3, 2012. http://www.infraeurope.com/news/spain-suspends-feed-tariffs-receives-eu-criticism/4963.

North Carolina Solar Center. Database of State Incentives for Renewables & Efficiency (DSIRE). http://www.dsireusa.org/ (accessed February 5, 2013).

Deloitte. “European Wind Services Study,” August 2012.

Executive Office of the President. “The President’s Climate Action Plan,” June 2013. http://www.whitehouse.gov/sites/default/files/image/president27sclimateactionplan.pdf.

Fessenden, Karl. Written testimony before the U.S. House Ways and Means Subcommittee on Trade, September 20, 2012. http://waysandmeans.house.gov/UploadedFiles/Lake Testimony unwatermarked for clerks 9.2 1.2012 TR.pdf.

First Solar. “First Solar Sells 550-Megawatt Desert Sunlight Solar Farm, among World’s Largest.” Press release, September 30, 2011. http://investor.firstsolar.com/release detail.cfm?releaseid=610018.

Frankfurt School of Finance and Management (FS) and United Nations Environment Programme (UNEP). *Global Trends in Renewable Energy Investment 2012*. Frankfurt, Germany: FS, 2012.

Herndon, Andrew. “NRG Gets U.S. Guarantee for $2.6 Billion Rooftop Solar Plan.” Bloomberg, June 22, 2011. http://www.bloomberg.com/news/print/2011-06-22/u-s-offers-guarantee-for-2-6-billionnrg-rooftop-solar-program.html (accessed February 4, 2013).

Inside U.S. Trade. “Danish Minister Says Time Is Ripe to Pursue Green Goods, Services Deal,” May 3, 2012.

———. “India, Argentina Cool to Using APEC Green Goods Deal for WTO Work,” November 15, 2012.

Kim, Joy Aeree. “Facilitating Trade in Services Complementary to Climate-friendly Technologies.”

International Centre for Trade and Sustainable Development (ICTSD) Issue Paper No. 16. Geneva: ICTSD, 2011.

Marantis, Demetrios. Office of the United States Trade Representative. Statement to the United States Senate Committee on Finance. The President’s 2013 Trade Policy Agenda, hearing, March 19, 2013.

McBee, David. *Solar Panel Installation in the U.S.* IBISWorld Industry Report OD4494, October 2012.

Organisation for Economic Co-operation and Development (OECD). "Services Trade Restrictiveness Index (STRI): Construction, Architecture and Engineering Services." TAD/TC/WP(2012)13. Working Party of the Trade Committee, Paris June 6–7, 2012.

Pancratz, Howard. "Vestas' Colorado Factories to Build Components for Canadian Project." *Denver Post*, April 9, 2013.

Peszko, Grzegorz, and Janina Ketterer. "Local Content Requirements for Renewable Energy: An Unnecessary Evil." *European Bank for Reconstruction and Development (EBRD) Blog,* November 23, 2013. http://www.ebrdblog.com/wordpress/2012/11/local-content-requirementsfor-renewable-energy-an-unnecessary-evil/.

Renewable Energy Policy Network for the 21st Century (REN-21). *Renewables 2012: Global Status Report*. Paris: REN-21 Secretariat, 2012.

———. *Renewables 2013: Global Status Report*. Paris: REN-21 Secretariat, 2013.

Solar Energy Industries Association (SEIA). "Solar Investment Tax Credit (ITC)," n.d. http://www.seia.org/policy/finance-tax/solar-investment-tax-credit.

Steenblik, Ronald, and Massimo Geloso Grosso. "Trade in Services Related to Climate Change: An Exploratory Analysis." OECD Trade and Environment Working Papers 2011/03. OECD: Paris, 2011. http://www.oecd-ilibrary.org/trade/trade-in-services-related-to-climate-change 5kgc5wtd9rzw-en.

Tulacz, Gary J. *The Global Sourcebook*. Engineering News-Record (ENR), December 10, 2012.

U.S. Department of Energy (DOE). Energy Information Administration (EIA). "What Is Renewable Energy?" May 25, 2012.
http://www.eia.gov/energyexplained/index.cfm?page=renewable home#tab1.

———.*International Energy Outlook 2011*. Washington, DC: EIA, September 2011.

———. "Biomass for Electricity Generation," n.d. http://www.eia.gov/oiaf/analysispaper /biomass/ (accessed February 6, 2013).

———. International Energy Statistics Database (accessed February 8, 2013).

———. "Our Projects." https://lpo.energy.gov/?page id=45 (accessed February 5, 2013).

U.S. Department of Energy (DOE). National Renewable Energy Laboratory (NREL). "Renewable Portfolio Standards," updated November 7, 2012.
http://www.nrel.gov/tech deployment/state local activities/basics portfolio standards.html.

U.S. Department of Energy (DOE). Loans Programs Office. "Deployment: How the Government Delivers Clean Energy and U.S. Jobs Today," August 2011.

———. "Feed-In Tariffs," updated November 7, 2012.
http://www.nrel.gov/tech deployment/state local activities/basics tariffs.html.

U.S. International Trade Commission (USITC). *Renewable Energy Services: An Examination of U.S. and Foreign Markets*. Publication No. 3805. Washington, DC: USITC, October 2005.

World Trade Organization (WTO). "Energy Services." Background Note by the Secretariat. S/C/W/311, January 12, 2010.

———. "Energy Services: Overview of Classification Issues." Informal Note by the Secretariat. JOB/SERV/94, March 5, 2012.

SECTION 3. SOLAR PV INDUSTRY

INTRODUCTION

U.S. and global solar photovoltaic (PV) installations rapidly increased during 2007–12, resulting in a significant rise in demand for associated PV services.[1] In 2011, the value of the U.S. services market associated with new PV installations totaled approximately $3.1 billion, and the value of the global market was about $34 billion. The United States has several thousand firms currently providing PV services, ranging from small firms installing a few residential PV systems per year to large construction firms and project developers. The majority of the firms currently providing services in the U.S. market are U.S.-based companies, and imports accounted for less than one-third of combined installation and project development services in 2012. U.S. firms are also actively exporting services globally, with the level of exports expected to rise as firms increase their presence in foreign markets, the U.S. industry matures, and markets such as Canada, Latin America, and South Africa expand. However, U.S. exports through the end of 2012 accounted for only a small share of the global market.

Overview of PV Services Segments

PV is currently the most widely deployed solar technology, both in the United States and globally. PV systems convert sunlight directly into electricity for on-site use or for distribution through the electric grid. The main components of PV systems are modules (also commonly referred to as panels), which are composed of cells that use crystalline silicon (c-Si) or another photosensitive material, such as cadmium telluride (CdTe), to convert sunlight into electricity.[2]

PV modules can be either ground-mounted or rooftop-mounted and are connected to an inverter, which converts the direct current (DC) generated by the system to alternating current (AC).[3] Most PV systems are connected to the grid and are installed in one of three market segments—residential, nonresidential, and utility (box 3.1).

BOX 3.1. Photovoltaic (PV) market segments

Grid-connected market: The grid-connected PV market is segmented into the residential market, nonresidential market, and the utility market:

- *Residential:* Residential systems are installed at homes and are usually connected to the grid, with homeowners using grid energy when solar electricity generation is not sufficient to meet demand and often feeding energy back into the grid when the system generates excess electricity. Where "net metering" is available, the power generated by homeowners and fed back into the grid can be used to offset utility charges for grid power used at other times. In the United States, the average size of a 2011 residential installation was 5.7 kilowatts (kW).

- *Nonresidential:* These are systems installed at commercial, industrial, government, and similar buildings and sites. They are larger than residential installations and, like residential installations, often use net metering. In the United States, the average size of a 2011 nonresidential installation was about 116 kW.
- *Utility:* In utility systems, electricity is generated either (1) by independent power producers (IPPs) and sold on the wholesale electricity market, or (2) by utilities. This electricity is then sold to households, commercial, and industrial users. Installations greater than 1 MW accounted for 92 percent of annual utility installations in 2011.

Off-grid market: In addition to the grid-connected market, there is a small off-grid market that includes a variety of uses such as homes not connected to the grid, industrial applications (e.g., lights and signs along highways), consumer goods, and military applications. There is also a small off-grid market, but that is not covered in this study, since it accounts for only a small share of PV installations.

Source: Installation size from Sherwood, *U.S. Solar Market Trends 2011*, August 2012, 6–7; Sherwood, *U.S. Solar Market Trends 2010*, June 2011, 7. For an overview of the electric industry, see the EIA website, "Electric Industry Overview 2007," http://eia.gov/cneaf /electricity/page /prim2/toc2.html (last accessed July 12, 2013).

Residential and Nonresidential Services

In the residential and much of the nonresidential market,[4] customers generally contract with installers or third-party owners—firms that lease systems to customers or sell customers the power generated by a system (box 3.2)—to install a system on-site.[5] These firms manage the provision of each major service associated with the installation, including (1) site assessment and design; (2) permitting, net metering agreements, and rebate applications; (3) financing (for some systems), including leases and power purchase agreements (PPAs)[6] from third-party owners;[7] (4) installation, involving the physical mounting of the system on the roof or ground, attaching the inverters and wiring, and connecting the system to the grid; and (5) operations and maintenance (O&M). Firms may provide all of the services associated with the installation directly, or they may contract out certain services to other firms that specialize in one or more of these services. Only a few third-party owners are also installers, and even those that do install may only do so in certain market segments; most third-party owners subcontract with installers to build the system.

BOX 3.2. Financing and ownership models for residential and commercial solar

Residential and commercial customers generally have several options when installing a PV system onsite, including (1) buying the system, (2) leasing the system, and (3) signing a power purchase agreement (PPA). The firms that offer leases and PPAs are generally referred to as third-party owners or third-party financiers.

Leasing: In this model, a company installs a PV system at a residential or commercial site and leases the system to the customer, who uses the electricity generated by the system.

PPA: In a PPA model, a company installs the PV system at a residential or commercial site and owns, operates, and maintains the system. The residential or commercial customer agrees to purchase the electricity generated by the system from the company over a long-term time frame.

For customers that want to buy their own PV systems, some communities offer property-assessed clean energy (PACE) programs. In these programs, a local government raises money by issuing a bond, or through other means. Property owners install a PV system using this money, paying it back over time through an additional amount on their property tax or utility bill. Customers may also use more traditional financing methods, such as bank loans, to pay for their systems.

Sources: SolarCity, "Form S-1," October 5, 2012, 47, 99; North Carolina Solar Center, DSIRE website, http://www.dsireusa.org/solar/solarpolicyguide/?id=26 (accessed March 4, 2013); SEIA website, http://www.seia.org/policy/finance-tax/third-party-financing (accessed March 4, 2013).

The residential and nonresidential markets include a broad range of installation sizes, from a few kilowatts (kW) to a megawatt (MW) or more. The extent and complexity of services provided, therefore, varies by installation size and individual site factors. For example, for a residential installation, site assessment may involve visiting the house or viewing it online, determining the system size by entering data into an online solar calculator, and designing a system to fit on the roof. For large nonresidential installations, on the other hand, higher-level engineering services may be required to design an optimized system. Similarly, a residential rooftop installation typically involves a small crew that installs the system in one or two days, and an electrician to connect the system to the grid and possibly supervise the installation. A large nonresidential installation may involve grading the site, fencing it, and installing a large number of pylons in the ground, as well as installing centralized inverters and a transformer, all of which may involve much more equipment and possibly a wide range of subcontractors. O&M services also vary considerably by project type. For a residential installation, services typically include monitoring the system and responding to component failures, whereas nonresidential O&M services may include more preventative maintenance.

Utility-scale Services

The development, construction, and operation of a utility project involves a number of steps. These may vary depending on factors such as whether the project is ground-mounted or roof-mounted, or whether it is on private land or public land (the latter choice may extend the project development process). However, three main services involved are project development; engineering, procurement, and construction (EPC); and O&M.

These services will be the focus of the discussion on the utility segment in this section. The following is a broad overview of these services:

- **Project development:** Project development includes a range of tasks, such as the initial site assessment, acquiring rights to the land, negotiating PPAs and interconnection agreements, permitting the project, conducting environmental impact assessments, securing financing for the project, conducting outreach to the local community, and managing the project and EPC contractor.

- **EPC:** The EPC services associated with utility projects include tasks such as fencing, grading, installing the pylons, mounting and connecting the modules, installing and connecting the wiring to the inverter, building the substation, and installing the transformer.
- **O&M:** In the utility PV sector, O&M involves monitoring the system, on-site and/or remotely. All O&M firms perform corrective/reactive maintenance (e.g., responding to equipment problems), and some firms perform preventative maintenance (e.g., cleaning, physical inspections, inverter maintenance) or condition-based maintenance (which bases maintenance priorities/tasks on the data generated through system monitoring).[8]

PV services and the associated Consumer Products Classification (CPC) codes are indicated in table 3.1.

Table 3.1. Services related to development of solar energy

		PV Services	
CPC code	Description	Residential/nonresidential	Utility
8675	Certain related scientific and technical consulting services	Site assessment	Project development-related services (e.g., site assessment, environmental impact assessment)
861, 862, 863, 8672, 8673, 9312, 93191, 932	Certain professional services, including engineering and integrated engineering services	System design	System design
633, 8861–8868	Maintenance and repair of equipment, except transport-related equipment	O&M (monitoring and corrective maintenance)	O&M (monitoring, corrective maintenance; possibly preventive or condition-based maintenance)
865	Management consulting and related services	Permitting, net metering agreement, etc.	Project development-related services (e.g., project management, PPA negotiations, interconnection agreements, permitting)
511-518	Construction and related engineering services	Installation of PV system (physically mounting the system, attaching the inverters and wiring, and connecting to the grid)	Engineering and construction (e.g., grading, fencing, installing pylons and modules, making electrical connections, substation construction)

Source: Compiled by USITC.

U.S. Market for PV Services

Market Size

Demand for installation-related services, as measured by annual U.S. PV installations,[9] increased from 160 MW in 2007 to 3,313 MW in 2012 (figure 3.1). All sectors of the market grew substantially during this period, but most of the increase was driven by the nonresidential and utility sectors. In 2012, the market was concentrated in a few states, with California, Arizona, and New Jersey accounting for a combined 65 percent of annual installations.[10]

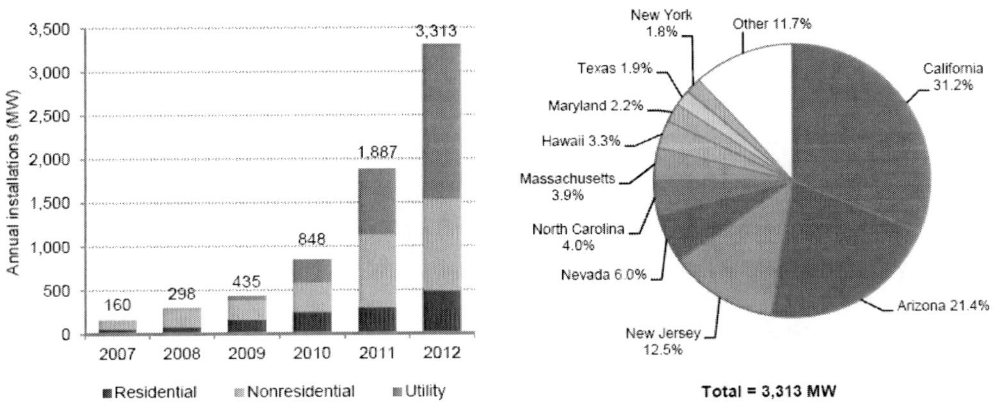

Source: GTM Research and SEIA, U.S. Solar Market Insight Report: 2012 Year-in-Review, 2013, 5–6.

Note: GTM does not include concentrating photovoltaic (CPV) systems in their PV installation data, though this segment only accounted for about 1 percent of installations in 2012.

Figure 3.1. Annual U.S. PV installations, 2007–12 (left), and installations by state in 2012 (right).

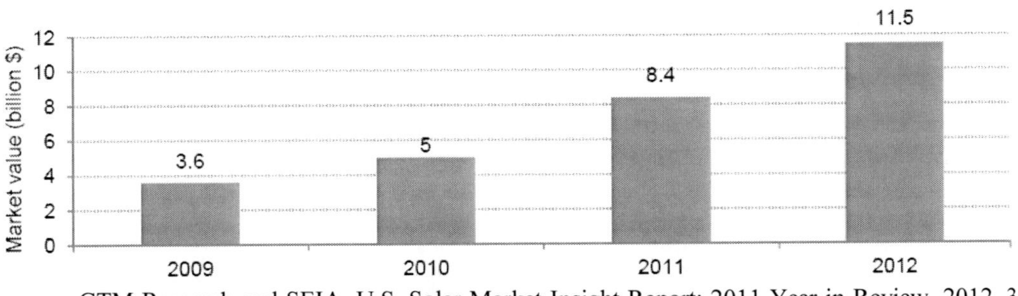

Sources: GTM Research and SEIA, U.S. Solar Market Insight Report: 2011 Year-in-Review, 2012, 3; GTM Research and SEIA, U.S. Solar Market Insight Report: 2010 Year-in-Review, 2011, 2; GTM Research and SEIA, U.S. Solar Market Insight Report: 2012 Year-in-Review, 2013, 2.

Notes: Market value based on MW installed. GTM does not include concentrating photovoltaic (CPV) systems, in their PV installation data, though this segment accounts for a small share of the market.

Figure 3.2. U.S. PV market value increased by more than 200 percent from 2009 to 2012.

The value of the U.S. market is also growing rapidly, with the value of installed systems (including both goods and services, but excluding O&M) increasing from \$3.6 billion in 2009 to \$11.5 billion in 2012 (figure 3.2).[11] The increase in the value of the market has not been as rapid as the increase in MW installed due to declining system prices within each market segment and a change in the mix of installations; utility installations, which have a lower price per watt, now account for a large share of the market.[12]

In 2011, services accounted for an estimated $3.1 billion (37 percent) of the $8.4 billion goods and services market (excluding O&M) for new installations.[13] The value of installations in each of the market segments are discussed below:

- *Residential market*: Services accounted for about $0.9 billion of the $1.9 billion residential goods and services market in 2011 (figure 3.3).[14]
- *Nonresidential*: Services accounted for an estimated $1.5 billion of the $4.0 billion nonresidential market.[15]

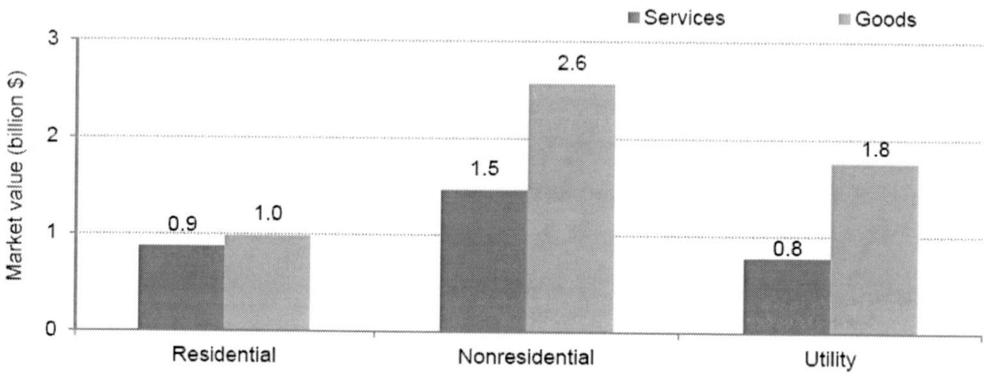

Sources: See text and footnotes 14–16.
Notes: See text and footnotes 14–16.

Figure 3.3. Services accounted for more than 30 percent of the PV installation market across all segments in 2011.

- *Utility*: Services accounted for an estimated $0.8 billion of the $2.5 billion utility market in 2011.[16]

In addition to the services associated with new installations, there is a market for ongoing O&M for all installed systems. The value of the O&M market in the United States, for all market segments combined, was likely around $0.1 billion in 2011.[17] While this is small compared to the value of services associated with new installations, the growing size of the installed PV base offers an increasing opportunity for O&M service providers.[18] Furthermore, there is anecdotal information about quality problems with some module brands, which may result in increasing demand for services to repair these systems.[19]

The amount of financing flowing to distributed installations is also on the rise. In the residential sector, third-party ownership of PV systems has quickly increased and now accounts for more than half of residential installations in some states.[20] In California, for example, third-party ownership increased from less than 10 percent of installations under the California Solar Initiative (CSI) in 2008 to almost 70 percent in 2012 (figure 3.4).[21]

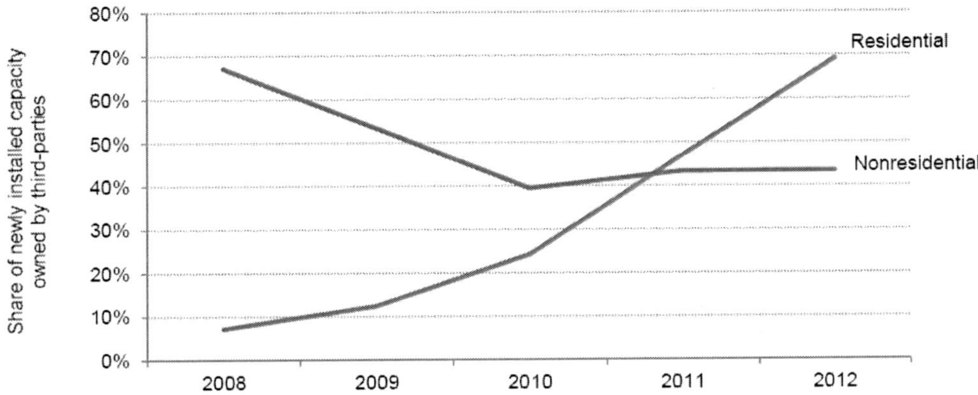

Source: CSI working data set, January 2, 2013 (accessed January 9, 2013).

Figure 3.4. Third-party ownership of residential PV systems in California has rapidly increased, while nonresidential ownership was flat during 2010–12.

Greentech Media estimated that the third-party financing market was valued at $1.3 billion in 2012.[22]

In the nonresidential sector, available data indicate that third party-ownership is not growing as quickly as a share of the market. In California, the share of newly installed third-party-owned systems fell from 2008 to 2010 and has subsequently remained relatively flat. However, the actual capacity installed in 2012 that was owned by third parties increased by 66 percent during 2008–12 due to the rapid growth in the size of the market. In addition, because of the larger size of the nonresidential market, the volume of third-party-owned residential systems installed in 2012 (in MW) exceeded that of nonresidential systems by only 13 percent.[23] The role of financing in the residential and nonresidential sectors is likely even higher than these data indicate, as they do not account for more traditional forms of financing, such as loans.

Factors Affecting U.S. Supply and Demand for PV Services

The key drivers of the U.S. PV market vary by segment, but are generally the declining price of PV installations and state and federal government policies. The market drivers are discussed in more detail below:

- *Declining cost:* The median price of distributed PV installations in the United States declined by more than 40 percent during 2002–11 (figure 3.5).[24]
- *Renewable portfolio standards (RPSs):* RPSs have been one of the main drivers of PV installations. Twenty-nine states and the District of Columbia have RPSs, which require a certain percentage of electricity to come from renewable sources by a particular date, while eight states have renewable portfolio goals. Sixteen states and the District of Columbia have specific requirements that a share of this electricity come from solar and/or other distributed energy sources.[25]
- *Other state policies:* States have implemented a number of laws, including rebates and tax incentives, that have helped spur PV installations.[26] For example, the CSI provides rebates for customers of investor-owned utilities in California, and Hawaii

offers a 35 percent tax credit (with a maximum of $5,000 for each 5 kW installed).[27] Net metering rules also play a role in enabling the growth of distributed generation.[28]

- *Federal government policies:* Another driver of PV market growth has been federal government policies. One of the primary federal government incentives for PV is the investment tax credit (ITC), which is a tax credit equivalent to 30 percent of the cost of a solar installation. For projects completed or under construction during 2009–11, developers could also opt to receive a payment equal to the amount of the ITC rather than taking the credit. [29] Many solar projects are also eligible for five-year accelerated depreciation and first-year bonus depreciation.[30]
- *Other:* A number of other factors are also contributing to the PV market growth. The increasing range of financing options may increase the deployment of PV by removing the barrier of high up-front costs, making financing easier to access and enabling some customers to use tax credits more easily.[31] Environmental concerns can help to motivate the installation of PV systems, though they are generally less important than cost.[32] Other factors include relatively short construction times, the ability to locate solar projects close to demand, the desire to minimize risks related to fossil fuel price fluctuations, excellent solar irradiation in many areas, and utilities' increasing experience and familiarity with solar and renewable energy technologies.[33]

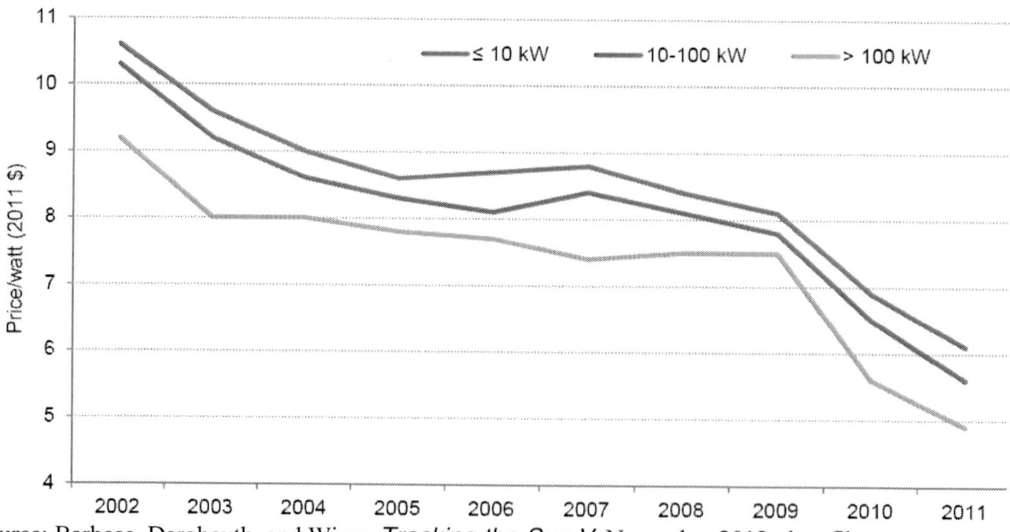

Source: Barbose, Darghouth, and Wiser, *Tracking the Sun V*, November 2012, data file.

Figure 3.5. The price of PV installations has declined more than 40 percent since 2002.

U.S. Industry Trends

Residential and Non-residential Services

The U.S. residential PV installer industry is highly fragmented and competitive.[34] More than 2,000 solar installers are active, with the top five firms combined accounting for less than one-third of the national market.[35] Many of the residential installers active in the United States are U.S.-based small and medium-sized enterprises, though there are also residential solar installation businesses belonging to larger firms, such as roofing company Petersen-Dean and Mainstream Energy's REC Solar.[36] Most firms installed less than 100 kW in the residential sector in 2012. Only about 37 firms installed 2 MW or more, but these firms accounted for more than half of all residential installations (by watts) in 2012.[37] Some residential installers are active in the U.S. nonresidential market segment, though many focus only on the residential sector. [38] Overall, the leading residential installers in the United States in 2012 appear to be U.S.-based companies SolarCity (14 percent of installations), Verengo Solar (5 percent), Trinity Solar (3 percent), RevoluSun (3 percent), REC Solar (2 percent), and Sungevity (2 percent) (figure 3.6).[39]

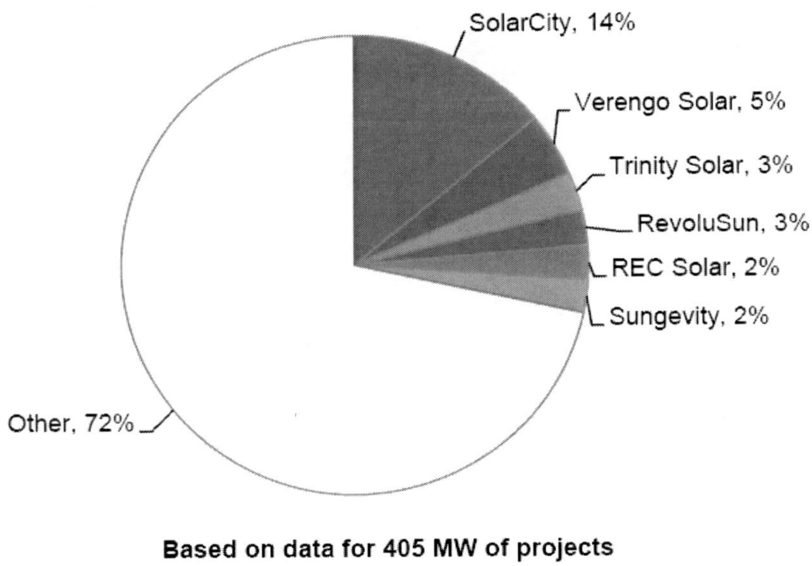

Source: Data compiled by USITC.

Figure 3.6. Approximate U.S. market shares of residential installers, 2012.

The nonresidential solar installer industry is also fragmented, with more than 1,000 nonresidential installers active in 2012. This is fewer than in the residential sector, but the top five combined accounted for less than 25 percent of installations (by watt).[40] Within the nonresidential installer industry there are also many small and medium-sized businesses, but the leading firm appears to be SunPower, a subsidiary of France-based Total that is also a large module manufacturer and service provider. CSI data indicate that for installations under that program, the number of nonresidential solar installers substantially increased during 2008–12, and the market share of the top 5 installers decreased slightly. [41] In 2012, SunPower had approximately 9 percent of the nonresidential market, followed by U.S.-based companies SolarCity (6 percent), SunEdison (3 percent), Borrego Solar Systems (2 percent), and Chevron Energy Solutions (2 percent) (figure 3.7).[42]

A number of residential and/or nonresidential installers are seeking to expand to new states or increase their competitiveness in existing states by opening new offices, acquiring other installers, undertaking mergers, and franchising.[43] SunEdison, a leading utility project developer, is entering the residential market in several states, signaling even more intense competition for installers. [44] The advantages that some large installation companies may have, such as the ability to provide financing and leverage lower equipment prices, along with their acquisitions of local firms, has led some industry representatives to predict significant consolidation among residential and nonresidential installers.[45] However, to this point there has not been a broad consolidation in the industry. In Massachusetts, for example, the number of residential installers approximately doubled from 2010 to 2012.[46] In California, the number of residential installers peaked in 2010, but the number of nonresidential installers continues to increase, and the number of residential installers remains well above the 2008 level.[47]

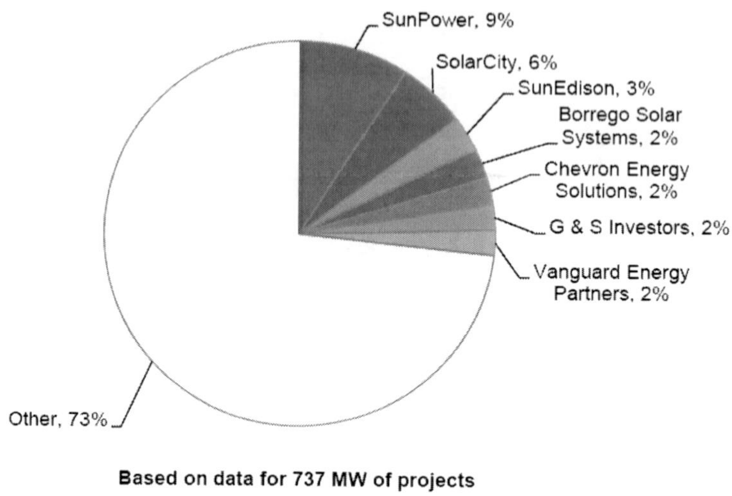

Source: Data compiled by USITC.

Figure 3.7. Approximate U.S. market shares of nonresidential installers, 2012.

A number of factors may account for the lack of significant consolidation to this point, including continued expansion of the PV market, the ability of solar installers to access financing from third parties and/or their continuing ability to compete for customers that prefer to own systems, and the ability of residential installers to adapt by, for example, contracting with larger firms. [48] Further, barriers to entry are low, [49] and the most prominent theme that emerges from a review of installers active in California is not consolidation but turnover, with a significant number of firms entering and exiting the industry each year.[50]

Third-party ownership in the residential sector is fairly concentrated among a few firms, but competition in this sector is increasing as more firms enter to offer their own financing options.[51] In the first quarter of 2012, for example, six firms accounted for at least 90 percent of third-party-owned residential systems installed in California (figure 3.8).[52] Similarly, in Maryland four companies accounted for all third-party-owned residential systems in 2012.[53] According to Greentech Media, the firms that have raised the most money nationally to finance residential third-party-owned systems include SolarCity, SunRun, SunPower, Clean Power Finance, and Sungevity.[54]

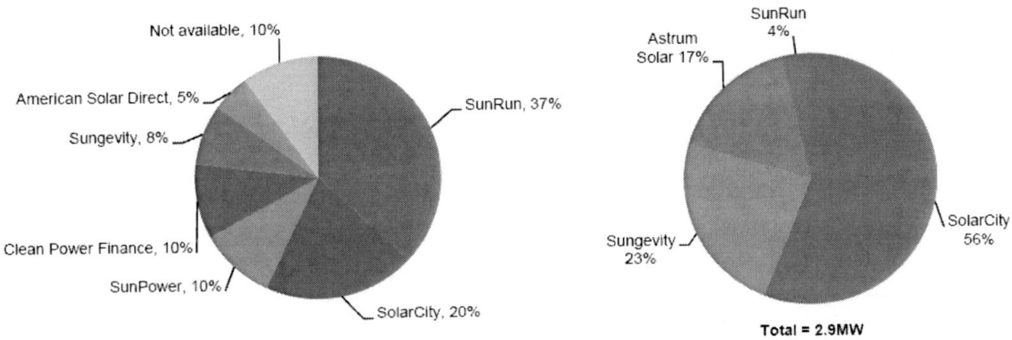

Sources: Wesoff, "Borrego Joins," July 31, 2012; Maryland Residential Clean Energy Grant Program, Historic Award Data Report Excel worksheet (accessed January 17, 2013).

Note: Total for California was not published by Greentech Media.

Figure 3.8. Residential third-party ownership industry was fairly concentrated in California (left) in the first quarter of 2012 and in Maryland in full year 2012 (right).

Utility-scale Services

The utility PV project development industry is composed of (1) firms whose primary business is project development (e.g., Strata Solar);[55] (2) firms that are engaged in both producing equipment (e.g., modules) and developing projects (e.g., First Solar, SunPower, and SunEdison);[56] (3) unregulated entities related to major utility companies (e.g., Sempra); [57] (4) other independent power producers (IPPs) [58] (e.g., Cogentrix, Constellation Energy, and NRG Energy); (5) utilities (e.g., Portland General Electric); and (6) other firms (e.g., a subsidiary of Alexander and Baldwin, a Hawaii real estate company).[59] The largest share of the market, through 2011, was accounted for by firms that also produce PV equipment, including SunEdison,[60] SunPower, and First Solar.[61] For these companies, low prices for equipment and related inputs have led to a significant change in how they view project development. While many of them expanded into project development, at least in part, to provide a path to the market for their equipment, they increasingly view project development as an area of profit; in one case, a firm now views the equipment side of its business as supporting project development.[62]

Utility project development is more highly concentrated among a smaller number of firms than is the case for residential or nonresidential installations, with the top 5 utility project developers accounting for 59 percent of U.S. projects completed in 2012 (figure 3.9). The top two project developers in 2012 were likely U.S.-based firms First Solar and Sempra, followed by three firms based outside of the United States—SunPower (France), EDF (France), and GCL Solar Energy (China), and then U.S-based SunEdison.[63] However, project development data can be highly influenced by the timing of project completion, so firms' positions in annual rankings can fluctuate significantly. In 2007–11, the leading large project developers, in descending order, were likely SunEdison, SunPower, First Solar, and Sempra, which accounted for a combined 28 percent of large projects completed.[64] More than 60 firms completed U.S. solar PV utility projects in 2012, but this only captures a portion of the industry, as there are other firms actively developing projects in the United States.[65]

The utility PV market has only recently emerged as a large market segment, and as firms have entered this market segment, they have employed a number of different approaches. The services provided may vary by company or even by project. Project developers may perform

EPC services on projects they develop themselves, but also on projects developed by other firms. The developers offering in-house EPC services tend to be PV- specific companies (whether manufacturers/developers or solely developers), while IPPs and unregulated entities related to utilities rarely provide EPC. Most of the project developers that do provide EPC services subcontract some or all of the actual construction. Large contractors usually do not develop projects, with some providing all three EPC services, others only construction services. Some firms handle all of the construction in-house, but many subcontract some of the construction.[66]

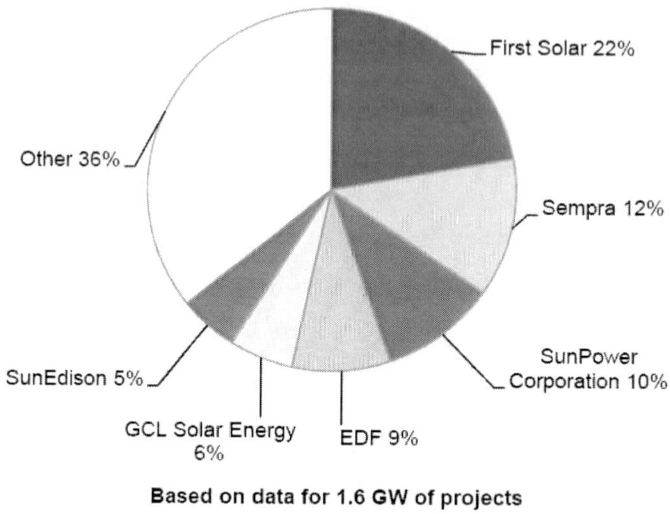

Source: Data compiled by USITC.

Note: U.S.-based firms (First Solar, Sempra, and SunEdison) are in blue, and firms based outside of the United States (SunPower, and EDF, GCL Solar Energy) are in orange.

Figure 3.9. Approximate U.S. market shares of PV utility project developers, 2012.

First Solar was the leading EPC provider for projects completed in 2012, with other leaders including Bechtel, Hanwha Q-Cells, juwi, Quanta Services, Strata Solar, SunPower, Swinerton, and Zachary Holdings.[67] In the utility sector, there is significant competition for small construction projects, but much less for large projects.[68] Price is an important competitive factor, but firms selecting a contractor also consider experience, innovation, timeliness, and ability to add value to the project.[69]

O&M services are often provided by the project developer, EPC contractor, or IPP. The leading providers of such services for projects completed in 2012 included Con Edison, EDF, First Solar, Hanwha Q-Cells, juwi, NRG, Strata Solar, SunEdison, SunPower, and Swinerton.[70] However, there are also independent providers of these services. True South Renewables, for example, reports that it provides O&M for more than 200 MW worth of installed power, including utility PV projects and distributed facilities.[71]

Services Employment

U.S. employment in the solar industry—including concentrated solar power and solar heating and cooling—totaled about 119,016 in 2012, according to The Solar Foundation, up from 105,145 in 2011. Of this total, installation had the largest share (57,177 employees, 48 percent), followed by manufacturing (29,742 employees, 25 percent), sales and distribution (16,005 employees, 13 percent), project development (7,988 employees, 7 percent), and other positions (8,105 employees, 7 percent). Employment in installation, sales and distribution, and other positions all increased from 2011 to 2012, including an increase of more than 10,000 in installation.[72]

Global Market for PV Services

Market Size

Global PV installations rapidly increased during 2007–12, rising from 2.5 GW to about 29.1 GW (figure 3.10). While Europe continues to account for the majority of demand, most of the demand growth since 2010 has been driven by markets outside of Europe, particularly the United States and the Asia-Pacific. The largest markets in 2012 were Germany (7.6 GW, 26 percent of installations), China (3.5 GW, 12 percent), Italy (3.4 GW, 12 percent), and the United States (3.3 GW, 11 percent).[73]

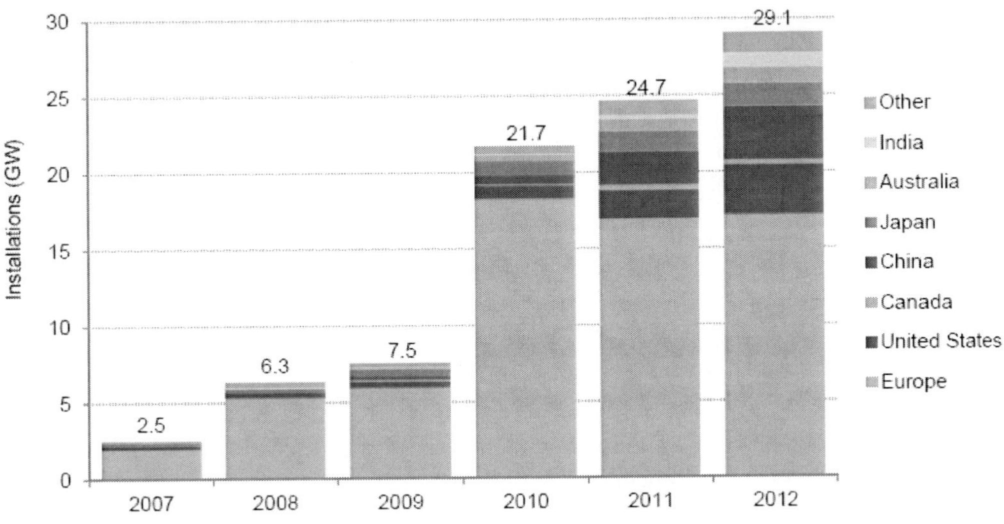

Sources: EPIA, Global Outlook for Photovoltaics until 2016, May 2012, 19, 66–67; EPIA, Global Market Outlook for Photovoltaics 2013–2017, 2013, 16, 18, 31; GTM Research and SEIA, U.S. Solar Market Insight Report: 2012 Year-in-Review, 2013, 5; JPEA, Statistics on Shipments, February 26, 2013, 5; METI, "Announcement Regarding the Present Status," March 13, 2012; Watt and Passey, "PV in Australia 2012," May 2013, viii; Energy Foundation and CREIA, China Solar, April 2013, 5–6.

Notes: Estimates of the size of the global PV market in 2012 range from 29 GW to 32 GW. The data included here are based primarily on data from EPIA. For 2007–11, data for Europe are estimated PV installations. For 2012, data are grid-connected capacity. In addition, for 2012 USITC staff updated data for several non-European countries, including Australia, China, Japan, and the United States.

Figure 3.10. Annual global PV installations rapidly increased during 2007–12.

The global PV market (i.e., the value of new installations), including goods and services, was an estimated $92 billion in 2011, with the services share accounting for approximately $34 billion. Italy was the largest services market ($9.8 billion in terms of services associated with new installations, 29 percent of the total services market),[74] followed by Germany ($5.1 billion, 15 percent), the United States ($3.1 billion, 9 percent), and Japan ($3.0 billion, 9 percent).[75] The services market has increased substantially over time. Clean Edge estimated the value of the entire market, including goods and services, at $20.3 billion in 2007, implying a services market of less than $10 billion based on the value of the module and inverter markets at that time.[76]

Factors Affecting Global Supply and Demand for PV Services

The primary factors affecting global demand for renewable energy services are price and government policies.

- **Price:** The price of PV installations is declining, thereby increasing the cost competitiveness of PV systems in both developed and developing countries. Prices in Germany, for example, declined 66 percent from the second quarter of 2006 to the fourth quarter of 2012.[77]
- **Market growth policies:** Government policies aimed at directly reducing the price of PV electricity, or mandating its use or purchase, have had a significant impact on the global PV market. Feed-in tariffs (FITs), which guarantee the purchase of renewable energy at a set price for a period of time, are widely used in Europe and, more recently, in countries such as China and Japan. Many countries now require that a certain percentage of electricity generation come from renewables, and some countries have tender processes to procure solar or other forms of renewable energy. However, sudden shifts in FIT levels, often in response to a surge in installations, have also led to significant contractions in demand in key markets.[78]
- **Market-enabling policies:** The implementation of policies that make PV installations practical, such as net metering, also drive market growth. Fourteen countries have national net metering policies, while other countries have such policies at the subnational level.[79]
- **Other:** Market drivers also include factors such as rising electricity demand, particularly in developing countries.[80]

Global Industry Trends

The global large-scale[81] project development, EPC, and O&M industries are highly fragmented and tend to be dominated by firms from the three largest markets, Europe, China, and the United States:

- **Project development:** Among project developers, the top 10 firms combined accounted for less than one-third of the market during 2012. U.S. firms had a relatively strong presence near the top of the rankings, with First Solar ranked first, SunEdison third, and Sempra sixth. France-based EDF was the second-largest developer globally, with China-based Talesun Solar and China Solar ranked fourth and fifth.[82]

- **EPC:** The leading global EPC firms (for projects over 10 kW) in 2012 were U.S.-based First Solar and SunEdison, with steady foreign project activity and expanding U.S. domestic activity propelling them to the top spots (table 3.2). Chinese and European firms are also strongly positioned in the top 15, with Chinese firms accounting for 6 of the top 15 EPC firms and European firms for 5 of the top 15. However, the market share of the leading global EPC firms is relatively small, including only 2 percent for leading firm First Solar and a combined 24 percent for the top 30 firms.[83]

Table 3.2. U.S., European, and Chinese firms were the leading global EPC firms in 2012

Company	Headquarters	2012 Rank	2011 Rank
First Solar	United States	1	3
SunEdison	United States	2	4
BELECTRIC	Germany	3	1
China Power Investment Corp.	China	4	2
juwi	Germany	5	6
Enerparc	Germany	6	15
EDF Energies Nouvelles	France	7	13
TBEA SunOasis	China	8	7
GD Solar	China	9	10
Jiangsu Zhenfa New Energy	China	10	11
Larsen & Toubro	India	11	>15
SunPower	France[a]	12	5
China Guangdong Nuclear Development	China	13	9
Swinerton, Inc.	United States	14	>15
Shanghai Solar Energy Co., Ltd.	China	15	>15

Source: IMS Research, "First Solar Ranked," March 27, 2013.

Notes: The ranking is based on MW of projects completed. First Solar installed more than 500 MW and SunEdison 390 MW, according to IMS Research.

[a]SunPower is based in the United States, but a majority of its shares were acquired by France-based Total in 2011 and 2012.

- **O&M:** Available data indicate that there are many firms providing O&M services, including many of the EPC firms and project development firms discussed above.[84] It is probable that among the leading providers are firms like Germany-based BELECTRIC and juwi, and U.S.-based First Solar and SunEdison.[85]

Trade and Investment

Imports and Exports

In the United States, residential systems tend to be installed primarily by domestic firms, so U.S. imports of residential installation services are relatively limited. [86] Firms performing U.S. residential installations, including most firms that installed at least 2 MW in 2012, are generally based in the United States. [87] The majority of U.S. nonresidential systems were also installed by U.S.-based companies, but there is substantial participation in this sector by foreign firms, and imports accounted for more than 10 percent of nonresidential installations in 2012. The leading nonresidential installer in the United States is SunPower, a subsidiary of France-based Total. A number of other foreign firms installed nonresidential systems in the United States in 2012, including Germany-based Conergy and Gehrlicher, and Portugal-based Martifer.[88]

Third-party ownership of U.S. residential projects was primarily by domestic firms.[89] In California, in the first quarter of 2012, for example, the only known imports were the 10 percent of the market accounted for by SunPower.[90] In Maryland, all third-party financing was provided by U.S.-based firms.[91] SunPower is likely one of the top providers of residential third-party financing nationally, and there are other non-U.S.- based firms in this sector, but there are no data indicating that any non-U.S. firm other than SunPower accounts for a significant share of the market.[92] In the nonresidential sector, comprehensive data on third-party financing are not available, though there are at least some imports from SunPower and other firms.[93]

Utility-scale projects in the United States have the most foreign participation among the three sectors, but the majority of services in this sector are also provided by U.S. firms. U.S. firms developed more than 65 percent of the PV projects completed in 2012, and were the EPC supplier/contractor for almost 80 percent of the projects completed—with EPC responsibilities for another 9 percent split between a foreign and domestic firm. U.S.-based firms were also contracted to provide O&M services for more than 60 percent of the projects completed. Another 13 percent of O&M services were split between a foreign and a domestic firm, but were expected to be solely provided by the U.S. firm by the third year of plant operation.[94]

In the residential and small nonresidential market segments, U.S. firms are starting to expand into markets outside the United States, though the value of exports is likely relatively small. Sungevity, for example, has expanded into Australia and the Netherlands.[95] SolarCity also has long-term plans to expand internationally, and has established a presence in Ontario, Canada.[96] However, since data on services exports are not available, it is difficult to compare trends over time.

Existing U.S. project developers are increasingly exporting project development services, with some expanding in Canada and many firms looking to expand beyond the foreign markets in which they have traditionally been active.[97] U.S. firms are increasingly interested in providing project development services in emerging PV markets like Chile and South Africa, and have large project pipelines in some of these countries.[98] For example, of SunEdison's 827 MW backlog of projects (projects with a signed offtake agreement such as a PPA, as of February 2013), 18 percent (150 MW) were in Canada and 16 percent (132 MW) were in emerging markets.[99] First Solar has shifted its longterm strategy to focus on markets in "the Americas, Asia, the Middle East, and Africa" where it expects electricity demand to increase and PV to be price competitive.[100] Other firms target one specific market, such as Penn Energy Renewables, which concentrates on the Ontario market.[101] Some U.S. project developers also provide EPC services in foreign markets for other project developers, as well as O&M services after project completion. While the increase in exports by existing U.S. firms is significant, and many of these firms do have growing foreign project development activity, their foreign installations remain only a small share of the global market.[102]

U.S. firms that provide only EPC and/or construction services[103] primarily export to Canada. For example, White Construction completed two 24 MW projects in Canada, and is the EPC contractor for two 10 MW projects currently under construction.[104] Similarly, Swinerton was contracted to provide EPC services for 30 MW of the Waubaushene project in Canada, and Signal Energy worked on a 30 MW project in Ontario.[105] Some firms are looking to further expand their operations in Canada, while others are looking at emerging markets like the Middle East.[106] When combined with exports of EPC services by the project developers discussed above, U.S. EPC exports are significant. However, the presence of U.S. firms in the global market should be kept in context—U.S. firms accounted for only 3 of the top 15 global EPC firms in 2012, as indicated in table 3.2 above.[107]

U.S.-based global firms like CH2M Hill and Black & Veatch also provide a variety of services worldwide in support of the project development process. For example, for an 80 MW project in Canada, Black & Veatch provided "design and engineering review, estimation of solar production, support for contract negotiation, construction monitoring and assistance with commissioning and performance testing."[108] For a 55 MW project in Thailand, Black &Veatch performed the "overall design and constructability review, operational analysis and detailed design review," as well as provided "information about procurement, construction, construction management, and consulting and planning."[109] Similarly, CH2M Hill provides a range of services, including "site evaluation, permitting, design, engineering, procurement, construction, startup, and commissioning services" and has provided project management in markets such as India.[110] More narrowly focused firms are also exporting globally. For example, 3TIER, which offers detailed solar forecasting and analysis, provides services in a number of foreign markets.[111] True South Renewables, which has an office in Ontario, Canada, provides final testing and inspection before commissioning of new plants, including for a recently completed 10 MW project in Ontario, and offers O&M at existing plants.[112]

Investment

There has been substantial foreign investment in the U.S. PV services industry. The largest transaction was France-based Total's 2011–12 $1.5 billion acquisition of 66 percent of the shares of U.S. manufacturer and service provider SunPower.[113] Several foreign equipment providers have acquired U.S. project developers, reflecting the trend toward vertical integration into the downstream market noted earlier. The largest, in terms of value, was Japan-based Sharp's $305 million acquisition of U.S. project developer Recurrent Energy in 2010.[114] Similarly, China-based LDK Solar acquired 70 percent of Solar Power, Inc., in 2011 for $33 million, and OCI Co., based in the Republic of Korea (Korea), acquired CornerStone Power Development.[115]

U.S. firms have acquired other U.S. domestic service providers, but have generally not used acquisitions to enter foreign markets. Among the few foreign acquisitions was First Solar's purchase of Chilean project developer Solar Chile in early 2013, and residential services firm Sungevity's investment in a Dutch firm in 2011.[116] SunEdison purchased German firm Business Institute Solar Strategy GmbH in 2009, which was developing 38 MW of PV projects at the time of the acquisition.[117]

Trade Barriers

PV firms indicated that local-content requirements, which mandate the local sourcing of goods and/or services, are the most significant trade barrier in the industry. Local-content requirements exist or have been used in markets such as South Africa and Ontario, Canada.[118] India implemented local-content requirements for projects constructed using c-Si products as part of its national solar program, the Jawaharlal Nehru National Solar Mission (JNNSM), though these rules do not necessarily apply to projects constructed under state solar programs. Batch 1 of the first phase of the JNNSM required locally sourced modules, while batch 2 required both c-Si cells and modules to be sourced locally.[119] Other countries, such as France, Greece, Italy, and Turkey, provide incentives for using local or regional content rather than mandating the use of local content.[120] For example, Turkey's FIT[121] is up to 50 percent higher for systems using local content.[122] In Brazil, local content is not mandated for PV projects, but firms that want to access lower financing rates from Brazil's national development bank, Banco Nacional de Desenvolvimento Econômico e Social (BNDES), have to meet the bank's local-content requirements.[123]

Local-content requirements for PV projects have been or are the subject of World Trade Organization (WTO) dispute settlement cases. In May 2013, the WTO Dispute Settlement Body adopted a WTO Appellate Body report and panel report (as modified by the Appellate Body) finding that the Canadian measures were inconsistent with the Agreement on Trade-Related Investment Measures (TRIMs Agreement) and the General Agreement on Tariffs and Trade (GATT) 1994.[124] Subsequently, Ontario's Minister of Energy directed the Ontario Power Authority not to procure any additional large projects. The Minister indicated that the government intends to replace the FIT program for large projects with a competitive procurement process, and that the Ministry intends to pursue legislation to bring the FIT program into compliance with the WTO ruling.[125] The United States filed a request for dispute settlement consultations with India on February 6, 2013.[126] In regard to European Union (EU) member states, China filed a request for dispute settlement consultations with the EU, as well as EU member states Greece and Italy, on November 5, 2012.[127]

Some local-content incentives/requirements apply to goods, while others apply to both goods and services. In Ontario, Canada, for example, the rules specify that for projects over 10 kW in size, 60 percent of the content must be sourced from within Ontario. In calculating this percentage, both goods and services are taken into account: Ontario identifies the qualifying percentage for each major piece of equipment and for the major services associated with solar installation. Services associated with crystalline silicon (c-Si) projects account for only 22 percent of the project value and those associated with thin-film projects for 28 percent of the value, so local sourcing of equipment is necessary to meet the 60 percent local-content requirement.[128]

Since several of the major U.S. utility project developers are also equipment providers, the requirement to source equipment locally may affect their ability to also provide services in the market. First Solar, for example, only develops projects with its own equipment and does not have a manufacturing plant in Ontario. As a result, the company indicates that its systems do not meet local-content requirements, and since First Solar provides services only in association with its own equipment, it is also excluded from providing services in Ontario.[129] SunEdison contracted with Flextronics to assemble modules locally in order to meet Ontario's local-content requirements, but the ease with which companies can adapt to these requirements by establishing local production or sourcing local products varies by firm and technology type.[130]

Standards and certification issues can also serve as a barrier to PV module trade, and therefore to U.S. exports of services. One example, according to the Office of the U.S. Trade Representative (USTR) and industry representatives, is Korea's certification requirement for PV modules.[131] Korea's PV standards cover all types of PV modules and, therefore, all types of modules can be sold in the Korean market. However, in order to qualify to be used in government programs, modules must be certified by the Korea Energy Management Corporation (KEMCO), and certification standards only exist for c-Si modules and amorphous silicon (a-Si) thin-film modules. Korea recently conducted an environmental assessment of copper indium gallium (di)selenide (CIGS) and cadmium telluride (CdTe) thin-film modules, and did not find significant concerns with CIGS modules. As a result, Korea is considering developing a standard for CIGS modules, though a Korean government representative indicated in 2012 that this would likely take two years. However, the Korean assessment found environmental concerns with the cadmium used in CdTe modules and is not planning to develop a standard for CdTe— though industry representatives and the U.S. Trade Representative's *2013 Report on Technical Barriers to Trade* state that there were methodological issues with the Korean environmental assessment. Certification rules in Korea thus effectively prevent U.S. firms that make CdTe from selling equipment, and therefore services, in much of the Korean market.[132] Until certification standards are developed, the same is true of firms that make CIGS modules.

The lack of harmonization of PV standards globally can also negatively affect trade in PV modules, with testing and certification requirements for entering new markets resulting in significant costs for manufacturers.[133] In addition, while in some countries certification by international testing bodies may be allowed, testing by organizations within the country may aid product acceptance in the local market.[134]

Country and Regional Profiles

This section profiles the largest market in 2012 (Germany), the largest non-European markets excluding the United States (China and Japan), and the emerging Canadian and Latin American markets (figure 3.11). These markets exhibit a variety of different characteristics, as will be discussed below. For example, Germany is a large, established market where demand has been flat the last few years, and there is extensive services competition in the domestic market. Latin American markets are small, but have significant growth potential and provide substantial export opportunities for U.S. firms. Government policies contribute to demand growth in many of the markets, but in parts of Latin America PV is becoming price competitive even in the absence of policy support.

Canada

Canada has a growing PV market, with annual installations increasing from 5 MW in 2007 to 268 MW in 2012. Utility installations accounted for 62 percent of newly installed capacity in 2011. Canadian system installation businesses generated $656 million in domestic revenue in 2011, and employment in the Canadian PV services sector (including distribution, installation, utilities, and government, other than research and development) was 2,210 employees. Demand in Ontario, the province that accounts for the majority of Canadian installations, is primarily driven by policies within the province.[135] In 2006, Ontario started the Renewable Energy Standard Offer Program (RESOP),which set a price for solar energy in the province and led to a number of utility-scale solar projects.[136] In 2009, Ontario replaced this program with the FIT Program [137] for projects over 10 kW, and the microFIT Program for projects 10 kW or less.[138]

U.S. firms are providing development, EPC, and O&M services in Canada, though the local content requirements discussed earlier are a barrier to entry for some of the firms that also produce equipment. U.S. project developers such as First Solar, Penn Energy Renewables, and SunEdison—some of which also provide EPA and O&M—are active in developing projects in Canada, though First Solar has not been awarded any contracts under the current Ontario FIT program.[139] Similarly, firms that do not engage in project development but provide construction and/or EPC services—such as Signal Energy, Swinerton, and White Construction—are active in the Canadian market, as noted earlier.

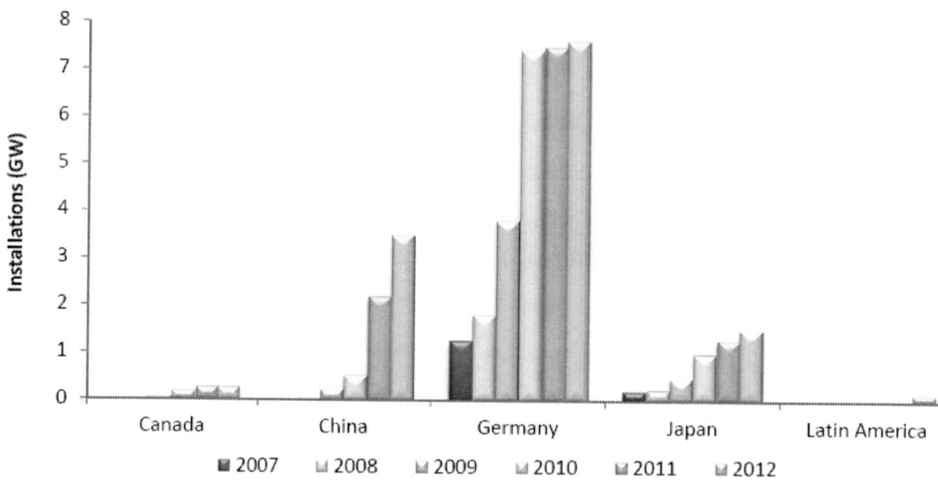

Sources: Poissant et al., "National Survey Report of PV Power Applications in Canada 2011,"June 2012, 8; EPIA, *Global Outlook for Photovoltaics until 2016*, May 2012, 66–67; EPIA, Global Market Outlook for Photovoltaics 2013–2017, 2013, 18, 31; BSW-Solar, "Statistic Data," February 2013, 1; JPEA, *Statistics on Shipments*, February 26, 2013, 5; METI, "Announcement Regarding the Present Status," March 13, 2012; Energy Foundation and CREIA, *China Solar*, April 2013, 5–6.

Note: Data for Latin America prior to 2011 were not published by EPIA.

Figure 3.11. Installations in profiled markets, 2007–12.

China

China's market is growing rapidly, increasing from 20 MW in 2007 to at least 3.5 GW in 2012, with one of the main drivers the implementation of a FIT in 2011.[140] The value of the services market (excluding O&M) totaled around $1.6–$2.5 billion in 2012.[141] The value of the services market is relatively small compared to other large markets, due to the low installation costs in China.

Project development and EPC services in China are primarily provided by Chinese firms, with many project developers providing in-house EPC in order to ensure quality. Chinese module suppliers are also increasingly providing EPC services in China.[142] There is, however, some participation in the Chinese market by foreign firms. German firm Wirsol Solar, for example, announced plans to develop PV projects in China with Suntech.[143] France-based SunPower formed a joint venture to manufacture and install its systems in China.[144] U.S.-based First Solar is awaiting approval to build a 30 MW project in China and has a memorandum of understanding to develop a 300 to 500 MW second phase of the project. [145]

U.S.-based Solaria, which is also an equipment producer, provides construction services for projects in China.[146]

Germany

Germany is the leader in global cumulative installed PV capacity, accounting for 35 percent of cumulative global installations through the end of 2011.[147] Annual installations increased from 1.3 GW in 2007 to 7.4 GW in 2010, but have been relatively flat since 2010 (figure 3.11). In 2011, installations totaled 7.5 GW and in 2012, installations totaled 7.6 GW.[148] The value of the German PV services market totaled an estimated $5 billion in 2011.[149] The main drivers of German demand are a FIT and the declining price of PV installations, which has continued to make system installation profitable and propel market growth despite cuts to FIT rates.[150]

Germany has a large services industry, with three of the top six global EPC services firms in 2012 based in Germany (BELECTRIC, juwi, and Enerparc).[151] It is likely that these three firms are also among the leading global O&M service providers, with more than 2 GW currently under management.[152] German PV installers also are the leading service providers in their home market. In 2010, the top 10 system integrators in Germany were all German-based firms.[153] German PV installers deliver some of the lowest installed costs in the world in their home market.[154] The industry, however, may face a period of consolidation if projections of German market contraction are accurate.[155] Of those top 10 system integrators in Germany in 2010, three have filed for bankruptcy protection, though two of these filings were also related to the equipment side of their businesses.[156] As a result of the market conditions in Germany, U.S. project developers and EPC firms are focusing on opportunities outside Germany, despite the significant size of the German market.

Japan

Japan was the first country to widely deploy PV systems, and was the world's fifth-largest market in 2012. The current incentive program for PV, a FIT that was implemented in July 2012, is contributing to further market growth. Japanese PV installations increased from 210 MW in calendar year 2007 to 1,559 MW in the first 11 months of the fiscal year ending March 2013. The Japanese market is dominated by residential installations, which accounted for 73 percent of installations during the first 11 months of the fiscal year. There have been many announcements of large projects in the utility sector, but it is unclear if this sector will be as big as expected and if demand in this sector will be sustainable, given Japan's land constraints.[157]

Japan's PV services market was valued at $3.0 billion in 2011. Japanese equipment makers are strongly positioned in the residential installation market, and there are home builders who install a large number of systems on new homes.[158] While domestic manufacturers also provide residential installation services, foreign module suppliers that enter the Japanese residential market typically do so through a local partner, usually either a distributor or installer.[159] The nonresidential sector includes participation by some of the same firms as the residential sector, though in this sector firms may contract out the EPC services rather than provide them in-house.[160]

Project development for the utility market in Japan is highly competitive and is currently dominated by domestic firms, which include subsidiaries of major utilities, PV manufacturers, and trading companies.[161] Foreign firms such as Germany-based juwi and

BELECTRIC, Spain-based Gestamp, and U.S.-based SunEdison are active in the market, typically entering through a joint venture or a partnership with a local firm, in part because local firms have a better understanding of Japanese practices, laws, and language.[162] Japan has a large domestic EPC industry, so these services are mostly provided by local companies.[163] A range of firm types currently provide O&M or are entering the market to provide these services, such as equipment providers and joint ventures between foreign project developers and Japanese firms.[164]

Latin America

Latin American markets are small at present, but substantial growth is taking place across many countries in the region, driven by factors such as declining PV equipment prices, rising electricity demand, the increasing cost competiveness of PV-generated electricity, excellent solar resources in many locations, and government policies. [165] While cumulative installed capacity was only about 91 MW at the end of 2012, the European Photovoltaic Industry Association (EPIA) forecasts substantial growth in the region.[166] Despite this potential growth, barriers to further PV deployment in the region—such as difficulties financing projects, a lack of sufficient policy support in some countries, and insufficient grid infrastructure—remain.[167]

The earliest entrants to project development in Latin America were mostly non-U.S. firms, though only a small number of projects were installed through the end of 2012.[168] However, U.S. project developers are increasing their presence in Latin America, and have initially been most active in Chile, where rising demand, excellent solar resources (especially in the Atacama Desert in northern Chile), and the cost competitiveness of PV-generated electricity make solar energy attractive. Among the U.S. firms active in developing projects in this market are AES, First Solar, and SunEdison. Some U.S. firms also plan to provide EPC and O&M services.[169]

U.S. project developers are also looking at other markets across the region, including Brazil, Central America, Mexico, and Peru.[170] Peru had the most installations in Latin America through the first quarter of 2013, with 84 MW completed in the fourth quarter of 2012 and the first quarter of 2013 combined, but up to this point large-scale project development in the Peru market has been done by Spanish firms. Mexico has a small distributed PV market, and a number of utility projects have been announced, but as in Peru, most of the companies involved are European firms.[171] Brazil recently introduced net metering, which may increase the size of the distributed market, though the industry continues to face challenges, such as high equipment prices. The Brazilian utility PV market has been slow to develop, reflecting factors such as difficulty competing with low wind and hydro prices, as well as the need to meet local-content requirements for BNDES financing. The market is growing, however, and prices are becoming more competitive. A number of large projects have been announced (with several related to World Cup venues), including a 1.1 MW project that will be built by U.S.-based developer SunEdison.[172]

References

Ahearne, Jack. "South African PV Market in Position to Promote Local Manufacturing Growth." *Renewable Energy World*, June 1, 2012. http://www.renewableenergy world.

com/rea/partner/first-conferences/news/article/2012/06/south-african-pv-market-in-position-to-promote-local-manufacturing-growth.

Alexopoulou, Maria. "Solaria Corporation to Construct Large-scale PV Project in China." *PV-Tech*, May 13, 2013. http://www.pv-tech.org/news/solaria corporation to construct large scale pv project in china.

Ali-Oettinger, Shamsiah. "China: Wirsol to Develop Projects with Suntech." *PV Magazine*, March 29, 2012. http://www.pv-magazine.com/news/details/beitrag/china--wirsol-to-develop-projects-withsuntech 100006267/#axzz2Uc1NkMT2.

Ardani, Kristen. "Benchmarking Soft Costs for PV Systems in the United States." Presentation for the ASES World Renewable Energy Forum, May 17, 2012. http://www.nrel.gov/docs/fy12osti/54689.pdf.

———. "Soft Costs in the U.S. Solar Markets–Survey Results." Presentation for Vote Solar, November 8, 2012. http://votesolar.org/wp-content/uploads/2012/11/NREL SoftCost Webinar.pdf.

Ardani, Kristen, and Robert Margolis. *2010 Solar Technologies Market Report*. U.S. Department of Energy, November 2011. http://www.nrel.gov/docs/fy12osti/51847.pdf.

Ayoub, Josef, Sylvain Martel, and Lisa Dignard-Bailey. "National Survey Report of PV Power Applications in Canada." Department of Natural Resources Canada. Prepared for International Energy Agency Co-operative Programme on Photovoltaic Power Systems, May 2008. http://www.iea-pvps.org/index.php?id=93.

B-GC. "Interview with Jared Haines, President, Mercury Solar Systems." B-GC, October 12, 2009. http://www.mercurysolarsystems.com/pdf/BGreen%20Collaborative%20'JH %20 Interview'%201 0-12-09.pdf.

Barbose, Galen, Naïm Darghouth, and Ryan Wiser. *Tracking the Sun V: An Historical Summary of the Installed Price of Photovoltaics in the United States from 1998 to 2011*. Lawrence Berkeley National Lab. LBNL-5919E, November 2012. http://eetd.lbl.gov/sites /all/files/publications/lbnl5919e-report.pdf.

———. *Tracking the Sun V: An Historical Summary of the Installed Price of Photovoltaics in the United States from 1998 to 2011*. Data file, Lawrence Berkeley National Lab, LBNL-5919E, November 2012. http://emp.lbl.gov/sites/all/files/lbnl-5919e-data.xls.

Berrill, Paul. "SunEdison Bumps up European Expansion with German Purchase." *Recharge*, January 22, 2009. http://www.rechargenews.com/solar/article1277749.ece.

BNEF database. http://www.bnef.com (accessed various dates). Bosworth, Melissa. "Split Ends." *Photon*, April 2012, 54–65.

Bosworth, Melissa, and Matthew Hirsch. "One Size Fits All." *Photon*, April 2011, 66–77.

California Solar Initiative (CSI). CSI working data set, January 2, 2013. http://www.californiasolarstatistics.ca.gov/current data files/.

Cameron, Christopher P., and Alan C. Goodrich. "The Levelized Cost of Energy for Distributed PV: A Parametric Study," n.d. http://energy.sandia.gov/wp/wp-content/gallery/uploads/LevelizedCostDistPV.pdf. (accessed March 5, 2013).

Canadian Solar. "Canadian Solar Announces Joint Venture and Distribution Deal with West Holding Ltd of Japan." News release, January 14, 2010. http://www.prnewswire.com/news-releases/canadian-solar-announces-joint-venture-and-distribution-deal-with-west-holding-ltd-of-japan-81475377.html.

Climate Change Business Journal. "Bechtel Applies Integrated Skills in Power Engineering to Utility-Scale Solar Market." Fall 2012, 1–4.

http://bechtel.com/assets/files/Bechtel Climate Change Journal Ivany Fall2012.pdf.

Cocke, Sophie. "RevoluSun Plans Mainland Expansion." *Pacific Business News*, February 11, 2011. http://www.bizjournals.com/pacific/print-edition/2011/02/11/revolusun-plans-mainland-expansion.html?page=all.

Commonwealth of Massachusetts. Executive Office of Energy and Environmental Affairs. RPS Solar Carve-Out Qualified Renewable Generation Units data. http://www.mass.gov/eea/energyutilities-clean-tech/renewable-energy/rps-aps/qualified-generation-units.html (accessed January 16, 2013).

Council on Energy, Environment and Water (CEEW) and Natural Resources Defense Council (NRDC). *Laying the Foundation for a Bright Future: Assessing Progress Under Phase 1 of India's National Solar Mission.* Interim report, April 2012. http://www.nrdc.org/international/india/files/layingthefoundation.pdf.

Crume, Richard, Hana Crume, and Mike Koshmrl. "Selling Solar." *Solar Today*, April 2011, 28–33.

Energy Foundation and Chinese Renewable Energy Industries Association (CREIA). *China Solar PV Policy Report: Differential Feed-in Tariffs*, April 2013.

Energy Information Administration (EIA). "Electric Industry Overview 2007." http://eia.gov/cneaf/electricity/page/prim2/toc2.html (accessed July 1, 2013).

———. *Solar Photovoltaic Cell/Module Shipments Report 2011*, September 2012. http://www.eia.gov/renewable/annual/solar photo/pdf/pv report.pdf.

Electric Power Research Institute (EPRI). "Addressing Solar Photovoltaic Operations and Maintenance Challenges." White Paper, July 2012. http://www.epri.com/abstracts/Pages/ProductAbstract.aspx?ProductId=000000000001021496.

Eleuteri, Stefano. "The Wait Is Over." *Photon International*, June 2011, 66–70.

Emerging Energy Research (EER). "New Study Highlights Utility Role behind US Photovoltaic Market." News release, December 3, 2009. http://www.emerging-energy.com/content/press-details/NewStudy-Highights-Utility-Role-Behind-US-Photovoltaic-Market/15.aspx.

———. "US Utility Solar PV Markets and Strategies: 2009–2020." Summary, November 2009. http://www.emerging-energy.com/uploadDocs/USUtilitySolarPV09ToC.pdf.

European Photovoltaic Industry Association (EPIA). *Global Outlook for Photovoltaics until 2016*, May 2012. http://www.epia.org/news/publications/global-market-outlook-for-photovoltaics-until2016/.

———. *Global Market Outlook for Photovoltaics 2013–2017*, 2013. http://www.epia.org.

First Solar. "First Solar Acquires Chilean Project Developer Solar Chile." News release, January 9, 2013. http://investor.firstsolar.com/releasedetail.cfm?releaseid=732358.

———. "Form 10-K." Annual report for Securities and Exchange Commission, February 28, 2011. http://investor.firstsolar.com/phoenix.zhtml?c=201491&p=irol-reportsAnnual.

———. "Form 10-K." Annual report for Securities and Exchange Commission, February 29, 2012. http://www.sec.gov/edgar.shtml.

———. "Form 10-K." Annual report for Securities and Exchange Commission, February 27, 2013. http://www.sec.gov/edgar.shtml.

German Solar Industry Association (BSW-Solar). "Statistic Data on the German Solar Power (Photovoltaic) Industry," February 2013.

http://www.solarwirtschaft.de/fileadmin/media/pdf/2013 2 BSW-
Solar fact sheet solar power.pdf.

Gestamp. "Gestamp Solar and Kankyo-Keiei to Partner for Roof-top Solar Projects All over Japan." News release, October 4, 2012. http://www.gestampsolar.com/en/media /information/gestampsolar-and-kankyo-keiei-partner-roof-top-solar-projects-all-over-japan.

Gibson, Robert. "Electric Utilities and Solar: Threat or Opportunity?" The GW Solar Institute 3rd Annual Symposium, "Solar Energy: A Critical Component of Meeting the Clean Energy Challenge," April 26, 2011, Washington, DC.

Government of France. Ministry of Ecology, Sustainable Development and Energy. "Mesures d'urgence pour la relance de la filière photovoltaïque française" [Emergency measures for relaunching the French photovoltaic supply chain], January 7, 2013. http://www.developpementdurable.gouv.fr/IMG/pdf/DP photovoltaique.pdf.

Government of Japan. Ministry of Economy, Trade, and Industry (METI). "Announcement Regarding the Present Status of Introduction for Facilities Generating Renewable Energy as of December 31, 2012." News release, March 13, 2012. http://www.meti.go.jp /english/press/2013/0313 02.html.

Government of Ontario. Ministry of Energy. FIT Program Directive. MC-2013-1450, June 12, 2013. http://powerauthority.on.ca/sites/default/files/MC-2013-1450- Direction RenewableEnergyProgram.pdf.

Greentech Media. "US Residential Solar Financing to Reach $5.7 Billion by 2016," February 11, 2013. http://www.greentechmedia.com/articles/read/us-residential-solar-financing-to-reach-5.7-billionby-2016.

GTM Research. "PV BOS Cost Analysis: Ground Mounted Systems," July 25, 2011. http://www.greentechmedia.com/articles/read/pv-bos-cost-analysis-ground-mounted-systems.

GTM Research and Solar Energy Industries Association (SEIA). *U.S. Solar Market Insight Report: 2012 Year-in-Review*, 2013. http://www.seia.org/research-resources/us-solar-market-insight-2012-yearreview.

———. *U.S. Solar Market Insight Report: 2011 Year-in-Review*. Executive Summary, 2012. http://www.seia.org/research-resources.

———. *U.S. Solar Market Insight Report: 2010 Year-in-Review*. Executive Summary, 2011. http://www.seia.org/research-resources.

———. "First Solar Ranked Largest Photovoltaic EPC in 2012 as European Companies Falter." News release, March 27, 2013. http://www.imsresearch.com/press-release/first_solar_ranked_largest_photovoltaic_epc_in_2012_as_european_companies_falter&ca t id=198&type=LatestResearch.

Hamm, Julie. "Why Are Utilities Increasingly Announcing New Solar Initiatives?" *Renewable Energy World*, September 2, 2008. http://www.renewableenergyworld.com/rea/news/article/2008/09/why-are-utilities-increasinglyannouncing-new-solar-initiatives-53426.

Hardcastle, Jessica Lyons. "Power Player." *Silicon Valley Business Journal*, August 31, 2012. http://www.bizjournals.com/sanjose/print-edition/2012/08/31/power-play.html.

Hering, Garrett. "A New Day in the New World." *Photon International*, April 2010, 10–18.

Herndon, Andrew. "OCI of Korea Purchases U.S. Solar Developer CornerStone Power." Bloomberg, January 18, 2011. http://www.bloomberg.com/news/2011-01-18/oci-of-korea-purchases-u-ssolar-energy-developer-cornerstone-power.html.

Herndon, Andrew, Christopher Martin, and Ehren Goossens. "Total to Buy 60% of SunPower for $1.38 Billion in Solar Bet." Bloomberg, April 29, 2011. http://www.bloomberg.com/news/2011-04- 28/total-to-begin-friendly-tender-for-up-to-60-of-sunpower-shares.html.

IHS. "Photovoltaic Industry to Enjoy Robust Installation Growth in 2013, but Revenue Dip Poses Challenges." News release, January 25, 2013. http://www.isuppli.com /Photovoltaics/News/Pages/Photovoltaic-Industry-to-Enjoy-RobustInstallation-Growth-in-2013-but-Revenue-Dip-Poses-Challenges.aspx.

IMS Research. "BELECTRIC Tops IMS Research PV System Integrator Rankings for 2011 but Chinese Companies Make Biggest Gains." News release, May 10, 2012. http://imsresearch.com/pressrelease/BELECTRIC Tops IMS Research PV System Integrator Rankings for 2011 but Chi nese Companies Make Biggest Gains.

Internal Revenue Service (IRS). *How to Depreciate Property*. Publication 946, February 15, 2013. http://www.irs.gov/pub/irs-pdf/p946.pdf.

International Energy Agency (IEA). Photovoltaic Power Systems Programme (PVPS). *Trends in Photovoltaic Applications: Survey Report of Selected IEA Countries between 1992 and 2012*. Report IEA-PVPS T1-19:2010, 2010. http://www.iea-pvps.org/index.php?id=32.

———. *PVPS Report: A Snapshot of Global PV 1992–2012*. Report IEA-PVPS T1-22:2013, 2013. http://iea-pvps.org/index.php?id=32.

Itron, Inc., and Kema, Inc. *CPUC California Solar Initiative 2009 Impact Evaluation*. Final Report. Submitted to Southern California Edison and California Public Utilities Commission Energy Division, June 2010. http://www.cpuc.ca.gov/PUC /energy/Solar/eval09.htm.

Jäger-Waldau, Arnulf. *PV Status Report 2012: Part I*. JRC Scientific and Policy Reports. EUR 25749 – 2012, October 2012. http://iet.jrc.ec.europa.eu/remea/pv-status-report-2012.

JPEA. *Statistics on Shipments of Solar Cells and Modules, Fiscal 2012 Third Quarter and Calendar Year* [in Japanese], February 26, 2013. http://www.jpea.gr.jp/pdf/t130226.pdf.

Kann, Shayle. "The U.S. PV Market—Today and Tomorrow." Presentation at Solar Power International 2011, Dallas, TX, October 18, 2011.

Kennedy, Danny. "The Changing Marketplace—How Mergers and Acquisitions are Shaping the Competitive Landscape in the Solar Industry." Panel presentation, Solar Power International 2012, Orlando, FL, September 13, 2012.

Krulewitz, Andrew. "Si Se Puede! Latin American Solar Markets Poised for Growth." *Greentech Media*, January 24, 2013. http://www.greentechmedia.com/articles/read/Si-Se-Puede-Latin-AmericanSolar-Markets-Poised-for-Growth.

Krulewitz, Andrew. "Who Reigns Supreme in Residential Solar?" *PV News* 31, no. 2 (February 2012): 1, 11.

Kyocera. "KYOCERA Establishes New Company to Operate 70MW Solar Power Plant in Japan." News release, July 31, 2012. http://global.kyocera.com/news/2012 /0706_ssll.html.

Lacey, Stephen. "SunPower to Acquire PowerLight." *Renewable Energy World*, November 15, 2006. http://www.renewableenergyworld.com/rea/news/article/2006/11/sunpower-to-acquirepowerlight-46567.

LDK Solar. "LDK Solar Agrees to Acquire 70% of Solar Power, Inc. for $33 Million." News release, January 6, 2011. http://investor.ldksolar.com/phoenix.zhtml?c=196973&p=irol-newsArticle&ID=1513453&highlight.

Lighthouse Solar. "The Lighthouse Solar Franchising Company Is Recruiting Building Industry Professionals with a Fresh Approach to Franchising." News release, February 17, 2010. http://www.reuters.com/article/2010/02/17/idUS190921+17-Feb-2010+BW 20100217.

Maryland Energy Administration (MEA). Residential Clean Energy Grant Program. Historic Award Data Report. Excel worksheet (accessed January 17, 2013). http://energy.maryland.gov/Residential/cleanenergygrants/index.html.

Matz, Michael D. "Going National." *Photon*, April 2011, 42–44.

———. "Ten Things Consumers Should Know about Solar Panel Certification." *Photon*, February 2012, 70–71.

Maui News. "Hawaii Solar Companies Join into One Entity," February 4, 2012. http://www.mauinews.com/page/content.detail/id/557805.html.

Mehta, Shyam, and Andrew Krulewitz. "The Direction of the Vertically Integrated Downstream." *PVNews* 30, no. 3 (March 2011), 1, 10–11.

MEMC. "Form 10-K." Annual report for Securities and Exchange Commission, February 29, 2012. http://www.sec.gov/edgar.shtml.

———. "Form 10-K." Annual report for Securities and Exchange Commission, March 1, 2013. http://www.sec.gov/edgar.shtml.

———. "Fourth Quarter 2012 Results Conference Call." Presentation, February 13, 2013. http://phx.corporate-ir.net/External.File?item=UGFyZW50SUQ9MTcxNDY3fENoaWxkSUQ9LTF8VHlwZT0z &t=1

———. "MEMC Completes Acquisition of SunEdison." News release, November 23, 2009. http://www.memc.com/index.php?view=investors&c=106680&p=irol-news.

———. "MEMC and Its SunEdison Subsidiary Join Flextronics to Celebrate Creation of Approximately 400 New Jobs Supporting Solar Module Production." News release, July 18, 2011. http://news.flextronics.com/phoenix.zhtml?c=235792&p=irolnewsArticle&ID=1586191&highlight.

———. "SunEdison and CAP Sign Agreement to Build 100MW Solar PV Plant in Chile." News release, January 31, 2013. http://www.memc.com/index.php?view =investors&c=106680&p=irol-news.

———. "SunEdison and Petrobras Sign an Agreement to Build One of the Largest Solar PV Plants in Brazil." News release, April 18, 2013. http://www.sunedison.com.

———. "SunEdison Closes R2.6 Billion (US$314 Million) in Funding for 58 MW (AC) in South Africa Solar Projects." News release, November 28, 2012. http://www.sunedison.com/wps/portal/memc/aboutus/newsroom/pressreleases.

———. "SunEdison First to Announce Confirmation of Ontario Solar FIT Content." News release, April 15, 2011. http://www.cansia.ca/sites/default/files/sunedisondomestic contentletter apr15final.pdf.

———. "Announcement Regarding the Present Status of Introduction of Facilities Generating Renewable Energy as of February 28, 2013." News release, May 17, 2013. http://www.meti.go.jp/english/press/2013/0517 02.html.

Mints, Paula. "US Supply/Demand and Market Growth 2012 to 2014." Presentation at Intersolar China, "Global PV Markets: The Americas/Mena Region/Asia," December 12, 2012.

Mortenson. "Solar Energy Industry and Construction Trends," January 2012. http://www.mortenson.com/Resources/Images/15287.pdf.

Mourgelas and Associates. "Mourgelas Greek Law Update," April 2012. http://www.mourgelas.gr/img/x2/collection/pdfs/Greek-lawupdate/NewsletterApril2012 1335951127.pdf.

Murphy, Andrew. "Private Sector Taking Up the Clean Energy Challenge." Panel discussion, GW Solar Institute Third Annual Symposium, "Solar Energy: A Critical Component of Meeting the Clean Energy Challenge," Washington, DC, April 26, 2011.

Nakayama, Michio, and Shigeru Sato. "Sekisui Expects Solar-Powered House Sales to Surge." Bloomberg, July 10, 2009.
http://www.bloomberg.com/apps/news?pid=newsarchive&sid=axgOwRScgUFE.

National Renewable Energy Lab (NREL). "Technical Study Addresses a Key Challenge to Harmonizing U.S. and International PV Module Standards." NREL Highlights. NREL/FS-5200-54528, July 2012. http://www.nrel.gov/docs/fy12osti/54528.pdf.

New York State Energy Research and Development Authority (NYSERDA). PowerClerk database. http://nyserda.powerclerkreports.com/Default.aspx?ReportId=1 (accessed February 1, 2013).

Nielsen, Stephan. "Brazil Promoting Rooftop Solar Market with New Policies." Bloomberg, April 19, 2012. http://www.bloomberg.com/news/2012-04-18/brazil-rules-let-consumers-trade-renewablepower-to-utilities.html.

Nielsen, Stephan. "Local Content Rule Thwarts Solar Development in Brazil." Bloomberg, August 9,
2012. http://www.renewableenergyworld.com/rea/news/article/2012/08/local-content-rule-thwarts-solar-development-in-brazil.

North Carolina Solar Center. Database of State Incentives for Renewables and Efficiency (DSIRE). "Renewable Portfolio Standard Policies," February 2013. http://www.dsireusa.org/documents/summarymaps/RPS map.pdf.

———. "Renewable Portfolio Standard Policies with Solar/Distributed Generation Provisions," February 2013. http://www.dsireusa.org/documents/summarymaps/Solar DG RPS map.pdf.

NPD Solarbuzz. "PV Installed Base in US Offers New Revenue Prospects for Systems Providers." News release, September 11, 2012. http://www.solarbuzz.com/news/recent-findings/pv-installed-baseus-offers-new-revenue-prospects-systems-providers.

———. "Solar Photovoltaic Demand in 2012 Falls Short of 30 GW Mark, Reports NPD Solarbuzz." News release, February 21, 2013. http://www.solarbuzz.com/news/recent-findings/solarphotovoltaic-demand-2012-falls-short-30-gw-mark-reports-npd-solarbuzz.

Ontario Power Authority. "Feed-in Tariff Contract (FIT Contract)." Version 2.1, December 14, 2012. http://fit.powerauthority.on.ca/sites/default/files/page/FIT Contract Version 2.1.pdf.

Ontario Power Authority. "Feed-in Tariff Program: FIT Rules." Version 2.1, December 14, 2012. http://fit.powerauthority.on.ca/sites/default/files/page/FIT Rules Version 2.1.pdf.

Pernick, Ron, Clint Wilder, and Trevor Winnie. *Clean Energy Trends 2012*. Clean Edge, March 2012. http://www.cleanedge.com/sites/default/files/CETrends2012 Final Web.pdf.

Podewils, Christoph. "The Air Is Getting Thinner." *Photon International*, November 2012, 54–57.

Poissant, Yves, Joseph Ayoub, Jennifer Hiscock, and Lisa Dignard-Bailey. "National Survey Report of PV Power Applications in Canada 2011." Prepared for International Energy Agency Co-operative Programme on Photovoltaic Power Systems, June 2012. http://www.ieapvps.org/index.php?id=146.

PROINSO. "PROINSO Supplies 100kW to Several Net Metering PV Projects in Brazil." News release, January 29, 2013. http://www.proinso.net/pub/doc/File/ingl/NPPROINSO brasil28012013.pdf.

PV News. "Strategies for the U.S. Utility-Scale PV Market." *PV News* 29, no. 8 (August 2010), 1, 12.

Real Goods Solar. "Form 10-K."Annual report for Securities and Exchange Commission, March 15, 2012. http://www.sec.gov/edgar.shtml.

Recurrent Energy. "Sharp Corporation to Acquire Recurrent Energy for Up to $305 Million." News release, September 21, 2010. http://www.recurrentenergy.com/sites /default/files/100921%20-%20Press%20Release%20-%20Sharp%20Corp%20to%20 Acquire%20Recurrent%20Energy.pdf.

Renewable Energy Policy Network for the 21st Century (REN21). *Renewables 2012: Global Status Report*. Paris: REN-21 Secretariat, 2012. http://www.ren21.net/Portals/0/documents/Resources/GSR2012 low%20res FINAL.pdf.

Rosell, Alejandro Diego. "Varying Rates." *Photon International*, December 2012, 28–31.

Ross, JP. "Expanding Residential Solar Markets." Presentation at Solar Power International 2011, Dallas, TX, October 20, 2011.

Sciaudone, Christiana. "AES Gener Unveils Plans for Up to 220MW of Desert PV in Chile." *Recharge*, November 26, 2012.

Sheppard, Michael. "Latin America: The Next Locus for Photovoltaic Growth." IHS iSuppli, January 21, 2013. http://www.isuppli.com/photovoltaics/marketwatch/pages/latin-america-the-next-locus-forphotovoltaic-growth.aspx.

Sherwood, Larry. "Big Time for Solar." *Solar Today* 26, no. 5 (July/August 2012): 40–42.

———. "Market Prospects for Solar in North America—New Data and Emerging Trends." Presentation at Solar Power International 2011, Dallas, TX, October 18, 2011.

———. "Utility-scale Installations Lead Growth." *Solar Today,* July/August 2011, 30–32.

———. *U.S. Solar Market Trends 2011*. Interstate Renewable Energy Council, August 2012. http://www.irecusa.org/wp-content/uploads/IRECSolarMarketTrends-2012-Web-8-28-12.pdf.

———. *U.S. Solar Market Trends 2010*. Interstate Renewable Energy Council, June 2011. http://irecusa.org/wp-content/uploads/2011/07/IREC-Solar-Market-Trends-Report-revised070811.pdf.

Shimogawa, Duane. "Haleakala, Solar Wave Set to Merge." *Pacific Business News*, February 3, 2012. http://www.bizjournals.com/pacific/print-edition/2012/02/03/haleakala-solar-wave-set-to-merge.html.

Sills, Ben, and Christopher Martin. "U.S. Utilities to Double Solar Investments Annually." Bloomberg, November 30, 2010. http://www.bloomberg.com/news/2010-11-30/utilities-solar-spending-todouble-each-year-in-u-s-to-2015-gtm-says.html.

Solarbuzz. "China's Feed-in Tariff Policy Stimulates 14 GW Photovoltaic Project Pipeline." News release, October 10, 2011.

———. "World Photovoltaic Market Grew to 27.4 Gigawatts in 2011, Up 40% Y/Y." News release,

March 19, 2012. http://www.solarbuzz.com/news/recent-findings/world-solar-photovoltaic-market-grew-274-gigawatts-2011-40-yy-0.

SolarCity. "Form S-1." Filing for Securities and Exchange Commission, October 5, 2012. http://www.sec.gov/edgar.shtml.

SolarPlaza. "Mexico More Attractive Due to Falling Prices of PV Modules," February 18, 2013. http://solarplaza.com/article/mexico-more-attractive-due-to-falling-prices-of-pv.

Solar Electric Power Association (SEPA). *Photovoltaic Incentive Programs Survey: Residential Participant Demographics, Motivations and Experiences*. SEPA Report 06-09, November 2009. http://www.solarelectricpower.org/media/114222/sepa%20 customer%20incentive%20program% 20survey%2011.18.09.pdf.

The Solar Foundation. *National Solar Jobs Census 2012: A Review of the U.S. Solar Workforce*, November 2012. http://thesolarfoundation.org/research/national-solar-jobs-census-2012.

Solar Frontier. "Solar Frontier and Yano Industry Connect 3.3MW of Power to the Western Japan Grid." Solar Frontier, March 21, 2013. http://www.solar-frontier.com/eng/news/2013/C017764.html.

Solar Industry. "Forecast: Many German PV Installers to Fade from Market This Year," February 27, 2013. http://www.solarindustrymag.com/e107 plugins/content/content.php? content.12190.

———. "Jinko Solar Planning 200 MW PV Project In China." December 20, 2012. http://www.solarindustrymag.com/e107 plugins/content/content.php?content.11782.

Steenblik, Ronald, Takenori Matsuoka, and Jim Hight. "Facilitating Trade in Selected Climate-ChangeMitigation Technologies in the Electricity Generation and Heavy-Industry Sectors." OECD Global Forum on Trade, Trade and Climate Change, Paris, June 9 and 10, 2009. http://www.oecd.org/tad/envtrade/42886254.pdf.

Stephens, Brandee. "Solar Plant's Production Costs Are Now Competitive." *Solutions*, no. 1, 2011.

http://solutions.bv.com/solar-plant%E2%80%99s-production-costs-are-now-competitive/.

Stuart, Becky. "SunPower Completes Tenesol Acquisition." *PV Magazine*, February 1, 2012. http://www.pv-magazine.com/news/details/beitrag/sunpower-completes-tenesol-acquisition 100005651/#axzz2VoqerPGU.

SunPower Corp. "Form 10-K." Annual report for Securities and Exchange Commission, February 25, 2013. http://www.sec.gov/edgar.shtml.

———. "SunPower Corporation and Toshiba Extend Long-Standing Partnership through 2018." News release, December 4, 2012. http://us.sunpowercorp.com/about/newsroom /pressreleases/?relID=135997.

———. "SunPower Signs Joint Venture Agreement in China." News release, December 3, 2012. http://us.sunpowercorp.com/about/newsroom/press-releases/?relID=135949.

Sunsong, Chris. “Latin America and Caribbean PV Demand Growing 45% Annually Out to 2017.” *Renewable Energy World*, January 2, 2013. http://www.renewableenergyworld.com/rea/news/article/2013/01/latin-america-and-caribbeanpv-demand-growing-45-annually-out-to-2017.

Sunvalley Solar. “Form 10-K.” Annual report for Securities and Exchange Commission, April 16, 2012. http://www.sec.gov/edgar.shtml.

Trabish, Herman K. “Emerging Solar Strategies Part 4: SunEdison and the Next Big Thing.” Greentech Media, November 23, 2012. http://www.greentechmedia.com/articles/read/Emerging-SolarStrategies-Part-4-SunEdison-and-The-Next-Big-Thing.

True South Renewables. “True South Renewables to Commission 10 MW Solar Power Project.” News release, December 13, 2011. http://truesouthrenewables.com/press/?p=42.

U.S. Treasury Department. “Payments for Specified Energy Property in Lieu of Tax Credits under the American Recovery and Reinvestment Act of 2009,” April 2011. http://www.treasury.gov/initiatives/recovery/Documents/GUIDANCE.pdf.

U.S. Trade Representative (USTR). *2013 National Trade Estimate Report on Foreign Trade Barriers*, March 2013. http://www.ustr.gov/sites/default/files/2013%20NTE.pdf.

———. *2013 Report on Technical Barriers to Trade*, April 2013. http://www.ustr.gov/sites/default/files/2013%20TBT.pdf.

Watanabe, Chisaki. “Toshiba, SunEdison to Build Solar-Power Stations in Japan.” Bloomberg, April 19, 2013. http://www.bloomberg.com/news/2013-04-19/toshiba-sunedison-to-build-solar-powerstations-in-japan.html.

Watson, Farley, and Williams. “Greece: Energy Briefing,” September 2012. http://www.wfw.com/Publications/Publication1130/$File/WFW-GreekEnergyBriefing2012.pdf.

Watt, Muriel, and Rob Passey. “PV in Australia 2012.” Executive Summary. Prepared for International Energy Agency PV Power Systems Programme, May 2013. http://iea-pvps.org/index.php?id=93.

Wesoff, Eric. “Borrego Joins OneRoof, SolarCity and Sunrun in the Third-Party Solar Cash Boom.” Greentech Media, July 31, 2012. http://www.greentechmedia.com/articles/read/Borrego-JoinsOneRoof-SolarCity-and-Sunrun-in-the-Third-Party-Solar-Cash-B.

———. “First Solar’s Q1: Steady Guidance, Profitable, Action in Chile and China.” Greentech Media, May 6, 2013. http://www.greentechmedia.com/articles/read/First-Solars-Q1-Steady-GuidanceProfitable-Action-in-Latin-America-and.

———. “What Really Motivates Consumers to Install Residential Solar?” Greentech Media, March 23, 2011.

Wissing, Lothar. “National Survey Report of PV Power Applications in Germany.” Prepared for International Energy Agency Co-operative Programme on Photovoltaic Power Systems, July 2012. http://www.iea-pvps.org/index.php?id=146.

Woodlawn Associates. “Solar Installation Effectiveness.” September 10, 2012. http://www.woodlawnassociates.com/wp-content/uploads/WA-SIE-Extract.pdf.

Woody, Todd. “California's Sungevity Takes Stake in Dutch Solar Startup.” Forbes, November 17, 2011. http://www.forbes.com/sites/toddwoody/2011/11/17/californias-sungevity-takes-stake-in-dutchsolar-startup/.

———. “Solar Industry Anxious over Defective Panels.” *New York Times*, May 28, 2013. http://www.nytimes.com/2013/05/29/business/energy-environment/solar-powers-dark-side.html?pagewanted=all& r=0.

World Trade Organization (WTO). Committee on Technical Barriers to Trade. "Minutes of the Meeting Held on March 24–25, 2011." G/TBT/M/57, May 26, 2011.

———. Committee on Technical Barriers to Trade. "Minutes of the Meeting Held on June 13–15, 2012." G/TBT/M/57, September 18, 2012.

Xu Honghua, Charlie Doum, Wang Sicheng, and Lv Fang. "National Survey Report of PV Power Applications in China 2011." Prepared for International Energy Agency Co-operative Programme on Photovoltaic Power Systems, August 15, 2012. http://www.iea-pvps.org/index.php?id=146.

Yonan, Alan Jr. "State Unveils New Rules for Solar Tax Credits." *Honolulu Star-Advertiser*, November 9, 2012. http://www.staradvertiser.com/news/breaking/178152371.html?id=178152371.

SECTION 4. WIND ENERGY SERVICES

Introduction

Overall, the global wind energy services market grew during 2007–12. The United States was second only to China in terms of newly installed wind energy capacity in 2011, followed by Germany, Spain, and India. However, many major markets are facing fluctuating demand for new wind energy installations and for wind development and construction services due to factors such as weak economic conditions and inconsistent government policies.[1] At the same time, competition has increased as demand has risen in the operations and maintenance (O&M) services market. Weak turbine sales have led a growing number of firms, particularly original equipment manufacturers (OEMs), to expand their provision of services, while a growing number of wind farm owners are now seeking services to maintain their recently installed turbines.

The global wind services market remains fragmented, with limited cross-border trade. Evidence suggests that the majority of trade involves firms in one country establishing a commercial presence abroad. For example, while a wide variety of firms provide wind energy services in the U.S. market, the largest are subsidiaries or affiliates of foreign multinationals. Due in part to their long history in the wind energy sector, European multinationals have the largest presence in the global market.

Overview of Wind Energy Services Segments

Wind energy services are typically provided in conjunction with the development and operation of utility-scale wind farms.[2] The life cycle of a wind farm can be broken down into three distinct stages, with each stage requiring a unique combination of wind services (figure 4.1). Although the bulk of services (by value) are provided during the project development phase, O&M services are provided for the longest time, often up to 20 or more years. Wind energy services and their respective Consumer Products Classification (CPC) codes are presented in table 4.1.

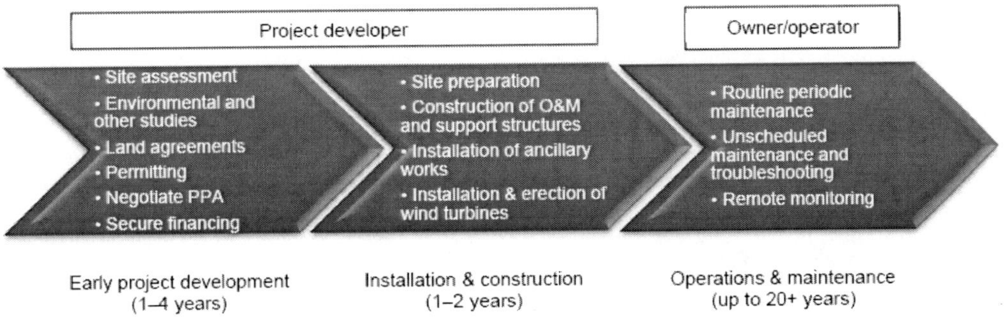

Source: Compiled by USITC.

Figure 4.1. Wind energy project development phases.

Table 4.1. Services related to provision of wind energy

CPC Code	CPC description	Wind-specific services
861	Legal services	• Preparation, documentation, and drawing up of legal documents pertaining to issues such as land agreements, federal filing requirements, and other documentation as needed • Legal advisory services
8673	Integrated engineering services	• Fully integrated engineering services provided for turnkey wind projects (i.e., project development services)
8672	Engineering services	• Advisory and consultative engineering services related to site location and design • Engineering design services related to the construction of a wind farm, including foundation, building structures, mechanical and electrical installations, and all necessary civil engineering • Advisory and technical assistance services throughout construction and installation to ensure work is in conformity with design and regulations
86753	Surface surveying services	• Geotechnical and geophysical evaluation of potential project site
865	Management consulting and related services	• Pricing policies, organization of distribution • Advisory, guidance, and operational assistance services related to safety • Risk management • Stakeholder and community relations • Environmental impact assessments and other studies as required (such as by regional transmission authorities)
62118	Sales on a fee or contract basis	• Electricity brokerage
511-518	Construction and related engineering services	• Site preparation, such as excavation, clearance, and foundation preparation • Construction of O&M facilities and other buildings as needed • Construction or bolstering of roads and foundations • Installation and erection of wind turbines

Table 4.1. (Continued)

CPC Code	CPC description	Wind-specific services
		• Installation of power cables and substations; development of other ancillary works • Crane rental
886	Repair services	• Operations and maintenance services
8676	Technical testing and analysis services	• Remote monitoring services • Testing of machinery, including turbines and generators

Source: Compiled by USITC.

Project Development, Construction, and Installation Services

Project development refers to the comprehensive planning and engineering services[3] provided for turnkey[4] wind projects. Developers compete on price, as well as the ability to identify and obtain permits for a high-quality wind resource and then find an appropriate buyer.[5] Some wind energy developers focus specifically on early development, overseeing the services related to finding and evaluating a potential site, before selling the project to another (often larger) developer for the construction and installation phases.[6] Although developers may have the ability to provide all the services required, they often retain only supervisory responsibility for most of them, which they contract out to firms that specialize in one or more of these functions. Decisions on which services to contract out are project specific and depend on factors such as location, demands of the project, and the developer's own level of expertise and willingness to take on risk.[7]

Engineering services are vital throughout the life cycle of a wind farm. In early development, once a potential project site has been identified, wind analysis is performed (a process that takes anywhere from six months to three years)[8] to calculate whether there is enough wind to make a potential project financially viable. [9] Surface surveying services—i.e., geotechnical and geophysical evaluations—assess the land to determine its suitability for bearing heavy turbine foundations and access roads. These services are provided by both turbine manufacturers/OEMs [10] and independent service providers (ISPs).[11] The leading resource assessment firms, though, tend to be niche specialty companies, often seeking competitive advantage through proprietary software that creates complex computer models.[12] Once a site has been selected, engineers begin project design, determining the optimal placement for all components, including wind turbines and building structures, as well as integrating the mechanical and electrical installations. Engineers may also provide advisory services and technical assistance throughout installation, providing due diligence for either the owner or bank.[13] Advisory services like these are typically provided by an experienced engineering, procurement, and construction (EPC) firm or an independent engineering firm. Additional engineering services may be needed once a project has been commissioned, in order to identify system failures or troubleshoot other problems.

After the resource assessment process has green-lighted a project site, management consulting and legal services are required for the process of siting and acquiring permits. Management consultants provide guidance on a number of issues, including risk management or community relations, as well as perform any studies that might be needed. For example, regional transmission organizations may mandate studies when developers apply for an interconnection agreement,[14] and environmental impact studies by an independent consulting firm may also be required, especially if the proposed site is located on public land.[15] Energy lawyers are necessary throughout the development process to handle the required documentation and contracts,[16] help with permitting, and provide advisory services throughout the development process.[17] Once a site has been chosen, legal services are particularly important to formalize land agreements.[18] Although legal services may be provided by in-house lawyers at an OEM or engineering and construction firm, there are a number of independent law firms that specialize in renewable energy law.[19]

Once a power purchase agreement (PPA) has been negotiated[20] and financing secured (box 4.1), construction firms prepare the site for assembly, before constructing and installing all components of the wind farm.[21] Overall, construction and installation of a wind farm is relatively quick, generally taking less than a year, compared to the lengthy process of development, which can last three to four years.

BOX 4.1. Financing a wind farm

Financing is a key element in the early stages of a wind energy project, particularly in terms of the services that are required. Both fossil fuel- and wind-powered energy-generating projects require significant upfront investment; however, because operating costs for wind projects are so much lower, properly financing the construction of the wind project becomes relatively more important.[a] It is the project developer who secures the financing, usually in the form of a construction loan that converts to long-term project financing.[b]

There are many variations of financing structures in the wind energy sector.[c] Larger developers are able to issue bonds to finance their projects, while smaller developers may require bank funding. The ability to issue bonds gives large developers a competitive edge, as bonds typically require lower interest rates than bank loans do.[d] Like developers, some OEMs offer financing. It is not clear how OEM financing compares to other third-party funding, but it is likely that they are able to use financing as a competitive advantage for their turbine sales, by offering favorable terms to developers as an incentive to use their equipment.[e] Before the financial crisis, some OEMs offered financing for all wind projects. Now, most are more fiscally conservative and finance only projects that use their turbines.[f] Finally, the role of export credit agencies has risen in emerging markets where financing for wind projects might not be readily available.[g] For instance, in 2011, the Export Import Bank of the United States (Ex-Im Bank) provided financing for a wind farm in Honduras and the subsequent expansion of the project, which used Gamesa turbines produced in the United States.[h]

In addition to the financial implications, the source of funding can also drive the development process. Typically, the more a project relies on third-party finance, the more outside verification is needed to demonstrate the economic benefit of the wind farm. Banks typically want studies from third-party consulting or geotechnical firms to demonstrate the economic benefit of a project, rather than accepting the project developer's assurances.[i]

[a] Harper, Karcher, and Bolinger, *Wind Project Financing Structures*, September 2007, i.
[b] Industry representative, interview by USITC staff, Washington, DC, January 23, 2013.
[c] For example, see "Description of Seven Financing Structures," in Harper, Karcher, and Bolinger, *Wind Project Financing Structures*, September 2007, ii.

[d] Industry representative, interview by USITC staff, Washington, DC, January 23, 2013; industry representative, telephone interview by USITC staff, January 25, 2013.
[e] Industry representative, interview by USITC staff, Washington, DC, January 23, 2013.
[f] Industry representative, telephone interview with USITC staff, January 18, 2013.
[g] O'Brian, "Wind Projects Look to Export Finance As New Markets Open," February 27, 2013.
[h] Ex-Im Bank, "U.S. Exports Will Expand Wind Farm in Honduras," March 5, 2013.
[i] Industry representative, interview by USITC staff, Washington, DC, January 23, 2013.

Wind projects typically use EPC contracts, in which a single firm is responsible for providing the necessary design, engineering, procurement, construction, commissioning, and testing.[22]

Although EPC contractors compete primarily on price, project developers also weigh a number of other factors, including technical quality, previous experience, availability, safety, and whether the construction firm accepts the developer's financing.[23] As noted earlier, once selected, the EPC contractors—typically engineering and construction firms or OEMs[24]—may then contract out most of the work while retaining overall responsibility, with the level of services subcontracted or provided by the contractor varying from project to project.[25] Projects typically employ local construction firms for basic construction and wind experts for specialized tasks, such as electrical infrastructure.[26]

O&M Services

Generally speaking, maintaining a wind farm includes an annual inspection of the entire wind system, as well as any maintenance required for planned or unplanned stoppages.[27] The cost of a turbine typically includes a warranty covering some or all of the turbine components for an average two to five years, which may be supplemented and continued by a service contract from either the OEM[28] or an ISP.[29] Increasingly important is the electronic monitoring of wind turbine performance,[30] which includes the use of sensors installed on location at the wind farm to gather data. The data are transmitted to a central monitoring site and analyzed by specialists to forecast when maintenance is needed based on usage, as well as to predict potential problems or breakages.[31]

U.S. Market for Wind Energy Services

Market Size

The U.S. wind energy market experienced rapid, albeit fluctuating, growth between 2007 and 2012. Wind energy capacity increased at an average annual rate of nearly 30 percent from 2007 through 2012, with cumulative installations growing from 16,700 megawatts (MW) to 60,007 MW (figure 4.2).[32] The United States was the world's largest wind energy market in 2012, in terms of annual installations, accounting for 29.3 percent of new global installations.[33]

The financial crisis, energy prices, electricity demand, and the status of the federal Production Tax Credit (PTC) contributed to fluctuations in the U.S. wind services market during 2007–11. Before the financial crisis in 2008, energy prices were reportedly high enough that wind projects required few incentives to attract buyers, driving growing development and installation through 2009. In 2009, energy prices fell, mainly due to declining natural gas prices, and the U.S. economy lagged due to the recession, both of which disrupted development of new projects. As a result, new installations fell in 2010.[34] Since then, development and construction of new projects have rebounded; new U.S. wind energy installations totaled 13,124 MW in 2012—up nearly 93 percent from 2011—as owners and developers raced to commission projects before the expected expiration of the PTC in 2013.

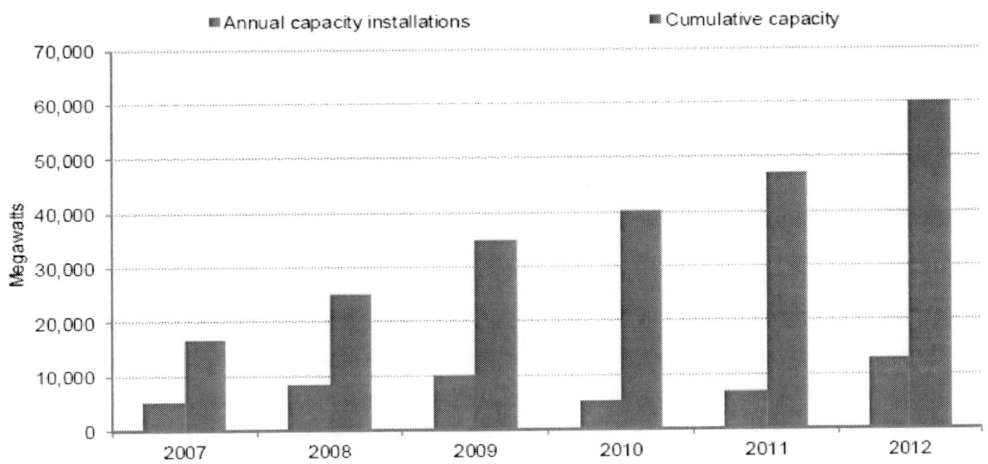

Source: AWEA, US Wind Industry Annual Market Report 2011, 2012, 9; AWEA website, "Industry Statistics," n.d. (accessed March 27, 2013).

Figure 4.2. U.S. annual and cumulative wind power capacity (utility-scale wind), 2007 through 2012.

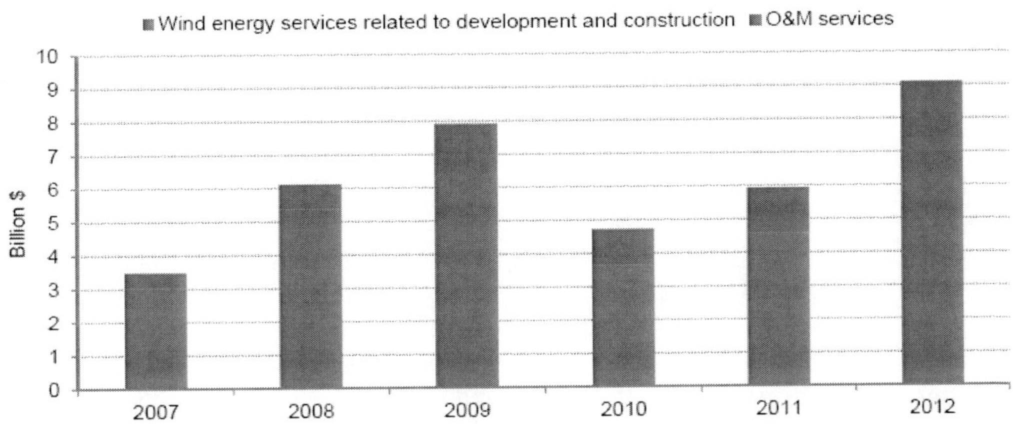

Sources: USITC staff calculations using data from Wiser and Bolinger, *2011 Wind Technologies Market Report*, August 2012; Black & Veatch, *Cost and Performance Data for Power Generation Technologies*, February 2012; and AWEA website, "Industry Statistics," n.d. (accessed March 27, 2013). EIA website, http://www.eia.gov/tools/glossary/index.cfm (accessed August 20, 2013).

Notes: Estimate of net capacity factor was not available for 2012. O&M value for that year was estimated using 2011 net capacity factor. The capacity factor is the ratio of the electrical energy produced by a generating unit for the period of time considered to the electrical energy that could have been produced at continuous full power operation duriing the same period.

Figure 4.3. U.S. wind energy services market: Estimated value of wind energy services.

Although the credit was renewed for 2013, given the uncertainty surrounding the decision as well as the relatively short timeframe, industry representatives expect that the PTC is unlikely to drive development of new projects in the near term. As a result, the number of new installations is forecast to be significantly lower in 2013.[35]

In 2012, wind energy installation services were estimated at $7.4 billion, or 32 percent of the broader market for wind power project installations (both goods and services).[36] By comparison, the U.S. O&M market was relatively small, at an estimated $1.7 billion in 2012, up from $467 million in 2007 (figure 4.3).[37] Not surprisingly, trends in the installation services market closely tracked trends in annual capacity additions.

Factors Affecting U.S. Supply and Demand for Wind Energy Services

Project Development, Construction, and Installation Services

In the United States, state-level policies and regulations, particularly renewable portfolio standard (RPS) programs, are primary drivers of new wind installations and hence of the wind development and construction market as well.[38] RPS programs typically require electricity providers to derive an increasing share[39] of energy from renewable energy sources, which creates a market for wind-generated energy and a demand for PPAs. Although RPS programs are not the sole driver of U.S. wind energy development,[40] many project developers cite them as the primary driver of new wind project development.[41] The programs also influence the geographic distribution of new installations. In 2011, 78 percent of new wind power capacity was installed in states with RPS programs.[42] Other state-specific factors that drive wind development include a favorable permitting environment,[43] the ease or availability of grid connection, and the political environment, which determines the level of community or government support for wind projects.[44]

Federal incentives such as the PTC further enhance the demand for wind energy created by RPS programs.[45] The PTC in particular is a significant driver of short-term demand for wind development and construction services. Initially established by the Energy Policy Act of 1992,[46] as of 2013 it provides a 2.3 cent tax credit per kilowatt hour (kWh) of energy generated.[47] By effectively reducing the price of wind-generated electricity and making it more competitive with conventional energy sources, the PTC encourages developers to initiate new projects, particularly in areas with high energy prices or strong wind resources.[48] However, because the PTC is not permanent, there tends to be a cycle of uncertainty that creates fluctuations in the market and reportedly impedes long-term planning.[49] Other federal level tax incentives that drove growth through 2012 include accelerated depreciation[50] and the option for either a 30 percent investment tax credit (both of which were extended through 2012) or a 30 percent cash grant in lieu of the PTC.[51] Some industry representatives note that such incentives as nonrefundable tax credits are difficult for small companies and foreign investors to use if they do not have a U.S. tax obligation equal to or greater than the tax credit.[52]

Finally, demand for project development and construction services is largely influenced by overall trends in the U.S. energy market. The effects of a sluggish economy, lower demand for electricity, and strong competition from other fuel sources, particularly natural gas, over the past five years have presented challenges to the U.S. wind industry.[53] Although new installations have resumed, electricity consumption has slowed since the recession, reducing demand for wind energy and thus for related installation services.[54]

O&M Services

The weak wind turbine market has been an important factor driving the supply of O&M services in the U.S. market. Turbine prices have dropped significantly since 2008, as competition among manufacturers has intensified and turbine demand has slowed in the wake of the financial crisis.[55] As a result, OEMs reportedly have been particularly aggressive in the O&M market, increasing the length of their warranties and service contracts. [56] There is growing demand among owners and operators for longer-term contracts, which OEMs are able to provide as they have become more confident in their equipment, as well as their ability to anticipate potential future failures and price the costs accordingly.[57] Additionally, the longer duration of service contracts provided by the OEM is an increasingly frequent condition of financing by lenders.[58] As a result, O&M services account for a significant share of some OEM's overall business; for instance, O&M reportedly accounts for 15 to 20 percent of global business for Vestas.[59]

On the demand side, the primary factor influencing the U.S. O&M market has been the growing focus among leading project owners and operators on maintaining existing assets,[60] rather than developing new projects.[61] Demand for O&M services is largely dependent on the age of equipment, and there is a growing need for services as many turbines come off warranty and are up for new service agreements.[62] Likely due to the aging of wind assets, a growing trend in O&M services has been a shift away from corrective and scheduled maintenance toward predictive and preventive maintenance, as owners and operators seek to avoid costly turbine failures and minimize time offline.[63] As a result, a growing number of O&M service providers offer remote monitoring services.

U.S. Industry Trends

Wind energy services are typically provided by either OEMs or ISPs. Although OEMs have traditionally focused on O&M services, a growing number have expanded their service offerings to also include full development and installation.[64] These expanded services offer a potentially strong revenue stream to supplement equipment sales as OEMs face an increasingly competitive equipment market.[65] ISPs, on the other hand, focus solely on the provision of services. They vary widely, ranging from small firms that specialize in a particular service, such as wind resource assessment, to large organizations that offer a full array of development, installation, and maintenance services. Unlike OEMs, ISPs also frequently own and operate wind farms, either by developing and constructing a project themselves or by purchasing a turnkey project from a project developer. As a result, particularly in the United States, many of the largest owner/operators provide their own O&M services.[66]

Throughout the wind services industry there has been a rise in vertical integration, and a growing number of OEMs, project developers, and large engineering firms offer end-toend services along the projected life cycle.[67] Despite this integration, most projects still involve multiple service providers, and even the largest project developers typically contract out specialized services, such as power systems engineering, to niche firms.

Services in Project Development and in Construction and Installation

Over the past decade, explosive growth in cumulative installed wind capacity in the United States drove concurrent growth in the project development market. A large number of firms offer project development services in the U.S. market, with the top five firms accounting for about 32 percent of installations in 2012 (table 4.2).[68] OEMs are reportedly providing increased competition to the largest project developers, by funding smaller developers who otherwise would not be able to compete against large firms such as Iberdrola.[69] As a result, the project development market has become successively less concentrated over the past five years. Whereas at the recent peak of development, in 2009, the 10 largest developers installed over 75 percent of new capacity, in 2011, the 10 leading developers accounted for less than 60 percent.[70]

Table 4.2. U.S. wind energy services: Top five project developers by capacity installed in 2012

Company	Country	Installed in 2012 (MW)	Percent of 2012 installations (%)
NextEra Energy Resources	United States	1,505	11.5
Iberdrola Renewables	Spain	716	5.5
EDF Renewables (formerly enXco)	France	658	5.0
Caithness Energy	United States	640	4.9
Duke Energy	United States	620	4.7
Other		8,992	68.5
Total		13,131	100

Source: Del Franco, "AWEA Reveals," April 11, 2013; AWEA, AWEA U.S. Wind Industry Annual Market Report: Year Ending 2012, 2013, 12.
Note: Numbers may not sum due to rounding.

Table 4.3. Top 10 EPC contractors in the U.S. wind power market, 2011

Rank	Company	Revenue (millions USD)
1	Mortenson Construction	725.8
2	Fluor Corporation	330.9
3	Aristeo Construction Company	66.0
4	Michels Corporation	57.6
5	Barton Malow Company	49.8
6	EMJ Corporation	46.4
7	The Boldt Company	46.4
8	Gemma Power Systems	43.2
9	Fagen Incorporated	38.9
10	Gray Construction	36.2

Source: ENR, *ENR Sourcebook*, September 17, 2012, 22.

In the United States, the bulk of the construction services required for a wind farm are typically provided by an EPC contractor.[71] Firms such as Mortenson Construction and Fluor Corporation (table 4.3) have expanded their EPC portfolios to include the wind power industry, and these two firms are now the largest EPC contractors in the United States. However, the wind industry's significant growth over the past decade has led to rising vertical integration, and firms in other segments of the wind industry, such as OEMs and specialty construction firms, are increasingly offering EPC services.[72] Firms have likely been motivated to enter the market by the profitable opportunities—EPC services reportedly account for the largest share of services costs associated with developing and building a wind farm.[73] Despite the growing number of entrants, the EPC market remains quite concentrated, with the top five companies accounting for over 75 percent of the market.[74] This is reportedly because larger construction firms tend to have an advantage in a number of important market factors such as safety records, reputation and experience, financial strength, and cost competitiveness.[75] Smaller developers, particularly those with less experience, often prefer to work with large EPC firms to help reduce the risk associated with a project.[76]

O&M Services

In the U.S. market, industry representatives report a wide variation in O&M strategies among wind farm operators. The majority of services tend to be provided in-house by asset owners or by OEMs, but there are O&M ISPs operating in the market as well.[77] For example, two of the leading U.S. service providers, NextEra Energy Resources and EDF Renewable Services, are wind farm operators; UpWind Solutions is the leading O&M ISP in the U.S. market.[78] The U.S. O&M market has seen consolidation among ISPs and a growing presence of OEMs since 2007. ISPs have been acquired by utilities or by other service firms seeking to expand their service offerings. As mentioned earlier, industry officials report that a growing number of OEMs are aggressively turning to the O&M market in an effort to boost profits.[79]

Competition in the O&M market varies depending on the services provided. Scheduled maintenance, which accounts for the majority of services required,[80] can typically be provided by most O&M firms, so competition is based on price. However, the highest priority for owners is typically unscheduled maintenance, which can take the turbine out of commission. In choosing an O&M provider for unscheduled maintenance, owners consider issues such as whether the service provider can offer remote monitoring to predict and prevent failures, if it has a procurement group to get spare parts, and whether it has personnel in the area or a hub nearby to reduce the cost of manpower.[81]

In general, according to one industry representative, it is often difficult for ISPs to compete against OEMs in the U.S. O&M marketplace because OEMs have become aggressive in negotiating service contracts in conjunction with turbine sales and have often shut third-party providers out of the market.[82] Additionally, OEMs have the competitive advantage of detailed knowledge of their technology and the ability to offer technology upgrades throughout the life of a turbine.[83] However, one industry representative noted that there are many basic services that ISPs can provide. Additionally, most ISPs offer the advantage of being able to work on multiple turbine brands.[84]

Global Market for Wind Energy Services

Market Size

Annual global installations of wind energy have grown rapidly over the past decade, increasing from 8.1 gigawatts (GW) in 2003 to 44.8 GW in 2012 (figure 4.4). In 2012, new installations reached record levels and were up 10 percent from the previous year—a change from the 2010–2011 period, which saw only moderate increases in demand. The United States was the largest market in 2012, with 13,124 MW in new installations (29.3 percent of all new installations), followed closely by China with 12,960 MW (28.9 percent), Germany with 2,415 MW (5.4 percent), India with 2,336 MW (5.2 percent), and the United Kingdom with 1,897 MW (4.2 percent) (figure 4.5).[85]

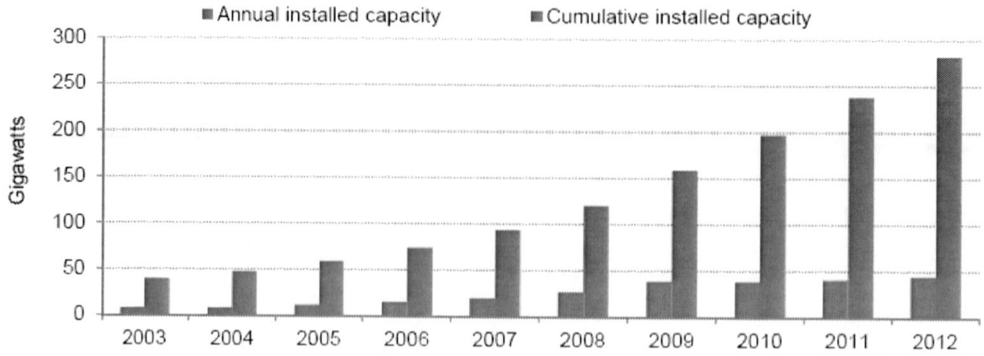

Figure 4.4. Global annual and cumulative installed wind energy capacity, 2003 through 2012.

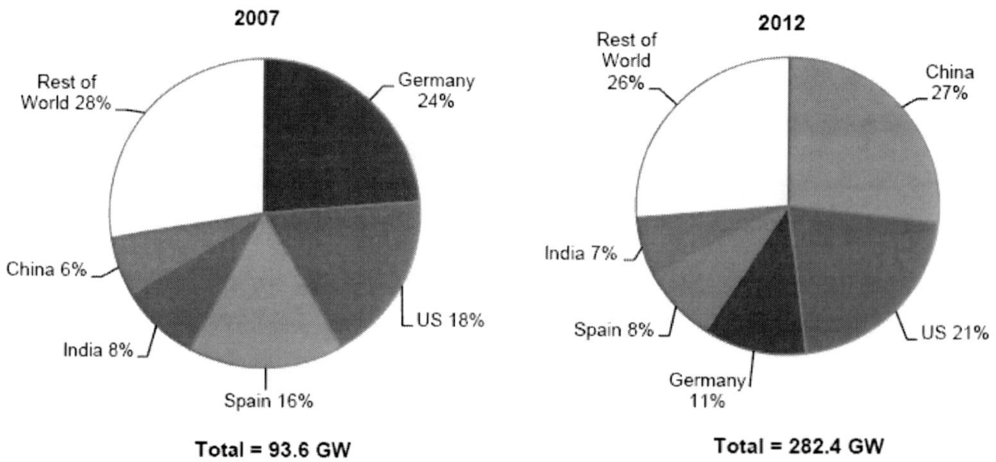

Figure 4.5. The distribution of cumulative wind energy capacity changed from 2007 to 2012.

The composition of the global wind energy market has changed substantially over the past five years, largely due to the growth in demand in China. China's share of cumulative installed capacity increased from 6 percent in 2007 to 27 percent in 2012, while Germany's decreased from 24 percent to 11 percent and Spain's from 16 percent to 8 percent. The United States' share shifted only modestly, increasing from 18 percent to 21 percent.[86] Additionally, the market has become less concentrated over the past decade; in 2012, the eight leading countries accounted for more than 80 percent of the global market, compared to just five countries with this share of the market 10 years ago.[87]

The global wind energy services market has grown as the value of global wind installations (including goods and services) rose from an estimated \$30.1 billion in 2007 to \$71.5 billion in 2011.[88] Wind services are estimated to have been \$23 billion in 2011, or 32 percent of the broader market for installations.[89] China and the United States together accounted for over half of the global market for wind development and installations services in 2011 (table 4.4).

Table 4.4. Top five wind energy development and installation services markets, 2011[a]

Country	Annual installed capacity, 2011 (MW)	Estimated value of services market (billion $)	Percentage of global services market[b] (%)
China	17,631	9.0–11.8	39–52
United States[c]	6,810	5.5–7.1	24–31
Germany	2,086	2.3–3.5	10–15
Canada	1,267	1.2	5
Spain	1,050	0.99	4
Total	40,629	22.7–30.5	100

Source: GWEC, Global Wind 2011 Report, 2012, 12; IEA, IEA Wind 2011 Annual Report, July 2012; IRENA, Renewable Power Generation Costs in 2012: An Overview, 2013, 32.

[a] Data were not available for India.

[b] Percentage is calculated based on the lower estimated value of services market ($22.7 billion).

[c] U.S. estimates may differ from those presented earlier due to slight variations in data sources. Estimates in the U.S. market section were calculated using data from AWEA and Wiser and Bolinger; however, for consistency across countries, estimates presented here were calculated using data from GWEC, IEA and IRENA.

The value of global O&M services also continued to rise steadily, as the volume of the global installed base of wind capacity has more than doubled over the past five years (figure 4.4), increasing demand for these services. Estimating the size of the global O&M market is more challenging, due to the limited availability of data. However, in 2011 the global O&M market was likely worth at least $7.2 billion.[90] Additionally, among the 14 countries for which specific O&M data are available,[91] the largest estimated O&M markets were found in developed countries, particularly Germany ($1.8 billion), the United States ($1.4 billion), and Spain ($1.1 billion). Despite the rapid growth in new wind installations in China and the fact that it has by far the most cumulative installed capacity globally, China's O&M market remains comparatively small at an estimated $917 million in 2011.[92]

Factors Affecting Global Supply and Demand for Wind Energy Services

Two distinct factors are reportedly driving demand for wind energy in the global market: the growing need for power, and the growing desire for environmentally friendly alternative power sources. Particularly in regions such as Latin America, where many countries face rising power demand and rely on expensive fossil fuels for energy, wind is a very competitive source of energy.[93]

In developing or emerging wind markets, such as in many countries in Latin America, there is typically strong demand for turnkey projects, as there is little or no domestic industry and markets lack the experience and skilled labor to support such projects. For example, in the Chilean market there is virtually no domestic industry, and the majority of services are provided by foreign firms.[94] As markets mature and service providers and customers gain experience, the demand for turnkey projects usually tapers off; however, foreign service providers often find new opportunities. For example, in Japan, wind service firms (both foreign and domestic) offer wind sector management services, which include diagnostic tests and reinstallation of existing wind projects to improve efficiency.[95]

Changes to government policies that support the development of renewable energy have had a direct effect on new installations and related demand for wind services, as illustrated by the recent slowdown in growth of new wind projects in many Western markets. For example, the size and steady growth of the European wind energy market is frequently attributed to long-term policies, such as feed-in tariffs (FITs)[96] or the European Wind Initiative,[97] which supported development of the wind industry.[98] However, in the wake of the economic downturn and budgetary concerns of the past several years, many European governments have reevaluated their renewable-energy legislation. For instance, both Ireland and the United Kingdom (UK) are seeking to cut government expenses by reducing incentives; Italy has reduced subsidies and introduced an additional tax on renewable companies; and in 2012, Spain suspended subsidies for new renewable energy projects.[99] It is expected that these policy changes will result in slower growth in new installations.

Global Industry Trends

As in the United States, utility companies globally are increasingly involved in project development and ownership, albeit on a smaller scale than industry leaders. This is particularly the case in China, where over half of the country's total wind capacity was owned by the five largest state-owned power generation corporations in 2011.[100] Utilities' increased presence is driven by growing pressure and incentives to increase the share of renewable sources in their portfolios.[101] This trend is seen in the largest markets around the world as utilities increasingly build, own, and operate large wind farms in China, Germany, Spain, the UK, and the United States.[102]

Overall, the global wind energy services market is fragmented along national borders. Typically, only the largest service providers have multinational operations.[103] Although most OEMs sell turbines worldwide, most do not provide services in markets where they do not have a commercial presence. Instead, in those markets they hire local firms to develop and install wind farms, as well as for O&M services.[104] Similarly, leading project developers are also multinational companies and although they typically solicit bids through an objective procurement process, they most frequently hire local service firms for new projects.[105] This is largely attributed to the regulatory or standards-based nature of many services, such as engineering and construction occupations, which may require local licensing or certification.[106] The exception to this is certain technology-based services such as data collection and analysis or remote monitoring, which are not site specific and may be easily provided across borders.[107] As a result, there are a number of multinational site assessment firms, the largest being GL Garrad Hassan (part of the Germany-based GL Group).[108]

Project Development, Construction, and Installation Services

Growth in global installations of wind farms over the past decade has created profitable opportunities for project developers. As a result of projected slow growth in their home markets, many European and U.S. service firms have sought to expand their presence in foreign markets, with a particular focus on Asia and Latin America.[109] Developers typically look for opportunities in politically stable countries with a consistent regulatory regime, strong contract law infrastructure, and demonstrated experience with successful PPAs.[110] Additionally, wind energy tends to be most competitive in markets, such as Central and South

America, that are heavily reliant on fossil fuels (typically diesel), which tend to be imported and very expensive.[111] European multinational service providers tend to be more active than U.S. firms in most regions of the world, possibly due to their greater experience.[112] German and Spanish firms in particular have shifted their focus to international activities and are actively targeting South America.[113] For example, Spain's Iberdrola has announced that Brazil will be the second-largest destination of renewable energy spending through 2014, after the UK. [114] Turbine manufacturers have also expanded their presence into emerging markets,[115] largely by building factories and providing services alongside equipment.[116]

The largest project developers have typically been established multinational firms from European countries, such as Iberdrola.[117] However, due to the growth of the Chinese market and the high level of U.S. demand in 2012, 5 of the 10 largest developers of new wind projects in 2012 were from China (all of which are state-owned enterprises), and 3 were from the United States (table 4.5).[118] EPC firms are less likely than project developers to be active outside their home markets. However, global wind EPC contractors (table 4.6) use their multinational presence to operate across multiple sectors including wind. Growth in global wind installations has facilitated the rise of EPC firms with a stronger focus on wind, such as Iberdrola Ingeniería y Construcción—the developer's construction and engineering branch.

Table 4.5. Top 10 global wind energy project developers, based on projects commissioned in 2012

Ranking	Company	Home country	Capacity developed (MW)
1	China Datang Corp.	China	1,526
2	NextEra Energy Resources	United States	1,215
3	Guohua Energy Investment Co.	China	1,142
4	China Huadian Corp.	China	1,084
5	China Longyuan Power Group Corp.	China	1,072
6	Gamesa Corp. Technológica SA	Spain	1,066
7	Iberdrola SA	Spain	1,012
8	Invenergy LLC	United States	905
9	Caithness Energy LLC	United States	845
10	China Huaneng Group Corp.	China	823

Source: Data compiled by USITC from BNEF database (accessed April 2, 2013).

Table 4.6. Top 5 international wind energy contractors, based on export revenue from wind energy construction, 2011

Rank	Company	Home country	Wind export revenue (millions $)
1	Fluor Corp	United States	330.9
2	Mortenson Construction	United States	212.7
3	Iberdrola Ingeniería y Construcción SA	Spain	132.7
4	Van Oord NV	Netherlands	61.7
5	Balfour Beatty PLC	UK	26.6

Source: ENR, ENR Global Sourcebook 2012, December 10, 2012.

Note: Export revenue is defined as revenue generated from projects outside each firm's home market.

O&M Services

Since 2007, the most striking trend in the global O&M market has been the rise in competition, which has created a buyer's market with lower costs and longer contract durations.[119] For example, Bloomberg New Energy Finance (BNEF) found that between 2008 and 2012, the price of full-service O&M contracts fell 38 percent, while the length of the average full-service O&M contract offered by OEMs increased from 4.5 years to 6.9 years.[120] Prices and contract length vary regionally due to factors such as the degree of competition within a market, labor costs, and the size of wind projects.[121] In China, where excess production capacity has created a surplus of wind turbines and created intense competition among manufacturers, wind project developers frequently demand warranty periods of up to 10 or even 20 years.[122] Higher labor costs are frequently cited as the driver of O&M costs in Europe, and are also likely a factor behind higher-thanaverage costs in emerging markets such as Brazil, Mexico, and South Africa.[123] O&M costs are lowest in the U.S. market, which enjoys lower labor costs than Europe, large projects that allow service providers to capture economies of scale, and strong competition that is due in part to the prevalence of in-house maintenance by large owner/operators.[124] However, despite falling prices, the global O&M market continues to grow as installed capacity rises.

Many firms provide O&M services in the global market. Leading providers include OEMs such as Enercon, Siemens, and Vestas, as well as ISPs—particularly large wind farm operators that provide in-house maintenance, such as Iberdrola.[125] O&M services markets vary substantially across countries. Compared to the United States, Europe tends to have smaller wind farms held by smaller owners typically not capable of providing their own O&M services. As a result, many of the European Union (EU) markets for O&M are dominated by OEMs and ISPs offering long-term service agreements.[126] For example, Germany's market is reportedly dominated by full-service contracts from OEMs, such as Enercon.[127] However, in markets with significant ownership by large operators and utilities, such as Acciona Energia and EDP Renováveis (EDPR) in Spain, there is a larger share of in-house maintenance supported by ISPs.[128] Similarly, in China—where over half of wind projects are developed by state-owned power groups— the majority of O&M services are performed in-house.[129]

With competition sharpening in the global O&M market, acquisitions of O&M service providers have increased as a growing number of firms seek to expand their service offerings and enter the O&M market. Similar to the trend in the United States, where utilities and operators have acquired wind energy service firms in order to self-perform maintenance, China's large state-owned utilities (which develop more than half of China's new wind projects) each have their own O&M groups.[130] In addition to this growing interest in self-performed maintenance by owners, component providers and other firms in the supply chain are increasingly acquiring service firms in an effort to offer specialized services.[131] In 2012, UK-based crane operator Ainscough entered the wind installation and maintenance market by acquiring Windcon, a UK-based provider of wind turbine erection and maintenance services.[132] A U.S. bearing manufacturer, Timken Co., similarly sought to diversify its services and enter the repair and maintenance market by acquiring a U.S. turbine maintenance company, Wazee, in early 2013.[133]

Trade and Investment

Data on U.S. and global trade in wind energy services are unavailable. Trade in wind energy services occurs predominately through commercial establishment (mode 3), with limited services provided cross-border (mode 1) or through movement of natural persons (mode 4).[134] The nature of wind energy services typically requires proximity to a project site, making remote provision difficult, if not impossible, for most services. However, certain technology-enabled services, such as wind resource assessment and remote monitoring of operations, are not location dependent and are increasingly provided cross-border. As a result, some wind service firms seeking to expand their global presence are diversifying by offering these services. One example is U.S. EPC firm Black & Veatch, which does not provide EPC contracting outside the United States, but does offer resource modeling and analysis services.[135] On the other hand, many industry representatives interviewed for this study emphasized the importance of local presence, and service firms often establish a commercial presence even for services that could be provided cross-border.[136] For instance, wind consultant GL Garrad Hassan, the global leader in wind resource assessment, has offices around the world in an effort to offer market-specific knowledge and support.[137]

Imports and Exports

The United States is a net importer of wind energy services, as many of the largest wind energy services firms in the United States are subsidiaries of foreign multinationals. GE is the leading supplier of turbines in the U.S. market and accounted for 38 percent of installations in 2012. However, only one of the other top 10 turbine suppliers in 2012 was a U.S.-based firm.[138] Similarly, some of the largest project developers in the United States are also U.S. subsidiaries of foreign parent firms (see table 4.2). In 2012, about one-third of newly installed wind capacity was developed by U.S. affiliates of foreign firms.[139] Development, construction, and O&M services provided by these firms to U.S. customers are considered to be imports in the form of affiliate sales (services provided by U.S. affiliates of foreign firms).

U.S. wind service firms, on the other hand, have a smaller global presence than some of the largest European multinational firms discussed above, and as a result, U.S. exports likely account for a smaller share of global exports of wind services than EU exports. Industry representatives indicate that a number of U.S. firms provide resource assessment services overseas.[140] For instance, AWS Truepower, which has been active in the Caribbean and Central America.[141] The largest U.S. project developers and EPC firms also operate internationally, particularly in Canada.

Canada is likely the largest export market for U.S. wind services. NextEra Energy Resources and Invenergy, which are U.S.-based, are both active in the Canadian market; U.S. EPC firm Mortenson Construction generates over 99 percent of its international revenue from wind operations in Canada.[142] U.S. wind services firms are reportedly attracted to the Canadian market due to its geographic closeness as well as the opportunities offered by significant planned MW capacity.[143] Other attractive markets for U.S.-based service providers include Mexico and other countries in Latin America, particularly for OEM turbine manufacturers that bundle service offerings with equipment sales.

Chinese OEMs are increasing their global presence and, as a result, Chinese exports of wind services are growing. For instance, Xinjiang Goldwind Science & Technology (Goldwind) is expanding internationally; currently, it generates 10 percent of its revenues overseas. [144] Chinese exports of wind-powered generating sets [145] increased from $78 million in 2007 to $467 million in 2012, with 39 percent of 2012 exports going to the United States.[146] In addition to established markets like the United States, Chinese OEMs are reportedly seeking to increase their presence in Latin America, including in Argentina, Bolivia, Brazil, Chile, and Ecuador, as well as in Africa.[147] Some Chinese firms reportedly have improved product offerings that, combined with low prices and the firms' access to low-cost financing, make Chinese turbines competitive in some of these markets.[148] Chinese service providers are also increasing their presence outside of their home market. For example, Chinese project developer HydroChina is developing wind power projects in counties such as Bolivia, Ethiopia, and Pakistan.[149]

Investment

The United States is considered to be a desirable market for foreign investors, as illustrated by the strong commercial presence of foreign wind service firms. European firms have sought investments in the U.S. market as a way to leverage their expertise in a market with expanding capacity.[150] European utilities in particular have invested in the U.S. market. For example, in 2007, Spain's Iberdrola acquired U.S. developer CPV Wind Ventures, as well as U.S. utility PPM Energy as part of its acquisition of Scottish Power.[151]

Foreign investment declined as growth in the U.S. wind industry slowed in 2010 and 2011.[152] However, it has since rebounded. One example is India-based Trishe Wind Energy, which acquired U.S. project developer National Wind LLC in 2012. Another is Australian private equity fund AMP Capital, which announced in February 2013 that it would invest $100 million in U.S. wind developer Capistrano Wind Partners.[153]

U.S. foreign direct investment in wind energy services has risen. U.S. firms are increasingly looking abroad to expand their presence in European and Asian markets, citing weakness in the U.S. market as motivation for their outward focus.[154] For example, in 2008, wind developer AES acquired a minority stake in a Chinese wind farm; in 2010, it acquired UK developer Your Energy and signed a deal for a majority stake in Polish wind developer 3E.

Trade Barriers

Restrictions on or barriers to trade in goods have frequently created challenges for wind energy service providers in the global market. Reportedly, among the most prominent of these challenges are local-content requirements, which mandate that a certain percentage of the wind turbines or labor required for a wind project must be sourced locally.[155] Since many OEMs are also leading service suppliers, even where the local-content requirements are focused specifically on equipment, these policies appear to be impeding trade in services.[156]

A number of localities and countries have local-content requirements in place. Reportedly, among the most problematic for U.S. wind service providers have been those in Ontario, Canada. Ontario's market provides significant export opportunities for U.S. service providers, but wind service firms must meet local-content requirements to qualify for the FIT. [157] Industry representatives indicate that while they try to meet these specifications, the requirements become particularly challenging and make the market unattractive when there are too few local suppliers to meet demand. This can result in less competition in the market.[158] Additionally, these requirements have reportedly influenced decisions on commercial presence, leading companies who would otherwise establish a presence in the

United States to establish a presence in Ontario instead in order to provide goods and services in both places.[159]

In May 2013, the WTO Dispute Settlement Body adopted a WTO Appelate Body report and panel report (as modified by the Appelate Body) finding that the Canadian measures were inconsistent with the TRIMs Agreement and the GATT 1994.[160] Subsequently, Ontario's Minister of Energy directed the Ontario Power Authority not to procure any additional large projects. The Minister indicated that the government intends to replace the FIT program for large projects with a competitive procurement process, and that the Ministry intends to pursue legislation to bring the FIT program into compliance with the WTO ruling.[161]

Brazil does not mandate the use of local content to operate, but local content must be used in order to receive low-cost financing for projects from BNDES, the national development bank.[162] In Brazil, one firm also noted that the lack of a reciprocal visa program with the United States limited the firm's ability to bring in U.S. technical experts to work on Brazilian wind farms. Instead, this firm brings in technicians from the EU, which does have such a visa program with Brazil.[163]

Certain markets, such as China and Latin America, were also identified as having restrictions on capital repatriation, and China has a joint venture requirement for market entry.[164] Otherwise, most industry representatives pointed to few formal trade barriers in wind services other than local-content requirements. Instead, many wind service firms face challenges related to specific economic or cultural characteristics of each market. For example, a number of service providers indicated that despite the potential opportunities, they had decided against establishing a presence in China due to perceived difficulties of doing business there as a foreign provider.[165] Other dissuading factors reportedly include the high level of competition from Chinese firms, the presence of intellectual property issues compounded by a weak court system, and a lack of transparency in the grid connection process.[166]

Country Profiles

This section profiles the wind energy markets in Brazil, Canada, and China. China was the largest market outside of the United States in 2012, with 12,960 MW in installations. Canada has a substantial wind energy market, with 935 MW installed in 2012, that provides significant export opportunities for U.S. firms. Brazil also has a rapidly growing wind market and was the largest market in Latin America in 2012, with annual installations reaching 1,077 MW in 2012.[167]

Brazil

Brazil is noted for having some of the world's best wind resources,[168] and its wind industry has been growing rapidly over the past five years, although its total installed capacity remains low relative to that of other countries. Cumulative installed capacity jumped from 247 MW in 2007 to 2,508 MW in 2012, and Brazil was the eighth-largest global market by new installed capacity in 2012. [169] This rapid expansion in wind development has been driven by the government's active support for development of the renewable energy sector to meet Brazil's growing demand for energy, as well as the competitiveness of wind relative to other energy sources in Brazil.[170]

The strong growth of wind projects, particularly since 2009, has attracted a number of wind firms to the market. Among project developers and construction firms, European wind service firms have a large presence in the Brazilian market, although Brazil's Renova remains the leading developer.[171] Among OEMs, the market is quite competitive, with substantial participation by foreign manufacturers, such as GE (based in the United States), IMPSA (Argentina), Enercon (Germany), Gamesa (Spain), and Vestas (Denmark).[172] In Brazil, OEMs typically do not participate in project development, but instead provide limited services related to logistics and the supervision of turbine erection or crane services, in addition to providing the majority of O&M services.[173]

As the Brazilian market has grown, a number of challenges have reportedly emerged. One is the inadequate grid, which prevents developers from selling surplus power on the unregulated market, as well as delivering power as contracted in PPAs.[174] As a result, in the future, grid connection will be the responsibility of the project developer, a change that will be reflected in higher PPA prices.[175] Moreover, PPAs for the regulated electricity market are awarded through reverse auction, a requirement that is also said to present a challenge for wind service firms. One industry representative indicated that not only is the process unfamiliar to foreign wind firms, but also it has pushed the price of electricity down while project costs are boosted by local-content requirements.[176] As mentioned earlier, local-content requirements are a prerequisite for funding from the national development bank, BNDES, and present a particular challenge for foreign OEMs.[177] However, these requirements also indirectly impact developers, as they can create bottlenecks in the supply chain. Nevertheless, due to its cost competitiveness, wind energy has won most of the energy auctions set aside for renewable energy.[178]

Canada

Canada has a robust and growing wind energy industry and is reportedly the United States' largest trading partner for wind energy services.[179] Total installed wind capacity has increased steadily over the past five years, growing 27 percent annually from 2007 through 2012.[180] Yet, the Canadian market—the ninth largest globally—remains far smaller than the U.S. market; Canada added 935 MW of new wind capacity in 2012 (compared to more than 13 GW in the United States) for a total of 6,200 MW of cumulative installed capacity.[181] Overall, however, the Canadian market exhibits more stable growth and fewer year-to-year fluctuations in installations than the U.S. market. Industry representatives attribute this to two factors. One is the existence of long-term policies supporting the development of wind energy—namely, the FIT program in Ontario and wind tenders in Quebec; the bulk of Canada's wind projects are located in these two provinces. The second factor is the greater transparency in Canada's planning process, which offers service providers information about future plans for wind energy.[182]

However, there is reportedly less competition in Canada's wind services industry than in the United States, and wind energy services in Canada tend to be more expensive, as developers are typically able to secure lucrative price agreements under provincial FITs, resulting in little downward pressure on prices in the supply chain.[183] Canada's more conservative investment environment reportedly means that fewer service providers, particularly smaller firms, are operating in the market. Canadian investors reportedly seek mature, financeable companies and as a result it is difficult for smaller firms to secure financing, particularly given the small market size.[184] In addition, industry representatives indicate that local-content requirements may have reduced the level of competition in the market, as noted earlier, because fewer firms have been able to meet these requirements and enter the market.[185]

Despite these issues, a number of foreign wind service firms operate in the Canadian market, attracted by its growth. In 2012, U.S. service firms developed nearly 20 percent of newly installed capacity, representing the largest share of foreign-developed projects. Services reportedly represent a significant opportunity for foreign firms, particularly EPC services.[186] Nevertheless, opportunities for project development remain somewhat limited, since the small market size and lengthy procurement process mean that many new projects already have PPAs awarded. As a result, many foreign developers are entering the market through acquisition.[187] For instance, in 2011, China's largest project developer, China Longyuan Power Group, purchased equity shares in an Ontario wind farm, and in 2010, EDP Renewables North America (the regional subsidiary of Spain-based EDPR) acquired new projects in Ontario.[188]

China

China's wind energy industry has experienced tremendous growth over the past five years.[189] Despite having less than 6 GW of installed wind energy capacity in 2007, China's installed capacity reached 75 GW by the end of 2012—overtaking both the United States and Germany.[190] However, the Chinese market faces a number of challenges.[191] The rapid expansion has created problems that have begun to moderate growth in annual installations: China has seen 26 percent fewer installations in 2012 than in 2011.[192] Factors in this trend reportedly include revised procedures for project approval by the National Energy Administration, which have delayed project commissioning; issues related to grid connection and curtailment of new wind projects;[193] delayed payments from the government; a general slowdown in China's economic growth (particularly in the industrial sector); and a decline in electricity consumption per unit of gross domestic product (GDP).[194]

China's wind energy services market is viewed as highly competitive, with the expansion of the wind energy industry driven largely by new project development carried out by state-owned enterprises (SOEs), with limited participation by OEMs or smaller ISPs.[195] Overall, SOEs undertake the bulk of investment in and construction of new farms, accounting for over 80 percent of installed capacity in 2012.[196] These groups typically provide all wind energy services in-house, from the design and construction of a project to the design and installation of turbines to O&M services.[197]

Few foreign firms participate in the Chinese services market, and those that do are limited to OEMs and some specialty ISPs. In 2011, foreign OEMs represented less than 15 percent of the market, in contrast to earlier years, when they controlled a majority of the Chinese market. Reportedly, many OEMs have left the market due to overcapacity and, to a lesser extent, protectionist policies.[198] Industry representatives indicate that foreign OEMs in China do not compete directly with their Chinese counterparts; instead of trying to compete on low equipment prices, they instead compete on the cost of energy (which reflects not just the initial purchase price, but also performance and reliability). As a result, these firms are generally in competition with other foreign firms in China, rather than with Chinese firms.[199]

Only a limited number of foreign ISPs (e.g., resource assessment or consulting firms) operate in the Chinese market, which they reportedly find difficult due to the market environment and the experience of their customer base. Foreign service firms are unable to compete directly with Chinese counterparts due to higher costs, and also reportedly find it difficult to differentiate their services from those that project developers can supply in-house.[200]

References

American Wind Energy Association (AWEA). "AWEA U.S. Wind Industry Annual Market Report 2012." May 2013. http://www.awea.org/learnabout/publications/factsheets /upload/AWEA-U-S-WindIndustry-Annual-Market-Rankings-2012.pdf.

———. *AWEA U.S. Wind Industry Annual Market Report: Year Ending 2012.* Executive Summary, 2013. http://awea.files.cmsplus.com/images/AWEA USWindIndustryAnnual MarketReport2012 Execu tiveSummary%282%29.pdf.

———. *Annual Market Report, Year Ending 2007*, 2008.

———. "Industry Statistics," n.d. (accessed March 27, 2013). http://www.awea.org/learnabout/industry stats/index.cfm.

———. *U.S. Wind Industry Annual Market Report 2011*, 2012.

———. *U.S. Wind Industry Fourth Quarter 2011 Market Report*, January 2012.

———. *Wind Energy Siting Handbook*, February 2008.

———. *Year End 2009 Market Report*, January 2010.

Anderson, Alan, and J. Britton Gibson. "Despite Uncertainty, Foreign Investors Eye U.S. Market." *North American Windpower*, April 2012, 1 and 28.

Anderson, Erin L., Greg D. Corbin, and Timothy L. McMahan. "The Law of Wind: Siting and Permitting Wind Projects," n.d. (accessed March 28, 2013). http://www.agmrc.org /media/cms/WindSitingandPermittingProjects 42DEBA672F7A7.pdf.

Associated Free Press. "China Energy Consumption Rises 3.9% in 2012," February 23, 2013. Bailey, Diane. "Building Boom Draws in Global Players." *Wind Power Monthly*, October 1, 2012.

Black and Veatch. *Cost and Performance Data for Power Generation Technologies*. February 2012. http://bv.com/docs/reports-studies/nrel-cost-report.pdf.

Bloomberg. "Eximbank to Lend Vietnam $1 billion for Mekong Wind Power Plant," October 18, 2011. Bloomberg New Energy Finance (BNEF). "Operations and Maintenance Price Index," October 23, 2012. BNEF Database. http://www.bnef.com. (accessed multiple dates).

Boyland, Kevin. "Gone with the Wind: After Years of Rapid Growth, Revenue with Slump with Fewer Projects." IBISWorld Industry Report OD4491: Wind Turbine Installation. December 2012.

Broehl, Jesse. "Operations & Maintenance – Industry prepares for Warranty Tipping Point." *Wind Power Monthly*, April 1, 2010.

Brown, Phillip. "U.S. Renewable Energy: How Does the Production Tax Credit (PTC) Impact Wind Markets?" Congressional Research Service Report for Congress R42576, June 20, 2012.

BTM Consult. *World Market Update 2011*, March 2012.

Canadian Wind Energy Association (CanWEA). "The Many Steps to Planning and Building a Wind Farm," n.d. (accessed January 16, 2013). http://www.canwea.ca/farms/planning e.php.

CanWEA. "Wind by the Numbers: Economic Benefits of Wind Energy," January 2013. http://www.canwea.ca/pdf/canwea-factsheet-economic-web-final.pdf.

ChinaDaily.com. "China's GDP Growth Eases to 7.8% in 2012," January 18, 2013. Cleantechinvestor.com. "Brazil: Wind Manufacturing Hub," August 15, 2011.

Connell, James and Dennis Grennan. "Who Needs an Owner's Engineer?" *Power Magazine*, March 1, 2011. www.powermag.com/business/who-needs-an-owners-engineer 3481.html.

Creed, Adrian, and Dr. Amir Kordvani. "The WTO Report on Local Content Requirements for Canadian Renewables Programme – MENA Implications." February 13, 2013. http://www.clydeco.com/insight/updates/the-wto-report-on-local-content-requirements-forcanadian-renewables-progra.

Cusick, Daniel. "With Congress at its Back, Wind Power Industry Pushes Ahead." Governors' Wind Energy Coalition, January 3, 2013. http://www.governorswind energycoalition.org/?p=4362.

Database of State Incentives for Renewables and Efficiency (DSIRE). "Renewable Portfolio Standard Policies," March 2013. http://www.dsireusa.org/documents/summarymaps/RPS map.pdf (accessed March 27, 2013).

David, Andrew. "Growth in Wind Turbine Manufacturing and Trade." USITC Executive Briefing on Trade, March 2009. http://www.usitc.gov/publications/332/Executive_ Briefings/USITC_EB_WindTurbines_David.p df.

David, Andrew and Dennis Fravel. "Wind Turbine Export Opportunities in Canada and Latin America." USITC Working Paper, No. ID-032, July 2012. http://www.usitc.gov/publications/332/working papers/ID-032 final.pdf.

Del Franco, Mark. "AWEA Reveals Numbers Behind U.S. Wind Sector's Record-Breaking Year." *North American Windpower*, April 11, 2013. http://www.nawindpower.com/e107 plugins/content/content.php?content.11362.

DLA Piper. "EPC Contracts in the Power Sector," 2011. http://www.dlapiper.com /files/Publication/18413b26-49b8-490e-acc6-3ff54faa55d7/Presentation/PublicationAttachment/1205e08d-e585-479d-ac17-42135efaf044/epccontracts-in-the-power-sector.pdf.

Engineering News-Record (ENR). *ENR Global Sourcebook 2012*, December 10, 2012. ENR. *ENR Sourcebook*, September 17, 2012.

European Wind Energy Association (EWEA). "WTO Rules Against Ontario Local Content Requirement for Renewables." Press release, December 20, 2012. http://www.ewea.org /pressreleases/detail/?txttnews%5Bttnews%5D=2003&cHash=257f985f5491 cfeb517a4bffd048795d.

Export Import Bank of the United States (Ex-Im Bank). "U.S. Exports Will Expand Wind Farm in Honduras." News release, March 5, 2013. http://www.exim.gov/newsandevents /releases/2013/Exports-Will-Expand-Wind-Farm-inHonduras.cfm.

Global Wind Energy Council (GWEC). *2012 China Wind Energy Outlook*, 2012.

———. *Annual Market Update 2011*, March 2012.

———. *Global Wind 2007 Report*, 2008.

———. *Global Wind 2011 Report*, July 2, 2012.

———. *Global Wind Statistics 2012*, February 11, 2013.

Goosens, Ehren. "AES Says U.S. Wind Market is Weak, Focuses on Europe and Asia." *Bloomberg New Energy Finance*, November 4, 2010.

Global Trade Information Services, Inc. (GTIS). Global Trade Atlas database, http://www.gtis.com (accessed July 9, 2013).

Harper, John, Matthew Karcher, and Mark Bolinger. "Wind Project Financing Structures: A Review and Comparative Analysis." Ernest Orlando Lawrence Berkeley National Laboratory publication no. LBNL-63434, September 2007. http://emp.lbl.gov/sites /all/files/REPORT%20lbnl%20- %2063434.pdf.

International Energy Agency (IEA). *IEA Wind 2011 Annual Report*. July 2012.

———. "Topic: Wind Power," n.d. (accessed September 13, 2012). https://www.iea.org/topics/windpower.

International Renewable Energy Agency (IRENA). *Cost Analysis of Wind Power*, June 2012.

———. *Renewable Energy Power Generation Costs in 2012: An Overview*, 2013.

Ireland, Beck. "Wind Service." *Electrical Construction and Maintenance*, November 2011.

Johnston, Adam. "China Reaches 50 GW of Connected Grid Wind Capacity; Expected to Top 140 GW by 2015." *Cleantechnica.com*, February 21, 2013. http://cleantechnica.com/2013 /02/21/chinareaches-50gw-of-connected-grid-wind-capacity-expected-to-top-140gw-2015/.

Jordan, Phil, and Cai Steger. *American Wind Farms: Breaking Down the Benefits from Planning to Production*. National Resources Defense Council, September 2012. www.nrdc.org/energy/files/american-wind-farms-IP.pdf.

Li, Junfeng, et al. China Wind Energy Outlook. n.d. http://www.gwec.net/wp-content/uploads/2012/11/China-Outlook-2012-EN.pdf.

Lima, Joao. "EDP Buys Horizon Wind from Goldman for $2.15 Billion." *Bloomberg*, March 27, 2007. http://www.bloomberg.com/apps/news?pid=newsarchive&sid=a0YT3J3T9 imA.

Logistics Business Review. "LM Wind Power Service & Logistics acquires Encore Power Services," January 13, 2011. http://www.logistics-business-review.com/news/lm-wind-power-servicelogistics-acquires-encore-power-services-130111.

Massy, Janice. "Banks Opting for Lazy Route could Increase Risk." *Wind Power Monthly*, October 17, 2010.

McGovern, Michael. "China puts Bolivia on Wind Map." *Wind Power Monthly*, March 11, 2013.

Morales, Alex and Ben Sills. "Spain Ejects Clean-Power Industry with Europe Precedent: Energy." *Bloomberg*, May 30, 2012.

Navigant Research. *World Market Update 2012*. A BTM Wind Report, March 2013.

Nielsen, Stephan. "BNDES Raises Local-Content Requirement for Brazil Wind Turbines." *Bloomberg*, September 13, 2012. http://www.bloomberg.com/news/2012-12-13/bndes-raises-local-contentrequirement-for-brazil-wind-turbines.html.

———. "Cheapest Wind Energy Spurring Renewables Deals: Corporate Brazil." *Bloomberg*, January 29, 2013.

———. "China Grabs Share in Latin America Wind with Cheap Loans." *Bloomberg*, November 20, 2012. http://www.bloomberg.com/news/2012-11-20/china-grabs-share-in-latin-america-wind-withcheap-loans.html.

———. "Wind Energy Makers Thrown by Local Supply Rules in Brazil." *Bloomberg*, March 18, 2013. http://www.bloomberg.com/news/2013-03-18/wind-energy-makers-thrown-by-local-supplyrules-in-brazil.html.

North, Will. "Ainscough Completes Windcon Deal." *Cranes Today*, October 24, 2012.

North American Windpower. "China holds spot as world's top wind energy market, on track to reach 100 GW by next year," February 4, 2013.

O'Brian, Heather. "Wind Projects Look to Export Finance as New Markets Open." *Wind Power Monthly*, February 27, 2013.

———. "Uncertainty Drives Italian Firms Abroad." *Wind Power Monthly*, December 23, 2011.

Ontario Power Authority. "General Information about the FIT and microFIT Programs." http://fit.powerauthority.on.ca/program-resources/faqs/general-information-about-fit-and-microfit-programs (accessed January 31, 2013).

Outland Energy Services. "Duke Energy Renewables Acquires Outland Energy Services," November 9, 2012. http://outlandenergy.com/duke-energy-renewables-acquires-outland-energy-services/.

Pernick, Ron, Clint Wilder, and Trevor Winnie. *2012 Clean Energy Trends*. March 2012.

Quilter, James. "AMP Capital to Invest $100 Million in U.S. Wind." *Wind Power Monthly*, February 12, 2013.

Qi, Wu. "China's Service Market Grows as Installations Fall." *Windpower Monthly*, July 1, 2012.

———. "China's Wind sector lost $1.6 Billion in 2012." *Windpower Monthly*, January 25, 2013.

———. "Debt Hampers Chinese Wind." *Windpower Monthly*, May 2013, 29.

———. "Hydrochina Wins 30MW Deal in Pakistan." *Windpower Monthly*, June 4, 2013. http://www.windpowermonthly.com/article/1184812/hydrochina-wins-30mw-deal-pakistan.

———. "Sany Wins 153 MW Ethiopian Contract." *Windpower Monthly*, May 20, 2013. http://www.windpowermonthly.com/article/1182941/sany-wins-153mw-ethiopian-contract.

Schoenberger, Robert. "Timken Buys Wazee Companies, Expands Wind Turbine Service Business." *Cleveland Plain Dealer,* January 2, 2013. http://www.cleveland.com/business/index.ssf/2013/01/timken buys wazee companies ex.html.

Scott, Mark. "A Spanish Blowout: The Iberian Giants of U.S. Wind Energy." *Spiegel Online International*, October 29, 2008. http://www.spiegel.de/international/business/a-spanish-blowoutthe-iberian-giants-of-us-wind-energy-a-587218.html.

———. "In Europe, Green Energy Takes a Hit from Debt Crisis." *New York Times*, November 13, 2012. http://www.nytimes.com/2012/11/14/business/energy-environment /in-europe-green-energy-takesa-hit-from-debt-crisis.html.

Sen, Sandip. "Wind Energy Majors Set up Low-Cost Plants in India." *Clean Biz.Asia*, February 23, 2011.

Sills, Ben. "Spain Halts Renewable Subsidies to Curb $31 Billion of Debts." *Bloomberg*, January 27, 2012.

Spatuzza, Alexandra. "Brazil – Lack of Grid Connections Keeps Wind Farms Offline." *Wind Power Monthly*, March 1, 2013.

———. "Turbine Makers Focus on Local Content Rules." *Wind Power Monthly*, September 26, 2012.

Tardieu, Pierre. "EWEA Trade Policy Views." Presented at ICSTD Conference, Bali, Indonesia, March 29, 2012. http://ictsd.org/downloads/2012/04/presentation-mr-tardieu.pdf.

Tegen, S., M. Hand, B. Maples, E. Lantz, P. Schwabe, and A. Smith. *2010 Cost of Wind Energy Review*. National Renewable Energy Laboratory Technical Report, NREL/TP-5000-52920. April 2012. http://www.nrel.gov/docs/fy12osti/52920.pdf.

U.S. Energy Information Administration (EIA). *Electric Power Annual 2011*. January 2013. http://www.eia.gov/electricity/annual/pdf/epa.pdf.

———. International Energy Statistics database (accessed August 19, 2013).

Weyndling, Richard. "Frustrated Spanish Wind Sector Looks Abroad to Stay in Business." *Wind Power Monthly*, June 28, 2011.

Windpower Engineering and Development. "Top Wind Power Developers (2012)." N.d. (accessed May 30, 2013). http://www.windpowerengineering.com/top-wind-power-developers-2012/.

Wind Power Monthly. "Balancing Act for a Hard Pressed Supply Chain." August 25, 2011.

———. "Canadian Growth Brings New Interest and Challenges." October 1, 2012.

———. "GE and Remote Turbine Monitoring," March 10, 2010.

———. "Operations and Maintenance," September 1, 2009.

———. "US O&M Spending to Reach $6 Billion by 2025," July 11, 2012.

Windustry.org. "Section 10: Tax Incentives." *Community Wind Toolbox*, n.d. (accessed January 30, 2013). http://www.windustry.org/community-wind/toolbox/chapter-10-tax-incentives.

———. "Section 13: Power Purchase Agreement." *Community Wind Toolbox*, n.d. (accessed March 24, 2013). http://www.windustry.org/community-wind/toolbox/chapter-13-powerpurchae-agreement.

Wiser, Ryan, and Mark Bolinger. *2011 Wind Technologies Market Report*. U.S. Department of Energy, Energy Efficiency and Renewable Energy. August 2012.

———. *LBNL Wind Technologies Market Report Data*. August 2012.

World Bank. "Power Purchase Agreements (PPAs)." PPP in Infrastructure Resource Center (PPPIRC), June 18, 2008. http://go.worldbank.org/477RW9M170.

World Trade Organization (WTO). "Dispute DS 412: Canada—Certain Measures Affecting the Renewable Energy Generation Sector."
http://www.wto.org/english/tratop_e/dispu_e/cases_e/ds412_e.htm (accessed July 2, 2013).

Xiao, Su. "Goldwind, Change is in the Air." *WWEA Quarterly*. Issue 4, December 2012.

Xinhua.net. "China's Largest Wind Power Developer Marches into Canadian Market," July 13, 2011. http://news.xinhuanet.com/english2010/china/2011-07/13/c 13983164.htm.

———. "SOEs Dominate China's Wind Power Projects," April 10, 2013. http://news.xinhuanet.com/english/china/2013-04/10/c 132298902.htm.

Yepez-García, Rigoberto Ariel, Todd M. Johnson, and Luis Alberto Andrés. *Meeting the Electricity Supply/Demand Balance in Latin America & the Caribbean*. The World Bank, September 2010. http://www.esmap.org/sites/esmap.org/files/REPORT%20LAC%20 Electricity%20Challenge%20 octubre%202010%20LESMAP%20FINAL.pdf.

SECTION 5. HYDROPOWER AND GEOTHERMAL SERVICES

The hydropower and geothermal electricity sectors show substantial growth potential in many countries. As electric power capacity in these sectors is added, opportunities for service providers will also increase. Service sectors that will substantially benefit include engineering, construction, and operations and maintenance (O&M). New projects in emerging economies, in particular, often provide export opportunities for U.S. service providers. This section will discuss both sectors in detail.

Hydropower Energy

Hydropower, a renewable resource and one of the oldest forms of electrical power generation, accounts for the largest share of electricity generated from renewable resources. Due to the environmental impact of large dams on rivers and surrounding regions, however, many policies that aim to provide incentives for renewable energy target only small-scale hydropower projects. Smaller projects are generally viewed as less environmentally detrimental because they are thought to present fewer environmental and social concerns and to need much less in the way of financing and startup costs, although smaller projects do not always have a smaller environmental impact.[1]

This section presents information related to the entire hydropower industry and related services and, where available, specific information for small hydropower. Note that there is no globally accepted definition of "small" hydropower, and project capacities for hydropower may range from several watts for the smallest individual installations to tens of gigawatts (GW) for the largest. According to one estimate, hydropower projects rated at capacities lower than 10 megawatts (MW) represent about 10 percent of global hydropower capacity.[2] In this section, discussion of small hydropower refers to projects of 50 MW capacities or less, unless otherwise noted.[3]

Overview of Hydropower-related Services

Services related to hydropower projects (tables 5.1 and 5.2) are supplied throughout the lifecycle of a project.[4] Services provided prior to project development broadly include resource assessment, site studies, and technical and economic feasibility studies. Services related to project development and construction include project design, construction, construction management, and quality control consulting services. For large projects that are expected to significantly impact the local population, social services such as public outreach, working with local communities and interest groups, and community relocation services can be significant.[5] Once construction has been completed and a project is operational, it requires services related to operations and maintenance (O&M). As noted by one company, post-development services run the gamut "from inspection to commissioning, from repairs to upgrades."[6]

The development cost of a particular hydropower project is generally split between civil engineering/construction costs, potentially including infrastructure needed to access the site, and the cost of electro-mechanical equipment. Equipment costs tend to be fairly predictable, while civil engineering costs vary significantly based on the specific site, design choices, and local costs for labor and materials. The costs of small hydropower projects tend to be higher per kilowatt-hour (kWh) on average than those for large hydropower projects. Plants with capacity below 1 MW tend to have significantly higher capital costs per kWh because civil works costs are significantly lower, so the cost of equipment accounts for a greater share of the overall project cost.

Many hydropower projects refurbish an existing plant, rather than develop a new site. These "life extensions" may upgrade original equipment to increase the capacity at an existing site, or they may add electricity generation capacity to existing dams that were originally built for non-electricity purposes, such as flood control. Rehabilitating or expanding an existing plant generally has the lowest costs per kWh, because much of the initial civil works development has been completed.[7] The cost of adding capacity to an existing hydropower site is estimated to be in the range of $400–$800/kWh.[8]

Most original equipment manufacturers (OEMs) in the hydropower field customize hydroelectric turbines for each site, and they have traditionally provided O&M services as part of the warranty on their equipment, or under contract after the warranty expires.[9] However, more companies now provide services on equipment provided by other firms, and more customers are willing to hire independent service providers to provide O&M services at the lowest cost. Nonetheless, a significant share of the services provided by OEMs is O&M services; one hydropower equipment manufacturer estimated that services accounted for 6–8 percent of its total revenue.[10]

For example, Germany-based Voith won contracts for two South American micro-hydroelectric projects in January 2013, and will supply both goods and services for the projects. Voith will supply and install a turbine, generator, automation systems, and all related electrical and mechanical parts for a small expansion of Brazil's 372-MW Santo Antônio do Jari hydropower project, adding about 3.4 MW of capacity to the site's overall capacity. Voith will also provide two small hydroelectric generators for a new 45.8 MW run-of-river project in San Miguel, Colombia. Voith will be responsible for the erection and commissioning supervision services for both projects.[11]

Table 5.1. Services related to the development of hydropower projects

CPC code	CPC description	Hydropower-specific services
511	Pre-erection work at construction sites	• Site access infrastructure • Time scheduling and construction phase planning
513	Construction for civil engineering	• Dam, reservoir, tunneling and canal construction • Grid connection
865	Management consulting and related services	• Project engineering, supervision, and quality management • Cost, time, and risk management • Community relocation and public outreach to NGOs and local communities •Financial consulting services, including electricity price forecasting, asset valuation, advice on green certificates (carbon credits), and technical and financial reviews for lenders

CPC code	CPC description	Hydropower-specific services
8672	Engineering services, including advisory and consultative services (86721), engineering design for civil works (86724), and engineering services during construction (86727)	• Field engineering and site supervision • Inspection and testing • Cost estimates for developers and owners, including planning, licensing and permitting • Feasibility studies, tender evaluations, and contract documents • Manuals and safety guidelines, technical specification documents • Geological investigations and mappings • Weir and dam type studies, powerhouse developments, siting and plant access analysis • Hydrological and hydraulic modeling services and data assessments covering cascade simulation, forecasting and warning systems, flood protection, and groundwater modeling and management • Earthquake assessment services for seismic design, modeling services, damage assessments and rehabilitation packages • Environmental services including monitoring, risk and impact assessments, resettlements, soil and groundwater investigations, permits, audits, and due diligence • Project development services, including construction drawings; sustainable water resource planning; dam, reservoir and water conveyance engineering; generation output modeling; power plant design and refurbishment; and transmission lines Specific to the United States: • Consulting services to assist with the Federal Energy Regulatory Commission (FERC) licensing or relicensing process
86739	Integrated engineering services for other turnkey projects	• Integrated processes and methods • Technical assistance and advice •Disaster recovery

Sources: Verdant Power, written submission to the USITC, March 1, 2013; lists of services from individual company websites and marketing publications. Central Product Classification codes were assigned by the Commission.

Note: Not all services will be needed on every project.

U.S. Market for Hydropower-related Services

There is little direct information on the size of the hydropower services market, either in the United States or globally. Instead, this section presents estimates of the size of the services market as a share of hydropower generation capacity.

Table 5.2. Post-development services related to operations and maintenance (O&M) for hydropower projects

CPC code	CPC description	Hydropower-specific services
8676	Technical testing and analysis services	• Testing of electrical machinery including turbines and generators
886	Repair services of machinery and equipment	O&M services • Visual inspection and maintenance testing • Generator rewinds and core restacks • Upgrades to existing equipment, to increase output and improve efficiency • Repairs of turbines, turbine welds, and other components Electromechanical and rehabilitation services • Turbines and pumps, generators, emergency and cooling systems, fire protection and HVAC, switchyards and substations, hydro mechanical equipment • Instrumentation and control as diagnostic systems • Water supply control regulators
86721	Advisory and consultative engineering services	• Dam monitoring services including geodetic surveys, dam safety assessments and seismic analysis • Turbine efficiency measuring
86739	Integrated engineering services for other turnkey projects	• Environmental monitoring • Contract management • On-site management and supervision

Sources: Written submission to the Commission, March 1, 2013; lists of services from individual company websites and marketing publications. CPC codes were assigned by the Commission.

Market Size

There are approximately 2,400 hydroelectric power plants of all sizes operating in the United States. Conventional (large and small) hydroelectric plants provided 74,800 MW of electricity, and pumped storage projects provided an additional 8,400 MW.[12] Of this total, 50 percent of generating capacity was federally owned, 25 percent was privately owned, and 25 percent was non-federal publicly owned.[13] Developers received licenses or exemptions from the U.S. Federal Energy Regulatory Commission (FERC) for 730 MW of new capacity between 2009 and 2012, suggesting the possibility of significant new hydropower development in the next five years.[14] As of January 2013, there were 878 existing small hydropower projects (50 MW or less) licensed by FERC, and 135 projects over 50 MW.[15] California has the largest number of projects overall, with the greatest overall generation capacity, while New York has the largest number of small hydropower projects.[16]

Electricity generation revenues for the U.S. hydropower industry were $6.1 billion in 2012, with revenue posting an average annual decline of 1.7 percent between 2007 and 2012. A major reason for the drop in revenues was the decline in the number of U.S. hydropower generating plants (figure 5.1). Additionally, droughts in the United States during 2007–09 reduced available water and thus reduced power generation in those years, although hydroelectricity generation rebounded in 2011.[17]

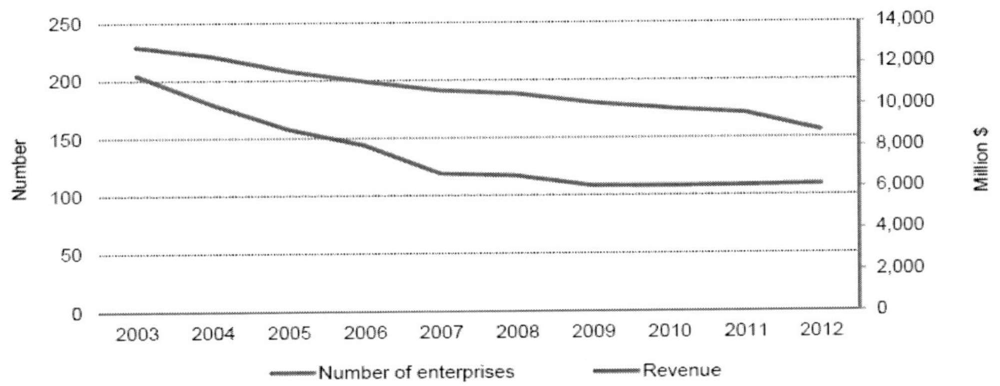

Source: Deonta Smith, "Hydroelectric Power in the U.S.," July 2012, 33.

Figure 5.1. U.S. hydropower revenue has declined since 2003.

Table 5.3. Estimated U.S. market for hydropower-related services based on existing large and small hydropower projects, 2010

Country/region	Hydropower capacity (total) 2009	2010	2009–10 growth	Estimated size of project development services market	Estimated size of fixed O&M services market	Estimated size of variable O&M services market
		MW			Million $	
United States	78,518	78,825	307	322	1,182	2,196
North America	164,571	165,064	493	518	2,476	4,598
World (all hydro)	887,378	917,544	30,166	31,674	13,763	25,560
U.S. as share of North America	48%	48%	62%	62%	48%	48%
U.S. as share of world (all hydro)	9%	9%	1%	1%	9%	9%
U.S. (small hydropower only, rated at ≤50 MW capacity)	826	826	0%	0	12	23
World (small hydropower only)	22,471	23,676	1,205	1,265	355	660

Source: BNEF; U.S. Energy Information Administration (EIA), "Annual Energy Review 2011," 2012; B&V, "Cost and Performance Data for Power Generation Technologies," February 2012, 35; and calculations by the Commission.

Notes: Figures for the market for project development services include estimates of EPC and owners' costs, and are based on the annual market for one year's new construction. Owner's costs primarily include services such as feasibility studies, environmental impact studies, resource assessments, and permitting and other legal costs. Figures for the market for fixed and variable O&M services are based on the cumulative reported capacity for each market. Figures for small hydropower projects are based on those in the BNEF database, which likely does not include all global small hydropower projects and should be seen as a base (floor) estimate. According to BNEF, there was no small hydropower development in the United States in 2010, so it is not possible to provide an estimate of the market size for new development services. Data for earlier years are not comparable. North America includes the United States, Canada, and Mexico.

Estimates of the size of the market for project development-related hydropower services are based on a constant share of total development costs for new hydropower projects, multiplied by the total of new capacity developed in the most recent data year.[18] As shown in table 5.3, the U.S. market for hydropower-related project development costs was estimated to be $322 million in 2010, and the combined market for fixed and variable O&M services was estimated at approximately $3.4 billion.[19]

Factors Affecting Supply and Demand for Hydropower Services

Hydropower services related to new project development and those related to O&M have different demand and supply drivers. Renewable portfolio standards (RPSs) and federal and state tax incentives are important drivers of demand for new renewable energy projects, including hydropower, and thus drive the market for hydro-related services.[20] However, many of the state incentives apply only to small hydropower projects, and under the federal incentives, only certain categories of hydropower projects are eligible.[21] For these reasons, the incentives have been more effective at promoting development of wind and solar power, and actually are considered to have been a factor in reducing the overall share of hydropower in the U.S. electricity market.[22]

The primary federal government incentive is the Investment Tax Credit (ITC), which offers a tax credit of 30 percent of a project's qualified investment costs.[23] Project developers also have the option of choosing to take the production tax credit (PTC) in lieu of the ITC. The PTC is based on the amount of electricity generated, and provides a credit of $11/MWh, as opposed to the ITC, which is based on investment costs. Twenty-four U.S. hydropower projects, all upgrades to existing facilities, qualified under the PTC in 2012. The federal government also permits project owners to use accelerated depreciation accounting to reduce their tax bills.[24] Both constructing hydropower projects and adding new capacity to an existing project often involve relatively high capital expenditures, so many hydropower developers may prefer the one-time, lump-sum benefits of the ITC over the production-based PTC. Relying on the ITC also removes the risks associated with production variability once the project is operational. Such variability is an important concern for hydropower, as electricity production can fluctuate significantly with changes in water levels.[25]

Many states have renewable energy incentive programs in place that only allow small hydropower projects to qualify, although the qualifying size varies by state. For most states, the qualifying capacity limit is set at either 10 MW or 30 MW. In New Hampshire, existing hydropower projects must be rated at or below 5 MW to qualify, while in Minnesota and Michigan, projects rated up to 100 MW may qualify. In New Mexico, any hydropower project brought into service after 2007 qualifies. In Ohio, projects do not face capacity limits, but must meet certain environmental standards. In addition to traditional hydropower, most coastal states include wave and tidal power in their definitions of eligible renewable energy technologies. Like the PTC and ITC, these state-level incentives have been significant drivers of new renewable energy capacity.[26]

Most hydropower projects in the United States are licensed and regulated by the FERC. The licensing process is a long and difficult one, which reduces the supply of new projects. Owners may choose from three different FERC licensing processes, but in all cases, the process averages 6–10 years from beginning to end, and is widely blamed for raising project development costs. Projects rated at capacities of 5 MW or less may be exempt from FERC licenses, although exempted projects are still regulated by state authorities.[27] All hydropower projects, both licensed and exempted, require environmental reviews, public participation, and agency consultation before development can proceed.[28] The FERC licensing process for small hydropower projects may face changes in the near future (box 5.1). The complexity of the process has created an entire category of legal and regulatory consulting service providers specialized in assisting project developers with the FERC licensing process.[29]

BOX 5.1. Changes to the FERC licensing process

Two bills changing the conditions for development of small hydropower were signed into law by President Obama in August 2013: H.R. 267, the Hydropower Regulatory Efficiency Act of 2013 (P.L. 113-23), and H.R. 678, the Bureau of Reclamation Small Conduit Hydropower and Rural Jobs Act (P.L. 113-24). Under the previous law, all hydropower projects were required to be licensed or exempted by FERC, but the new law changes that.

P.L. 113-23 allows FERC to exempt hydropower projects of up to 10MW that use existing canals, pipelines, or other conduits from its usual permitting process (under previous law, only projects up to 5 MW were exempt). It also waives license requirements for certain projects of 5 MW or less. According to hydropower industry representatives, the cost of the FERC licensing process is a high hurdle for small projects, sometimes surpassing actual project development costs. The measure also orders FERC to explore a two-year licensing process for hydropower development at existing non-powered dams and closed-loop pumped storage projects (significantly faster than the current process). In addition, it directs the Department of Energy to study pumped storage to support intermittent renewable energy sources like wind and solar.

P.L. 113-24 authorizes the development of small hydropower facilities at existing canals and other manmade waterways that are owned by the Bureau of Reclamation. The bill also streamlines the licensing process for some projects by removing them from coverage under the National Environmental Policy Act of 1969 (NEPA).

Sources: *HydroWorld.com*, "Hydroelectric Power Industry Leaders," August 12, 2013; *HydroWorld.com*, "Obama Signs Hydroelectric Power Bills," August 9, 2013; CRS, Bill Summary & Status, H.R. 267, CRS Summary; industry representative, telephone interviews with USITC staff, March 13 and April 4, 2013.

Sales of hydropower equipment for hydropower O&M services, particularly turbines, are an important demand factor because the equipment needs to be maintained and repaired. Most manufacturers of turbines initially provide maintenance and repair services under the warranty on their equipment.[30] Once the warranty period has passed, many hydroelectricity producers continue to receive these services from the same manufacturers; others may contract for O&M services with another manufacturer or an independent service provider. Leading manufacturers of hydropower turbines in the United States and globally, all of which provide O&M services, include General Electric, Siemens, Weir American Hydro, Voith, Andritz, and Alstom.[31] One firm which specializes in contract O&M work is NAES, headquartered in Issaquah, WA. It is an affiliate of Itochu Corporation (Japan).[32]

U.S. Industry Trends

The U.S. hydropower industry is mature, with few sites left for building new large-scale dams, and significant opposition to doing so.[33] Nonetheless, the federal government plans to significantly increase U.S. hydropower production by upgrading existing facilities, adding power-generating capacity to dams that currently have none, and developing small hydropower facilities.[34] In one example of a project that uses existing facilities to produce new power, U.S.-based American Municipal Power Ohio is building four new hydroelectric power plants along the Ohio River at existing dams and locks in West Virginia, Indiana, Kentucky and Ohio. Germany-based Voith will supply generators, turbines, and automation systems from its plant in York, Pennsylvania, generating 313 MW of renewable energy without adding environmental impact.[35] Another potential area of growth is the emerging technology of marine and hydrokinetic power (box 5.2).

Two leading service sectors related to hydropower project development are engineering design and construction services for new power plants. Table 5.4 presents the top five leading U.S. firms in each of the latter categories in 2012, and the firm's hydropower-related revenue for those activities.

Global Market for Hydropower-related Services

Market Size

Hydropower provided 16 percent (about 3,500 TWh) of the world's electricity in 2010. It supplied more electricity than nuclear power did (12 percent) and much more than wind, solar, geothermal, and other renewable sources combined (4 percent), but much less than fossil fuel plants (67 percent). China is the world's leading generator of hydroelectricity, followed by Brazil and Canada (table 5.5). The 10 countries listed in the table produce 70 percent of the world's hydroelectricity.[36] Hydropower added 25 GW of new power generation capacity in 2011, bringing the global total to 970 GW.[37] Hydropower facilities tend to be long-lived—some existing projects were commissioned as early as the 1870s.

Box 5.2. Renewable energy from marine and hydrokinetic power

Besides classic hydropower electricity generation, FERC also regulates projects using an emerging technology called marine and hydrokinetic power (MHK). MHK generates electricity from three types of sources: ocean tides, waves, and current; temperature gradients in the ocean; and free-flowing rivers and streams, but without using dams or barrages, as in traditional hydropower projects.

As of March 2013, FERC has issued one MHK commercial license, for the Reedsport OPT Wave Park in Oregon, and three pilot licenses. The pilot license process was developed in 2008 specifically to test the emerging MHK technology. Pilot licenses are issued for 10 years, and are initiated by a draft application with appropriate environmental analysis. All pilots must be small, short-term, removable projects located in an environmentally non-sensitive area. Licensees must be able to shut them down on short notice, and must remove them and restore the site before the end of the license term unless they obtain a new license.

The first pilot license for a tidal energy project was issued for Verdant Power's 1 MW-rated MHK project in New York's East River. As of March 2013, FERC had also issued 100 preliminary MHK permits (see table), which authorize permit holders to maintain rights to a particular site while conducting feasibility studies and before applying for a license. Most of the inland projects are located on the Mississippi River, with a few in the Great Lakes. The tidal projects are located in the Northeastern states, Washington state, and Alaska, while the wave permits are located in California, Oregon, and Alaska.

Summary of FERC preliminary permits for MHK power projects

	Number of permits	Capacity (MW)
Tidal	22	171.1
Wave	5	3,290.0
Inland	73	6,754.7
Total	100	10,215.8

Source: FERC, "Issued HydroKinetic Preliminary Permits," n.d.

Many services relevant to hydropower will also be relevant to MHK electricity projects, including engineering consulting and advising, environmental and engineering analysis, and legal and management consulting services. In addition, because MHK technology is so new, there is a relatively undeveloped market for resource assessment services to inform investors about the energy extraction potential of a particular water resource, along with environmental and economic aspects of a particular site. If this technology develops and becomes more widely used, U.S. firms that develop expertise in such resource assessment services are likely to generate additional market potential in related site development services and equipment sales, both for domestic sales and for exports.

Sources: FERC, "FERC Issues First Pilot License," January 23, 2012; Taylor, written submission to the Commission, March 1, 2013.

Table 5.4. Leading U.S. hydropower design firms and contractors, 2012

Design firms		Construction firms	
Company	Hydroplant revenue (million $)	Company	Hydroplant revenue (million $)
Aecom Technology	182.7	Kiewit	414.0
Tetra Tech	70.0	The Walsh Group	152.5
MWH Global	68.0	PCL Construction Enterprises	48.8
HDR	48.5	The Shaw Group	35.3
AMEC	33.1	The Cianbro Cos.	7.2

Source: ENR, "The Top Contractors in Power," September 17, 2012; ENR, "The Top Design Firms in Power," July 2, 2012.

Table 5.5. Top 10 global hydropower producers in 2010

Country	Hydroelectricity production (TWh)	Hydropower share of country's electricity generation (%)	Country share of global hydropower production (%)
China	694	14.8	20
Brazil	403	80.2	12
Canada	376	62.0	11
United States	328	7.6	9
Russia	165	15.7	5
India	132	13.1	4
Norway	122	95.3	3
Japan	85	7.8	2
Venezuela	84	68.0	2
Sweden	67	42.2	2

Source: IEA, Hydropower Technology Roadmap 2012, 2012, 10, and calculations by the Commission.

Note: Data includes all hydropower, not just small hydro. These numbers do not include electricity imports such as those from Paraguay's side of the Itaipu hydropower plant to Brazil, which represent almost half of that hydropower plant's generation (36 TWh).

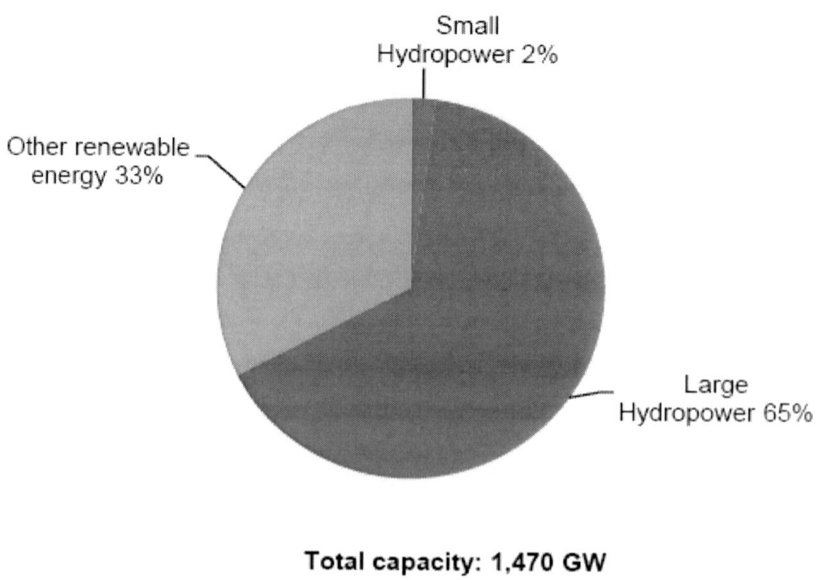

Source: REN-21, Renewables 2013, 2013, 94.

Figure 5.2. Hydropower accounted for the majority of renewable energy capacity globally in 2012.

Although the primary focus of this section is small hydropower, projects rated at capacities of less than 50 MW account for only a small share of all hydropower generation capacity (figure 5.2). Moreover, large hydropower projects are on the rise, as they often deliver the greatest returns in terms of power generation, especially in countries like China and Brazil where energy demand is rising faster than new capacity can be installed (box 5.3).

There were at least 2,600 small hydropower projects in operation (commissioned) globally at the end of 2012, with a total capacity of just over 29,400 MW, equal to less than 3 percent of all global hydropower capacity.[38] The average project size for reported small hydropower projects was 12.8 MW, but individual projects varied in size from 0.1 MW up to the cutoff of 50 MW.[39] Figure 5.3 shows newly commissioned small hydropower projects on an annual basis beginning in 2007.

BOX 5.3. Recent Developments in Large Hydropower

Large hydropower projects have raised social and environmental concerns given that they may change the natural course of rivers, lessen the supply of water available to traditional downstream users, and may flood wide areas for reservoirs behind new dams. For these reasons, and because large hydropower is often already price-competitive with existing fossil fuel power plants, government policies that subsidize and promote renewable energy generally apply only to small hydropower facilities. Hence, the discussion of hydropower in this report concentrates primarily on small hydropower facilities where possible.

Nevertheless, large hydropower facilities accounted for more than three-fifths of renewable energy generation capacity worldwide at the end of 2012 (see figure 5.2). Moreover, there are many large hydropower projects underway and in the planning stages; these projects certainly present potential opportunities for U.S. service providers. In addition, attitudes toward large hydropower may be changing, as electricity providers search for a way to expand access to electric power while remaining consistent with stated goals with respect to climate change. For example, the World Bank announced in May 2013 that it would once again begin funding hydropower projects after scaling back dramatically in the mid-1990s. The Bank stated that it sees large hydropower as the only way to meet its global development objectives without accelerating greenhouse gas emissions.[a]

China led the world in new capacity additions in 2012, as it has in other recent years, followed by Turkey, Brazil, Vietnam, and Russia. China installed 15.5 GW of new capacity, reaching almost 229 GW of installed hydropower capacity. China's three largest hydropower facilities are Three Gorges (22.5 GW), Xiluodu (13.9 GW when completed), and Xianjiaba (6.4 GW when completed). The Three Gorges achieved full capacity in July 2013, and reached a record output of 98.1 TWh in 2012. China targets a total of 290 GW of installed hydropower capacity by 2015.[b]

Turkey is increasing its hydropower capacity at a rapid rate to address chronic shortages of electricity and frequent power outages, adding about 2 GW in 2012, for a total of 21 GW installed. Brazil placed 1.86 GW of hydropower into operation in 2012, including 394 MW of reported small-scale (<30 MW) capacity, for a total of 84 GW. Construction continued on hydropower facilities around the country, including the 11.2 GW Belo Monte project, expected to be Brazil's second largest after the 14 GW Itaipu plant. Itaipu set another output record in 2012, matching China's Three Gorges at more than 98 TWh.[c]

Vietnam added at least 1.8 GW of new capacity in 2012, raising its total capacity to 12.9 GW. The country's Son La plant is reportedly the largest hydropower project in Southeast Asia, with a total capacity of 2.4 GW.[d]

In Canada, the 200 MW Wuskwatim plant was commissioned in Manitoba, and Hydro-Québec completed the 768 MW Eastmain 1-A powerhouse, with the 150 MW Sarcelle plant scheduled to come online in 2013. India added about 750 MW of hydropower capacity in 2012, of which 157 MW was categorized as small-scale (<25 MW), for a total of about 43 GW.

Ethiopia's Grand Renaissance Dam is under construction, with commissioning of the first phase to start in late 2013. The final project is expected to deliver 6 GW and to be the largest hydropower facility in Africa. Several transmission projects are also under way, to permit Ethiopia to export hydropower to neighbors in the Horn of Africa.[e]

[a] Hydroworld.com, "World Bank Announces Renewed Support for Large Hydropower," May 30, 2013; *South China Morning Post,* "World Bank U-turn Brings Hydropower in From Cold," May 11, 2013.
[b] REN21, *Renewables 2013*, 2013, 35.
[c] Ibid., 35.
[d] Ibid., 36.
[e] Ibid., 38.

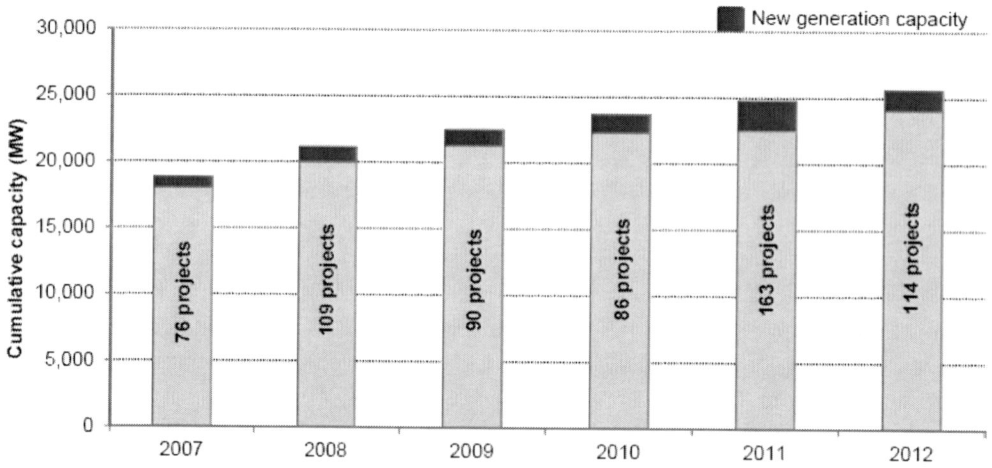

Source: BNEF database (accessed January 18, 2013).
Note: Includes projects reported by BNEF. May not include all existing projects.

Figure 5.3. Small hydropower generation capacity has steadily increased from 2007–12.

China and Brazil lead the rankings of countries with small hydropower projects, as they do for all hydropower, but Japan and India rank third and fourth, even though the latter two countries rank lower in the listing of total hydropower generation. China, Brazil, and India are rapidly adding new small hydropower capacity: 60 percent, 55 percent, and 40 percent, respectively, of each country's small hydropower capacity has been added since 2007. In contrast, less than 1 percent of Japan's small hydropower capacity has been added since 2007; Japan ranked first globally for small hydropower generation capacity in 2006.[40] Table 5.6 shows the top 10 countries by total small hydropower capacity in 2013.

Table 5.6. Largest small hydropower producers, by country and generation capacity, projects commissioned through February 2013

Country	Capacity (MW)	Number of projects	Average capacity per project (MW)
China	6,942.8	416	16.7
Brazil	5,339.1	414	12.9
Japan	3,613.6	487	7.4
India	2,622.0	327	8.0
Canada	1,180.7	75	15.7
Vietnam	934.5	72	13.0
United States	825.9	91	9.1
Chile	736.4	41	18.0
Australia	702.9	56	12.6
Colombia	587.7	48	12.2

Source: BNEF database (accessed March 1, 2013).

Table 5.7. Estimated market for hydropower-related services based on existing large and small hydropower projects, by region and selected country, 2010[a]

Country	Hydropower capacity (total) 2009	2010	2009–10 growth	Estimated size of project development services market	Estimated size of fixed O&M services market	Estimated size of variable O&M services market
		MW			Million $	
China	196,800	219,000	22,200	23,310.0	3,285.0	6,100.6
Brazil	79,291	80,703	1,412	1,482.6	1,210.5	2,248.1
Turkey	14,553	15,831	1,278	1,341.9	237.5	441.0
India	39,598	40,610	1,012	1,062.6	609.2	1,131.3
Japan	21,784	22,362	578	606.9	335.4	622.9
Canada	74,510	74,901	391	410.6	1,123.5	2,086.5
United States	78,518	78,825	307	322.4	1,182.4	2,195.8
Sweden	16,544	16,624	80	84.0	249.4	463.1
France	18,199	18,229	30	31.5	273.4	507.8
Venezuela	15	15	0	0.0	0.2	0.4
Russia	46,873	46,873	0	0.0	703.1	1,305.7
Norway	28,188	27,677	(511)	0.0	415.2	771.0
World (all hydro)	887,378	917,544	30,166	31,674.0	13,763.2	25,559.8
World (small hydropower only, rated at ≤50 MW capacity)	22,471	23,676	1,205	1,265.3	355.1	659.5

Sources: Capacity data: EIA, International Energy Statistics, http://www.eia.gov/cfapps/ipdbproject/IEDIndex3.cfm?tid=2&pid=2&aid=7; (accessed March 20, 2013); calculations by the Commission. Information on share of services is from B&V, "Cost Estimates and Performance Data," 2012, 34–5.

Note: Figures for the market for project development services include estimates of EPC and owners' costs, and are based on the annual market for one year's new construction. Owners' costs primarily include services such as feasibility studies, environmental impact studies, resource assessments, and permitting and other legal costs. Figures for the market for fixed and variable O&M services are based on the cumulative reported capacity for each market. Figures for small hydropower projects are based on those in the BNEF database, which likely does not include all global small hydropower projects and should be seen as a base (floor) estimate.

As was the case for the U.S. market, the Commission relied on the EIA for data on global hydropower capacity, and estimates from B&V for the share of services in a hydropower project to estimate the size of the global hydropower services market (table 5.7).[41] China was the largest market for EPC services in 2010, given its significant installation of new capacity that year, followed by Brazil, India, and Turkey. The market for O&M services (both fixed and variable) was dominated by China, Europe, Brazil, the United States, and Canada, reflecting those countries' well-established hydropower resources.

The International Energy Agency (IEA) estimates that annual O&M costs range from $5 to $20/MWh for new medium to large hydropower plants, and approximately twice as much for small hydropower.[42] The Commission's estimates of the O&M market are slightly higher than the high end of the range provided by the IEA. According to one industry representative, O&M costs are generally calculated as 1–2 percent of annual revenue, which depends on the plant load factor.[43] This factor is different for all sites and ranges from 30 to 70 percent for most projects.[44]

Factors affecting initial investment costs and the return on investment include the project scale, which can range from over 10,000 MW to less than 0.1 MW; the project location; the costs of site preparation and materials; the presence and size of existing reservoirs; the use of the power supplied, whether for baseload or peak load or both; and whether a project is designed to handle functions other than electricity generation, such as flood control, irrigation, or freshwater supply.[45]

Factors Affecting Global Supply and Demand for Hydropower Services

There is significant global development potential for new hydropower generation capacity, which would bring with it demand for hydro-related services. Global installed hydropower capacity has grown by an average of more than 24 GW annually in recent years, reaching close to 1,000 GW at the end of 2012. Total capacity is expected to reach 1,300 GW in 2017, also implying strong growth in the market for related services. Given the long lead times for developing hydropower projects, these figures represent capacity that is currently under construction and is virtually certain to come on line. Global hydropower capacity will increase by approximately 180 GW by 2020 if projects currently under construction proceed as planned, an increase of one-fourth of currently installed capacity. One-third of this increase is expected to come from China alone, followed by Turkey, Brazil, and India.[46]

Despite these promising numbers, hydropower developers must face diverse problems at different stages of their projects. These include the rising cost of construction, including increasing costs for some raw materials; increasing social and environmental opposition to new hydropower projects; and the costs of security for staff and property.[47]

Global Industry Trends

The IEA forecasts global installed hydropower capacity of 1,947 GW by 2050, nearly twice the current level but a less certain estimate than that for 2020. At that level, generation of hydroelectricity would near 7,100 TWh, doubling the power currently generated and substantially increasing the market for related services. Ten countries (China, the United States, Russia, Brazil, Canada, the Democratic Republic of the Congo, India, Indonesia, Peru, and Tajikistan) account for about two-thirds of global long-term hydropower potential, with emerging markets expected to supply the largest share of the growth.[48] In addition to new construction, many existing hydropower plants could increase their capacity by 5–20 percent by installing new and more efficient turbines. Refurbishment projects are often easier to carry out, for both technical and social reasons, and are completed faster than new plants.[49]

Trade and Investment

Imports and Exports

Data on U.S. trade in services related to hydropower generation are not available.[50] However, it is likely that services constitute the biggest part of the hydropower industry, and with an increasing number of firms developing projects in foreign markets, services trade may outpace hydropower equipment trade.

Independent power producers (IPPs) are the principal owners and developers of small hydropower projects; the development of large hydropower plants is generally a utility- or state-led effort.[51] Globally, Canada-based Brookfield Renewable Energy Partners is the largest IPP owner of small hydropower assets, controlling 103 U.S. projects in 11 states, for a total of 1,966 MW of capacity. Brookfield also owns other large and small hydropower assets in Canada and Brazil. In March 2013, the company completed the acquisition of 19 existing hydropower projects and eight upstream storage reservoirs in Maine and New Hampshire, collectively valued at about $760 million and with 351 MW of hydroelectric generating capacity. The acquisitions will bring Brookfield's assets in New England to nearly 1,300 MW of installed hydropower capacity.

Table 5.8 presents the leading international design and construction firms in hydropower for 2012, by revenue. Many of these companies have global subsidiaries that perform work in a variety of countries. Chinese firms are prominent on both lists, consistent with China's adding more hydropower capacity than any other country in recent years. Three U.S.-based firms are among the top hydropower designers (Aecom Technology, Tetra Tech, and MWH Global). Kiewit Corporation is the only U.S.-based firm among the top 10 hydropower contractors.

According to one industry representative, the global market leaders in hydroelectric turbine manufacturing, in order of market share, are Voith (Germany), Alstom (France), Andritz (Austria), Dongfang Electric (China), and Toshiba (Japan). The five companies together represent two-thirds of the global market, with Voith and Alstom each accounting for slightly less than 20 percent of the total.[52] Firms around the world reported increased sales in 2011, likely generating increased services and foreign trade revenues as well. For example, Dongfang Electric reported the production of 11 hydroelectric turbine generator sets totaling 2.5 GW in the first six months of 2011; Voith reported a sales increase of 6 percent and a dramatic increase in forward orders, up 81 percent over 2010. Other manufacturers, including Andritz, Alstom, IMPSA (Argentina), and Toshiba, also reported increased sales and/or backlogs in filling orders. Consistent with this demand, several large manufacturers have invested in new plants or in acquisitions of smaller firms. Alstom, for instance, opened a new equipment factory in India, expanded its turbine factory in Tianjin, China, and agreed with RusHydro on a joint manufacturing facility in Russia. Voith, which spent $106 million on research and development in 2011, has a new turbine component manufacturing facility in Brazil and a new workshop in Austria. Andritz continued to expand its Chengdu (China) facility for large hydropower components, and IMPSA is expected to open a new production facility in Brazil in 2012.[53]

Table 5.8. Top international contractors and design firms for hydropower plants

	Design firms				Construction firms		
Rank	Company	Home country	Revenue (million $)	Rank	Company	Home country	Revenue (million $)
1	Hydrochina	China	159.2	1	Sinohydro	China	1,687.0
2	Aecom Technology	United States	84.0	2	China Gezhouba Group Co.	China	806.7
3	Lahmeyer International	Germany	69.3	3	Construtora Norberto Odebrecht	Brazil	668.0
4	Poyry	Finland	54.0	4	Kiewit Corporation	United States	413.0
5	AF	Sweden	35.3	5	Dongfang Electric	China	345.7
6	Sinohydro Group	China	34.1	6	China National Machinery Industry	China	309.6
7	Tetra Tech	United States	30.0	7	C.M.C. Di Ravenna	Italy	199.9
8	Amec	United Kingdom	28.9	8	Astaldi	Italy	148.7
9	MWH Global	United States	27.0	9	OAS	Brazil	140.4
10	Fichtner	Germany	21.0	10	China International Water & Electric	China	117.4

Source: ENR Sourcebook, December 10, 2012.

Investment

In the United States, BNEF estimates financing of hydropower assets at $1.9 billion between 2008 and 2011. This figure was primarily attributable to American Municipal Power, which began construction on several plants totaling 300 MW of new capacity on the Ohio River in 2009–11, with total estimated costs of $1.7 billion for these new installations. [54] In the fourth quarter of 2012 alone, there were 13 greenfield small hydropower deals in the United States, collectively valued at $1.4 billion.[55] Global investment in small hydropower totaled $6 billion in 2011, a 59 percent increase over the previous year.[56] Globally, small hydropower was the only renewable energy sector to show report investment growth in 2012, compared with 2011, with a 17 percent increase in investment to $7.6 billion.[57]

The World Bank and other global development banks are important sources for financing both small and large hydropower projects in emerging markets (table 5.9). As of March 2013, the World Bank Group had $6 billion of lending to hydropower projects under development; one-fourth of these were projects to rehabilitate existing hydropower plants.[58] The World Bank often funds hydropower projects as part of financing syndicates that involve funding from one or more development banks, commercial banks, and local governments.[59] Most development banks have programs focused on funding sustainable development and renewable energy projects, including hydropower projects. Much of the related services work, including resource assessment, feasibility studies, environmental impact statements, engineering design, construction management, and quality control, is carried out by firms from the United States and other developed countries that can meet the qualification requirements of the funding entities.[60]

In addition to funding capital costs for particular projects, some global development banks also contribute equity financing to companies based in developing countries. In one example, in 2013 the Asian Development Bank announced its first equity investment of $30 million in a private, India-based renewable energy company—India-based NSL Renewable Power Private Ltd. The company will use the funds toward construction of the 100 MW Tidong run-of-river hydropower project in Himachal Pradesh and a 75 MW Chilarewadi wind plant in Maharashtra.[61] A representative of one development bank noted that hydropower projects can be particularly difficult to finance due to their social and environmental impacts; run-of-river projects were easier than others, presumably because they tend to have the lowest impacts.[62]

Trade Barriers

There are very few barriers specifically targeting hydropower service providers, though more general restrictions on investment or movement of workers may affect a firm's ability to operate in a given market. One of the barriers to exporting hydropower services reported by U.S. industry representatives was difficulty matching the financing terms that some foreign firms are able to offer to project developers. [63] A U.S. industry representative working on small hydropower projects in India also reported that in his experience, U.S. firms face particular scrutiny from the Indian government, compared with local firms, including frequent tax audits, difficulty moving money out of the country, and high tariffs (22 percent) on imported equipment. This situation has led at least one company to rely much more extensively on Indian-made equipment than they would otherwise.[64] In Indonesia, the government also imposes local-content requirements on hydropower based on the size of the project.[65]

Table 5.9. Selected small hydropower projects with financing contributed by global development banks

Country	Development Bank	Project details
Brazil	Inter-American Development Bank (IADB)	A government-backed rural electrification program called Luz Para Todos has installed small solar, wind, and hydropower projects in remote villages throughout Brazil since 2003. One particular small hydropower project is the 50kW Cachoeira de Aruã micro-hydro facility in the Amazonian state of Pará, which benefits 50 low-income households in a remote area where it would be difficult to bring in transmission lines from larger power generation facilities.
Honduras	IADB	Two small hydropower plants with a total installed capacity of 6.4 MW form part of the MIPYMES Verdes program, which has disbursed almost $7 million to small and medium-sized companies to develop projects to cut energy consumption, improve efficiency, or generate power from renewables.
Suriname	IADB	At the end of 2011, Suriname's rural electrification program received more than $25 million in loans and grants to fund electrification in rural villages in Suriname's interior. The project aims to install almost 700kW of solar photovoltaic capacity and about 2.7 MW of micro hydropower plants.

Table 5.9. (Continued)

Country	Development Bank	Project details
Haiti	IADB	In Haiti, one small hydropower project, the 48 MW Peligré plant, represents 45 percent of the country's total electricity capacity. However, the plant fails to operate at full capacity most of the time, due to lack of water during the dry season and aging equipment. In 2011, Haiti received a $20 million grant from the Inter-American Development Bank to rehabilitate Peligré. In 2012, France-based Alstom said it would provide turbines to restore the project to its full nameplate capacity.
Panama	IADB	In Panama, small hydropower is the largest destination for clean energy investment, with most of the $1.1 billion in such investment in recent years destined for the small hydropower sector.
Sri Lanka	Asian Development Bank (ADB)	In Sri Lanka, a $1.29 million credit line from the Asian Development Bank (ADB) Sustainable Power Sector Support Project will be used to rehabilitate and repair 19 micro-hydropower projects. The credits will be granted to private developers and will finance up to 100 percent of the total estimated costs for restarting the small hydropower sites. The 19 projects will add about 1.3 MW of combined capacity. Similar programs have existed in Sri Lanka since the mid-1990s, rehabilitating numerous small hydropower projects of 0.25 MW to 10 MW. These have added more than 253 MW of total hydroelectric capacity to Sri Lanka's grid.
Papua New Guinea	ADB	ADB approved a loan of $57.3 million to the local utility to fund renewable energy projects including run-of-river hydro.
India	ADB	India is developing four run-of-river hydropower projects for a total capacity of 856 MW, with $800 million in ADB funding.
Ecuador	China Development Bank	$680 million for four hydropower plants, rated at 21 MW, 276 MW, 50 MW, and 15 MW. Financed through an 8-year loan at 6.9 percent, with a 2-year grace period.
Romania	European Bank for Reconstruction & Development (EBRD)	$150 million to state-owned hydropower company Hidroelectrica S.A. in May 2011 for modernization work on an existing 210 MW project. The loan will pay for refurbishing six units, adding 30 years to the life expectancy of the plant.
Macedonia	EBRD	$8.2 million for development of four small hydropower plants with a total capacity of 4.1 MW.

Sources: Ingram, "Banks Provide Critical Support," November 1, 2011; BNEF, *Climatescope 2012*, 2012; development bank websites; *HydroWorld.com*, "Sri Lanka Micro Hydro Rehabilitation Projects," January 18, 2013.

Regional Profiles

Asia and Latin America are expected to experience the most rapid growth in hydroelectricity generation in coming years, leading to overall growth in those regions in the market for related services. In Africa and Europe, some growth in small hydropower projects is expected, but rapid growth of large hydropower projects is less likely.

Latin America

There was at least 8.4 GW of installed small hydropower capacity in Latin America and the Caribbean at the end of 2011, or 2.8 percent of total installed power capacity in the region.[66] Small hydropower accounted for a particularly large share of the renewable energy portfolio in certain countries, including Belize (39 percent), Honduras (9 percent), Peru (6 percent) and Ecuador (6 percent). With 241 MW of capacity, small hydropower represented 10 percent of total installed hydropower capacity in Guatemala, and had attracted $152.5 million in new capital at the end of 2011.[67] Brazil has a larger amount of installed small hydropower capacity (5,339 MW) but it represents a lower overall share of power (4 percent).[68] In Chile, there is a renewable portfolio standard (RPS) that includes small hydro, but only for projects rated at 20 MW or less. For large hydropower companies, such projects are not economically sound, but smaller companies are reportedly more interested in developing small hydropower projects in Chile.[69]

Brazil's 2009 national energy plan shows the country's small hydropower capacity doubling to 9 GW by 2030, and all hydropower almost doubling, relative to 2010, to 150 GW.[70] However, according to industry representatives, small hydropower projects are generally not competitive with either large hydropower or other renewable energy sources, particularly wind. In recent years, Brazil has held reverse auctions for renewable energy development, which set a cap for the price paid for energy. Small hydropower from newly developed projects has not been cost competitive compared with other renewables because the cost of the civil works (dam construction) has grown tremendously in recent years, now accounting for about 70 percent of project costs. Labor costs and concrete prices are very high in Brazil, as is construction risk, making it more difficult for small hydropower projects to recoup their investment. However, costs are significantly lower for new hydropower capacity coming online from retrofitting and modernizing existing dams.[71]

According to one industry representative, the future of small hydropower in Brazil will depend on government policies that support small hydropower, which is competitive only at rates of 140 reais/MWh ($61.50/MWh), compared with large hydropower, which is competitive at rates of 70 reais/MWh ($30.75/MWh).[72] Recently, new hydropower capacity has been split between new large hydropower projects (83 percent), modernization and upgrades of existing large hydropower projects (6 percent) and new small hydropower projects (10 percent).[73]

Brazil's National Bank for Economic and Social Development (BNDES) offers significantly lower interest rates than local commercial banks, making it the leading source of financing for project developers. However, BNDES financing generally carries a 60 percent local-content requirement and other restrictions.[74] To get the financing, some firms import equipment and parts and assemble them in Brazil. The project can start with 40 percent local content, but must have a plan to reach 60 percent. From 2003 to June 2008, BNDES contributed about $10.5 billion to 142 power generation projects, of which $6.7 billion went to hydroelectricity. In addition to BNDES financing, there are some federal tax benefits associated with wind and small hydropower related to the value-added tax (VAT) and income taxes, and some income tax incentives for small hydropower projects offered by the northern Brazilian states.[75]

Asia

Asia is expected to experience continuing growth in hydropower capacity through at least 2030, which will also lead to growth in the hydro-related services market. Countries with growing markets include Burma, Laos, Vietnam, Indonesia, and the Philippines.[76] In India, hydropower provided the majority of renewable electricity in 2012 and will continue to grow, with hydropower capacity expected to add 25 GW by 2030.[77] An Indian program to privatize small hydropower sites to encourage development has provided rare opportunities for small U.S. engineering services companies that do not have the resources to handle large hydropower projects. Such opportunities for small firms can be difficult to find in the United States.[78]

China has become a major market for hydro-related services. In 2011, its total installed hydropower capacity (large and small) was about 230 GW, with another 50 GW under construction.[79] The total was expected to grow to at least 300 GW by 2020, including 41 GW of pumped hydro, and possibly to 400 GW by 2030,[80] ensuring a growing market for hydropower-related services.

Chinese companies are also actively involved in exporting equipment and services for both small and large hydropower projects. Most Chinese firms active in hydropower globally have focused on providing EPC services, but a few are also providing O&M services. In general, management teams are sent from China to the foreign locations, but the companies tend to hire mostly local employees, consistent with the practices of large firms based in the United States and elsewhere when operating outside their home markets. Chinese firms have been particularly active in Asia and Africa, notably Pakistan, Vietnam, and Nigeria.

Foreign firms active in China's hydropower market include Alstom and Shanghai Fuji. Alstom Hydro China Co. is investing in the second phase of its hydropower technology center in Tianjin, which is expected to be the company's largest hydropower equipment factory when completed in mid-2013. Alstom has partnered with Chinese firms to work on hydropower projects in China, Ghana, and Iran.[81]

In Japan, small hydropower has been slow to expand, because smaller power stations have typically been considered to be a high-cost technology. Nonetheless, Japan still ranks third globally in total small hydropower capacity.[82] Japan enacted a strong feed-in tariff (FIT) to promote investment in renewable energy in late 2011. The FIT applies to small hydropower (capacity rated at less than 30 MW),[83] as well as other renewable energy technologies, and is encouraging the development of new small hydropower projects and related services. However, the FIT is likely to be less effective at promoting small hydropower than other renewables, because hydropower takes longer to design and is generally more difficult to implement than projects based on other renewable energy technologies.[84]

Africa

In parts of Africa, where regional instability may make it difficult to secure financing for the significant investments required for large hydropower projects, small hydropower may play an important role in alleviating power shortages. An increasing number of IPPs are actively developing small hydropower projects in Africa. Hydropower and other renewable energy sources are seen as a way to reduce dependence on fossil fuels in areas far from the existing power grid, such as in mining operations. Smaller hydropower projects are often easier to fund, have faster startup times, and have smaller social and environmental impacts than more traditional, large hydropower projects. Small hydropower plants are often used off-grid, or in small "mini-grids"; large plants tend to be connected to centralized grids in order to fully utilize their generation capacity.[85] South Africa's 20-year resource plan, introduced in 2012, calls for renewables to represent 38 percent of all new electricity capacity added through 2030; while the new capacity is expected to draw largely on wind and solar sources, the plan also envisions 2.6 GW of new hydropower.[86]

European Union

Recent EU regulations will have significant impacts on hydropower development. On one hand, the EU Water Framework Directive aims to turn rivers back to their original state as far as possible, with a focus on pollution reduction. In certain rivers, this will reduce hydropower generation capacity. On the other hand, to promote renewable energy, many EU countries have introduced economic support programs such as FITs that promote new hydropower projects. Some of these systems include smaller-scale hydropower, but most exclude large hydropower projects. The lack of harmonization between overall EU energy policy and EU water management policies reportedly serves as a significant barrier for future development of hydropower in Europe.[87] Table 5.10 highlights policies that support small hydropower in EU countries.

Table 5.10. Support policies for small hydropower in the EU, 2011

Type of policy	Relevant countries
Feed-in tariff (guarantees a certain price per kWh for power generated)	Austria, Bulgaria, Czech Republic, Estonia, France, Germany, Greece, Hungary, Ireland, Italy, Latvia, Lithuania, Luxembourg, Portugal, Slovakia, Spain, United Kingdom
Feed-in premium (FIP) (offers a premium above the average spot electricity market price)	Czech Republic, Denmark, Estonia, Netherlands, Slovenia, Spain
Quota obligation/green certificates (require electricity suppliers to show that a certain amount of electricity delivered to consumers comes from renewable energy)	Belgium, Italy, Lithuania, Poland, Romania, Sweden, United Kingdom
Investment grants (direct support to research and investment projects)	Belgium, Czech Republic, Finland, Greece, Luxembourg, Poland, Slovakia
Tax exemptions, deductions, or other fiscal incentives (permits project owners to deduct a percent of investment costs from taxable profits)	Belgium, Germany, Greece, Netherlands
Tendering (contracts offered by the government for designated projects)	Belgium, France

Source: ESHA, Small Hydropower Roadmap: Condensed Research Data for EU-27, 2013, 14–16.

Until 2012, there was considerable incentive for developers to build small hydropower plants in Italy, including a comprehensive FIT system for hydropower projects up to 1 MW. The success of the sector prompted communities to look into small hydropower projects and pursue joint public-private opportunities. Although this has encouraged some development, some observers believe that the incentives created a problem by overly encouraging development of small hydropower plants anywhere in the country. Starting in 2013, a new incentive system is in force in Italy that decreases all tariffs and is expected to reduce new small hydropower plant installations as a result.[88] In Spain, in an effort to reduce government deficits, new measures were approved in February 2013 to cut the revenue that renewable energy plants receive from the FIT. FITs for clean energy, as well as other power system costs, will now rise annually based on a reduced inflation index. In addition, generators will no longer be able to switch between a tariff and subsidized market rates. The measures were expected to save between €600 million ($775 million) and €800 million ($1.0 billion) in 2013.[89]

Geothermal Energy

Although about two-thirds of global geothermal energy is delivered in the form of direct heat,[90] this section focuses on the services related to the electricity generation segment of the geothermal energy market. The United States is the world leader in geothermal electricity production and installed geothermal power capacity.[91] Geothermal energy is an important source of U.S. renewable energy in the western United States, where there is a concentration of geothermal resources. However, geothermal energy production is limited by several factors, including geography, lack of access to transmission lines, and the relatively high development and capital costs associated with significant exploration and production risks. Indeed, geothermal power's installed costs are among the highest of all renewables.[92] Government policies have been key in helping defray some of these costs and facilitating the development of geothermal energy projects.

International demand for geothermal energy continues to grow at a healthy pace, creating a significant opportunity for U.S. suppliers of geothermal equipment and services. Indeed, the pace of investment growth in the global geothermal energy sector has been strong over the last five years. Emerging markets and developing countries represent most of the untapped growth potential. Given the highly competitive U.S. geothermal market, it is not surprising to find evidence of U.S. geothermal services companies expanding their operations to international markets. More U.S. companies may follow suit, especially to high-potential and low-barrier countries.[93]

Overview of Geothermal Energy Services

Geothermal services include a variety of activities provided by engineering and consulting firms, petrochemical exploration and development firms, and original equipment manufacturers (OEMs). Firms in this sector provide a wide variety of services, including exploration; field development; project engineering and analysis; field operations and project management; plant design; drilling engineering services (well design, drilling program design, onsite supervision, drilling program management, and engineering support); geology, geochemistry, and hydrogeology services; and training services.[94] Particular O&M-related services identified by one geothermal power operator include construction, rental of drilling rigs, other equipment rental, engineering services, and warehouse services.[95] Table 5.11

identifies the CPC codes associated with the services most relevant to the production of geothermal energy.

Table 5.11. Services related to development of geothermal energy

CPC code	Description	Geothermal services
8675	Certain related scientific and technical consulting services	Project development related services (e.g., geological surveying, site assessment, environmental impact assessment)
861, 862, 863, 8672, 8673, 9312, 93191, 932	Certain professional services, including engineering and integrated engineering services	System design (well design, drilling program design)
633, 8861–8868	Maintenance and repair of equipment, except transport-related equipment	Operations and maintenance (monitoring, corrective maintenance; possibly preventative or condition-based maintenance)
865	Management consulting and related services	Project development related services (e.g., project management, PPA negotiations, interconnection agreements, permitting)
511–518	Construction and related engineering services	Engineering and construction (e.g., drilling, onsite supervision, electrical connections, substation construction)

Source: Compiled by USITC.

U.S. Market for Geothermal Energy Services

Market Size

U.S. geothermal energy production and installed capacity grew moderately between 2007 and 2011, yet the geothermal sector continues to represent a very small share of U.S. renewable energy generation. Between 2007 and 2011, U.S. geothermal electricity production grew from about 15 billion kWh to 17 billion kWh, while total installed geothermal capacity increased from 2.2 GW to 2.4 GW.[96] Further, between 2007 and 2012, U.S. geothermal revenues are estimated to have grown by 17 percent annually reaching $2.4 billion in 2012.[97] Despite this growth, geothermal electricity production still accounted for only about 3 percent of total U.S. renewable energy generation and an even smaller share (0.4 percent) of total U.S. electricity generation in 2011. [98] In comparison, during the same period, production of wind energy increased by almost 250 percent to 119.7 billion kWh, [99] while solar power generation grew by over 200 percent, although from a low base.[100]

The cost of constructing geothermal power plants varies significantly depending on a number of factors, including resource temperature and pressure, location, drilling market, and the number and types of plants.[101] While the development and construction of geothermal power plants require large investments in capital, services also represent an important cost when it comes to establishing new plants and then operating and maintaining those plants.[102] As shown in table 5.12, the U.S. market for geothermal-related project development costs was estimated to be $34.3 million in 2010, and the market for variable O&M services was estimated at $594 million.[103]

Table 5.12. Estimated U.S. market for geothermal-related energy services based on existing capacity, 2010

Country	Geothermal power capacity			Estimated size of project development services market	Estimated size of variable O&M services market
	2009	2010	2009–10 growth		
		MW		Million $	
United States	2,381.9	2,405.0	23.1	34.3	594.3
North America	3,346.9	3,370.0	23.1	34.3	832.8
World	9,892.8	10,104.9	212.1	315.0	2,497.1
US share of North America	71.2%	71.4%	100.0%	100.0%	71.4%
US share of world	24.1%	23.8%	10.9%	10.9%	23.8%

Source: U.S. Energy Information Administration (EIA), "Annual Energy Review 2011," 2012; B&V, "Cost Estimates and Performance Data for Renewable Electricity Technologies," 2012; and calculations by the Commission.

Notes: Figures for the market for project development services include estimates of EPC and owners' costs, and are based on the annual market for one year's new construction. Owners' costs primarily include services such as feasibility studies, environmental impact studies, resource assessments, and permitting and other legal costs. Figures for the market for variable O&M services are based on the cumulative reported capacity for each market. According to B&V, fixed O&M costs for geothermal are zero, although other sources do quote fixed O&M costs for geothermal plants. Since the estimates in this section rely on B&V, we do not cite fixed O&M costs here. North America includes the United States, Canada, and Mexico.

Factors Affecting U.S. Supply and Demand for Geothermal Power and Related Services

Factors affecting the development of geothermal energy in the U.S. market are primarily found on the supply side, due to the relatively high development and capital costs of geothermal projects. These include advances in technology, government incentives and regulations, transmission access, project lead times, and the price of electricity and alternative fuels.

Recent technological advances have facilitated the expansion of the geothermal power industry to new areas. Traditionally, the expansion of geothermal power production was limited by geography, because deposits of high-temperature groundwater are required to produce geothermal electricity. However, new technologies like "binary" geothermal technology can use lower temperatures to generate electricity, which has allowed plants to be built in areas where geothermal development was previously unfeasible. In addition, enhanced geothermal systems (EGSs), although more costly, represent an emerging technology allowing use of geothermal energy resources not otherwise accessible.[104]

Government incentives and regulations have also played a role in the development of geothermal energy. Under the American Recovery and Reinvestment Act (ARRA or the Stimulus Bill), the Department of Energy's Geothermal Technologies Office made 148 awards for a total of $368.2 million, including cash grants, tax credits, and funding for research and development. These awards provided an unprecedented level of fiscal stimulus for the industry and are likely to lead to new private sector investment.[105] In addition, as discussed in section 2, more stringent renewable energy requirements and new laws and regulations at the state level are also expected to promote new investment in renewable energy and renewable energy services, including geothermal energy.[106]

Other factors affect the development of geothermal energy projects in different ways. For instance, the lack of access to transmission lines is an important factor limiting the growth potential of geothermal power plants, as most geothermal resources are not near existing transmission infrastructure.[107] In addition, geothermal power projects take significantly longer to complete than many other renewable energy projects, such as wind or solar farms (it can take 4–8 years to complete a geothermal power project). This time factor adds to project risk. The price of electricity and alternative fuels like natural gas can also affect the competitiveness of geothermal power producers. For instance, the recent trend towards low natural gas prices may constrain the ability of geothermal power plants to be competitive once the plant becomes operational. By contrast, rising electricity prices make renewable sources of energy more attractive to consumers and more profitable for geothermal power companies.[108]

U.S. Industry Trends

Geothermal energy production has traditionally been concentrated geographically. However, advances in technology are allowing geothermal energy production to expand to new states. For example, although California and Nevada currently account for about 97 percent of U.S. geothermal installed capacity,[109] Hawaii, Utah, Idaho, Alaska, Oregon, and Wyoming have also established operational geothermal energy plants. [110] The industry's geographical reach is widening as new methods such as binary geothermal technology, which uses lower temperatures to generate electricity, allow plants to be built in places where it was previously unfeasible.[111] In fact, in 2012 there were 130 confirmed projects at various stages of development spread out over 15 U.S. states, of which three-quarters were greenfield projects rather than expansions of existing ones. This is an increase from 123 confirmed projects under development in 2011, and represents about 1,800 MW of new planned capacity.[112]

Table 5.13. Top geothermal power companies in the United States by installed and operating capacity

Company	Installed capacity (MW)	Operating capacity (MW)
Calpine	1,310	725
Ormat Technologies	627	408
Terra-Gen Power	352	344
CalEnergy	329	300
Northern California Power Agency	220	108
Enel Green North America	65	65
U.S. Renewables Group	55	11
Nevada Geothermal	50	45
PacificCorp	32	32
Alterra Power Corp.	23	11

Source: Islandsbanki, United States Geothermal Energy Market Report, October 2011, 16.

Note: Some of the listed companies are affiliates of or foreign multinational companies. For example, Enel North America is affiliated with Enel Green Power of Italy; Nevada Geothermal and Alterra Power Corp. are Canadian-based companies.

There were 58 suppliers of geothermal energy in the United States in 2011. Although a few electric utility companies own and operate geothermal plants,[113] IPPs produced approximately 93 percent of all geothermal electricity in the United States in 2011.[114] The top two companies, Ormat Industries and Calpine Corporation, are expected to account for 35 percent of total industry revenue in the United States.[115] Table 5.13 presents the top 10 companies operating in the United States as measured by installed and operating capacity, while table 5.14 presents the top 10 as measured by geothermal MWs under development.

Table 5.14. Geothermal energy capacity under development, by company

Company	Capacity under development (MW)
Gradient Resources Inc.	1,035
Oski Energy	563
Ram Power Corp.	541
Nevada Geothermal	268
U.S. Geothermal	264
Alterra Power Corp.	231
Terra-Gen Power	200
Ormat Technologies	198
CalEnergy	159
Eureka Green Systems	150

Source: Islandsbanki, United States Geothermal Energy Market Report, October 2011, 16.
Note: Capacity under development includes geothermal projects in the exploration, pre-feasibility, and feasibility phases, as well as the design and construction phases.

There is some overlap between the two top 10 groups. Ormat Technologies, CalEnergy, and Terra-Gen are major players when it comes to owning and operating plants as well as developing new geothermal energy capacity. Ormat Technologies, Ram Power Corp., U.S. Geothermal, Cal Energy, Chevron, and Unocal Corp are among the most active U.S. geothermal companies operating outside of the United States.[116] The majority of these companies provide EPC and O&M services, but there are some like Ormat Technologies, which manufactures equipment as well.

Global Market for Geothermal Energy Services

Market Size

Comprehensive published data on the size of the global market for geothermal energy services are unavailable. However, the growth in global geothermal power production likely reflects a vibrant market for geothermal services. The United States is the world leader in geothermal energy production and installed capacity. In 2011, the U.S. geothermal power sector held 28 percent of the 11.2 GW in global installed geothermal power capacity. Worldwide, there were at least 24 countries generating electricity from geothermal power plants in 2012. [117] In addition to the United States, seven other countries, most situated in the Ring of Fire area of the Pacific, account for the bulk of the remaining global installed geothermal capacity: the Philippines (1.9 GW), Indonesia (1.2 GW), Mexico (just under 1

GW), Italy (0.8 GW), Iceland (0.7 GW), New Zealand (nearly 0.6 GW), and Japan (0.5 GW) (figure 5.4).[118]

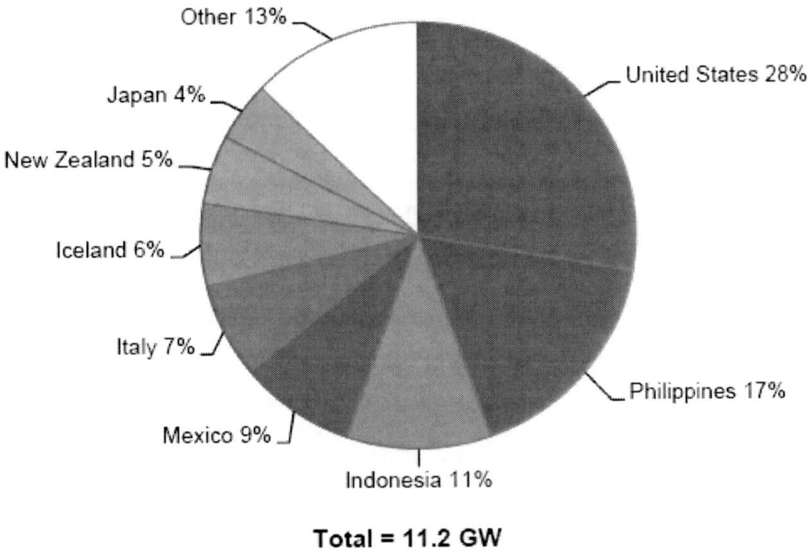

Source: REN21, *Renewables 2012*, 2012, 40.

Figure 5.4. The U.S. was the global leader in geothermal electricity generating capacity in 2011.

Global geothermal power capacity grew by a modest 136 MW in 2011. This growth came entirely from three countries: Iceland (90 MW), Nicaragua (36 MW), and the United States (10 MW). While growth was modest in 2011, it is expected to accelerate with projects under development in traditional markets, emerging markets, and new markets in developing economies, such as Rwanda and countries elsewhere in East Africa.[119] Indeed, most of the untapped growth potential lies in developing countries and emerging markets, especially those with significant proven untapped conventional geothermal resources, such as Chile.[120] By some estimates, global installed geothermal capacity is expected to grow by roughly 6–13 GW by 2021. [121] These long-run scenarios offer sizable opportunities for geothermal energy service providers.

As was the case for hydropower, the Commission relied on the EIA for data on global geothermal capacity, and estimates from B&V for the share of services in a geothermal power project to estimate the size of the global geothermal services market (table 5.15).[122] New Zealand and Italy installed the most new geothermal power capacity in 2010, so had the largest markets for services related to project development. The United States and the Philippines had the most installed geothermal capacity, and thus the largest markets for geothermal O&M services.

Factors Affecting Global Supply and Demand for Geothermal Services

Factors affecting the development of geothermal energy are primarily supply-side due to the relatively high development and capital costs of geothermal energy projects. Particularly important are advances in technology, which are making the development of geothermal energy production feasible in new countries. Government policies, including aggressive geothermal energy targets, renewable portfolio standards (RPSs), and feed-in tariffs (FITs), are also helping to facilitate the sector's growth, especially in emerging markets and developing countries.[123]

One particular problem for geothermal power, as opposed to solar and wind, is identifying the extent of existing geothermal resources. As in the case of oil and gas, developers must drill wells to verify that resources exist, and only a minority of the wells are successful, making the exploration process both risky and expensive. In one effort to overcome this problem, the government of Indonesia has instituted a geothermal energy funding facility which funds exploratory activities through local governments, aiming to provide potential investors and financing entities with better information about geothermal resources prior to the tendering process for specific development blocks. The program had received about $200 million as of 2012. Another program provides loan guarantees to certain geothermal energy project developers after the tendering process, as they actively develop new power projects.[124]

Table 5.15. Estimated market for geothermal energy-related services based on existing projects, by region and selected country, 2010

Region/country	Geothermal power capacity 2009	2010	2009-10 growth	Estimated size of project development services market	Estimated size of variable O&M services market
	MW			Million $	
New Zealand	633.0	731.0	98.0	145.5	180.6
Italy	700.0	750.0	50.0	74.3	185.3
United States	2,381.9	2,405.0	23.1	34.3	594.3
Philippines	1,953.0	1,966.0	13.0	19.3	485.8
Indonesia	1,189.0	1,197.0	8.0	11.9	295.8
Japan	535.0	537.0	2.0	3.0	132.7
Mexico	965.0	965.0	–	–	238.5
Iceland	575.0	575.0	–	–	142.1
El Salvador	204.4	204.4	–	–	50.5
World	9,892.8	10,104.9	212.1	315.0	2,497.1

Sources: U.S. Energy Information Administration (EIA), "Annual Energy Review 2011," 2012; B&V, "Cost Estimates and Performance Data for Renewable Electricity Technologies," 2012; and calculations by the Commission.

Note: Figures for the market for project development services include estimates of EPC and owners' costs, and are based on the annual market for one year's new construction. Owners' costs primarily include services such as feasibility studies, environmental impact studies, resource assessments, and permitting and other legal costs. Figures for the market for variable O&M services are based on the cumulative reported capacity for each market. According to B&V, fixed O&M costs for the geothermal sector are zero, although other sources do quote fixed O&M costs for geothermal plants. Since the estimates in this section rely on B&V, we do not cite fixed O&M costs here.

In contrast, unclear government policies, conflicting regulations, and bureaucratic problems in obtaining project permits have acted as roadblocks to new development in Indonesia, a country with aggressive targets and some of the world's largest geothermal energy resources.[125] Argentina, Ethiopia, Indonesia, Kenya, and the Philippines have among the most ambitious long-term geothermal energy targets, although it remains unclear whether those targets will be met. In Indonesia, for example, industry observers consider it unlikely that the country will meet its 2020 targets for new geothermal power production.[126] On the demand side, populations and economic growth are increasing the need for affordable and reliable energy in emerging markets. Other significant factors are access to power transmission, project lead times, and the price of electricity.

Global Industry Trends

The global geothermal power industry is dominated by multinational firms that provide geothermal energy services worldwide. The industry is relatively concentrated—the top 10 geothermal energy producers account for 64 percent of installed geothermal power capacity worldwide (table 5.16).[127] Chevron, a U.S. multinational energy producer, is reportedly the largest producer of geothermal energy. In 2012, Chevron's geothermal operations in Indonesia and the Philippines had a total capacity of 1,273 MW, and the company is assessing the prospects for additional development of geothermal energy production in both countries.[128]

Table 5.16. Companies owning geothermal energy capacity over 300 MW in 2010

Company	Country	Installed capacity (MW)	Country of operations
Calpine	United States	1,310	United States
Chevron	United States	1,087	Philippines and Indonesia
CFE	Mexico	958	Mexico
Enel Green Power	Italy	915	Italy and Latin America
Ormat	Israel	749	United States
EDC	Philippines	707	Philippines
Terra Gen	United States	337	United States
Contact Energy	New Zealand	335	New Zealand
Reykjavik Energy	Iceland	333	Iceland
CalEnergy Generation	United States	329	United States

Source: World Bank, ESMAP, *Geothermal Handbook,* 2012, 28. Handbook based on Bertani 2010.

Drilling accounts for a large share of the services revenue produced by geothermal energy development. The drilling techniques for geothermal energy are not the same as those used for oil and gas drilling, but the same type of equipment is often used in both industries. Specific services performed in connection with geothermal drilling include provision of drilling fluids; completions (putting pipe into the ground); servicing of drill bits; wireline and proliferating services; and testing of wells.[129]

Drilling companies that focus on oil and gas operations also perform drilling for geothermal projects, In fact, Schlumberger and Halliburton, two multinational firms associated primarily with oilfield services, dominate the geothermal drilling services market.[130] This involvement of oil and gas companies in geothermal project development contributes to greater geothermal production capacity and expands the size of the global geothermal market. On the other hand, the geothermal industry competes with oil and gas companies for access to drilling rigs. This competition sometimes reduces geothermal drilling, as it causes drill rig costs to rise to levels that are difficult for geothermal companies to pay. Within large companies involved in both geothermal and oil and gas exploration, there also may be competition for drilling rigs or other resources, requiring a balance between the two.[131]

Trade and Investment

Imports and Exports

The top international markets for geothermal energy services include Argentina, Chile, Iceland, Indonesia, Kenya, Mexico, New Zealand, Nicaragua, and the Philippines. These countries have the largest pipeline of geothermal projects and the largest amount of recently installed geothermal power plant capacity.[132] As noted previously, U.S. firms like Chevron, Schlumberger, and Halliburton are actively involved in foreign markets and contributing to U.S. exports of geothermal energy services. While comprehensive published data on trade in geothermal energy services are unavailable, evidence suggests that it is common for multinational companies that are engaged in geothermal services trade to set up affiliates abroad to facilitate access to the market.[133]

Investment

Global investment in geothermal energy grew from $1.4 billion in 2007 to $2.9 billion in 2011.[134] Of this, $1.8 billion went to developed countries and $1.1 billion to developing countries. [135] Development banks represent a major source of investment in new geothermal projects, particularly in developing or emerging markets that have potential geothermal energy sources.[136] One significant source of funding for geothermal power in emerging markets is the World Bank. Between 2007 and 2012, the World Bank lent $1.1 billion for geothermal energy projects, with annual funding growing from $73 million to $336 million over the period.[137] Much of this funding is directed toward purchases of geothermal services. As noted above, geothermal energy, unlike other sources of renewable energy, requires a substantial and risky exploration and drilling process to confirm the extent of the usable resource, and many private sector firms are not able to bear that risk.

In order to attract more funding to project development, in 2013 the World Bank announced the Global Geothermal Development Plan (GGDP), a $500 million geothermal investment fund focused on funding exploration and drilling in emerging markets. This new initiative expands on previous efforts by its global scope and its focus on test drilling. It will identify promising sites and leverage financing for exploratory drilling, to enable development of commercially viable projects. The initial effort will work to mobilize $500 million from donors, who can contribute through bilateral aid or through such existing World Bank programs as the Climate Investment Funds (CIFs) or the Global Environment Facility (GEF). The GGDP will be managed by the World Bank's long-standing Energy Sector Management Assistance Program (ESMAP). Donors will also be able to contribute by identifying viable projects.[138]

Trade Barriers

Important emerging markets such as Indonesia and the Philippines are, to varying degrees, considered to be fairly open to foreign geothermal energy development companies entering their countries. For example, Indonesia permits foreign geothermal energy development companies to own 95 percent of the project, with 5 percent of equity required to be owned by an Indonesian firm. Under the 2009 Electricity Law, Indonesia also has local content requirements that vary based on the size of the project.[139] Other countries have higher entry barriers; still others allow entry, but not at certain stages of the project. In Kenya's case, it is difficult to enter at the initial development stage, but once the steam field development is complete, then the barriers to entry are reportedly low.[140]

Regional Profiles

Southeast Asia (Indonesia and the Philippines)

Indonesia has enormous geothermal potential, as noted above, and has instituted an ambitious plan to expand geothermal power production by 2,000 to 3,000 MW by 2020. However, it is unlikely that the country will be able to meet those targets. There has been no new geothermal power development in Indonesia since 2003, even though a number of blocks were opened to exploration through a tender process (the second Fast Track Program or FTP2), beginning in 2010, following the promulgation of the 2009 Electricity Law.[141]

Industry observers cite several problems inhibiting new development:

- Confusion over the FIT rates that will be paid to project developers, and the government's reverse auction system that requires project developers to bid for the lowest acceptable tariff, and may force prices below viable rates;
- A tender process that awards exploration blocks to the lowest bidder, even though that bidder may not be qualified to carry out the work, or be able to obtain financing;
- A new geothermal regulation widely expected to be introduced in the second half of 2013, encouraging project developers to delay breaking ground for new projects;
- Regulatory conflict between government agencies. For example, projects licensed by the Ministry of Energy and Mineral Resources may have difficulty obtaining permits from the Forestry Ministry to build on forest land, or projects that were planned to rely on loan guarantees may not be able to meet requirements from the Ministry of Finance;
- Local protests may prevent developers from beginning construction.[142]

The Philippines will likely add new capacity as well, although less than 1,000 MW by 2020. The overall additions will depend on the respective governments' ability to mobilize new investment from private sector developers. Malaysia and Papua New Guinea are also likely candidates to increase their geothermal capacity in the next decade.[143]

East Africa

The East African Rift Valley is the region with the second-strongest geothermal potential after Asia, and Kenya is the region's leader in geothermal power development. The Kenyan government has taken concrete steps to encourage new development, including the creation of a dedicated public company, Geothermal Development Company (GDC), and is planning to add 2,000 MW of capacity by 2020. Based on 2011 project preparations, Djibouti and Ethiopia are also likely to increase their installed capacity by 50 to 200 MW in coming years.[144] As of March 2013, Tanzania, Burundi, Rwanda, Uganda, and Ethiopia were participating in a "Geothermal Compact," led by the World Bank and Iceland, which supports surface exploration studies and provides technical assistance toward developing geothermal resources. Zambia is expected to join the compact soon.[145] Rwanda is believed to have total geothermal resources of more than 700 MW, and in 2011, the Rwandan government announced plans to spend an estimated $935 million to develop 310 MW of new geothermal capacity by 2017.[146] Rwanda started exploratory drilling for geothermal resources near the Karisimbi volcano in April 2013. If tests are successful, a 10 MW pilot phase of the project will be constructed. Great Wall Drilling Company (China) won the contract for test drilling, a Rwandan firm will handle rehabilitation of the road to the drilling site, and Yashinoya Ltd. (Kenya) will handle civil works engineering. Exploratory drilling in Rwanda's Kinigi region, another area likely to hold significant geothermal resources, is expected to begin in 2014.[147]

Latin America

The countries most likely to develop new geothermal power projects in Latin America are Mexico, Costa Rica, Nicaragua, and El Salvador. Total additional capacity is expected to be in the range of 500 to 1,500 MW by 2020, depending on particular country policies and available funding. Other countries with geothermal resources and the potential for new development include Guatemala, Honduras, Panama, Colombia, Ecuador, Chile, Bolivia, Cuba, Haiti, and Dominica.[148]

Japan

Japan ranks eighth in installed geothermal energy capacity, but third globally in potential geothermal resources (its last large geothermal power plant was completed in 1999).[149] Japan has implemented an FIT and has at least partially deregulated the sector. Combined with the need for new power generation following the March 2011 Tohoku earthquake and tsunami, these actions are seen as encouraging the development of new geothermal energy plants.[150] Obstacles, however, remain. One of the main challenges in developing geothermal resources in Japan is that most of the best resources are in national parks and/or close to areas that rely on hot springs to support their local economy; in such areas, residents and businesses are concerned about the potential impact of geothermal power plants on the hot springs.[151] In addition, the development of large power plants typically takes 8 to 10 years, so it is unlikely that a significant number of new large power plants will come online in the next few years. Smaller plants can, however, be developed more quickly.[152] While no large projects have moved forward recently, U.S.-based equipment supplier Ormat has been active in the Japanese market. It supplied equipment in 2004 for at least one smaller project, a 2 MW plant, and is the supplier for an upcoming 1.5 MW project. Several other projects, both small and large, have been announced, but available information indicates that participation in these projects by U.S. firms is limited, at least so far.[153]

References

American Chamber of Commerce in China. White Paper, 2013.

Black & Veatch Corporation (B&V). "Cost Estimates and Performance Data for Renewable Electricity Technologies." Prepared for the National Renewable Energy Laboratory, 2012.

Bloomberg New Energy Finance (BNEF). "Global Trends in Clean Energy Investment," January 14, 2013.

———. "Global Trends in Renewable Energy Investment 2012: Chart Pack," n.d. (received via email from BNEF on January 10, 2013).

———. *Climatescope 2012: Assessing the Climate for Climate Investing in Latin America and the Caribbean,* 2012. http://www5.iadb.org/mif/Climatescope/2012/img/content/pdfs /eng/Climatescope2012- report.pdf.

———. "New Investment in Clean Energy Fell 11% in 2012." Press release, January 14, 2013. http://about.newenergyfinance.com/about/press-releases/new-investment-in-clean-energy-fell-11- in-2012-2/.

———. *Sustainable Energy in America 2013 Factbook,* January 2013. http://www.bcse.org/factbook/pdfs/BCSE BNEF Sustainable Energy in America 2013 Factbo ok.pdf.

Bloomberg New Energy Finance (BNEF) database. http://www.bnef.com (accessed various dates).

Bloomberg News. "Alstom Building Center in China to Expand Hydro Business," July 11, 2012. http://www.bloomberg.com/news/2012-07-11/alstom-building-center-in-china-to-expand-hydrobusiness.html.

BNEF Newswatch. "Asian Development Bank Invests in Indian Renewable Power Company." May 2, 2013.

Center for Climate and Energy Solutions, "Renewable and Alternative Energy Portfolio Standards," March 21, 2013. http://www.c2es.org/us-states-regions/policy-maps/renewable-energy-standards.

Cichon, Meg. "The Next Challenge for Geothermal: Turning the Corner toward Private Investment." RenewableEnergyWorld.com, April 10, 2013. http://www.renewableenergy world.com/rea/news/article/2013/04/the-next-challenge-forgeothermal-turning-the-corner-towards-private-investment.

———. "Rwanda Set to Start Major Geothermal Development." RenewableEnergy World.com, March 27, 2013. http://www.renewableenergyworld.com /rea/news/article /2013/03/rwanda-set-to-startmajor-geothermal-development.

Congressional Research Service (CRS). Bill Summary & Status, H.R.267—113th Congress (2013-2014), CRS Summary. http://beta.congress.gov/bill/113th/house-bill/267.

Crowe, Robert. "Japan's Geothermal Resources Get a Closer Look." *Renewable Energy World*, May 11, 2011. http://www.renewableenergyworld.com/rea/news/article /2011 /05/japans-geothermalresources-gets-a-closer-look.

EIA. *See* U.S. Department of Energy. Energy Information Administration.

Engineering News-Record (ENR). "The Top Contractors in Power," September 17, 2012.

———. "The Top Design Firms in Power," July 2, 2012.

Environmental Business Journal (EBJ). "Multilateral Development Banks Still Influential in Water and Now Renewables." EBJ 25, (2012), 33–38. http://ebionline.org/ebj-archives/1355-ebj-v25n06-07.

European Small Hydro Association (ESHA). *Small Hydropower Roadmap: Condensed Research Data for EU-27*, 2013. http://www.esha.be/fileadmin/esha_files/documents /publications/2013/SHPRoadmap_FINAL_Pu blic.pdf.

———. *State of the Art of Small Hydropower in EU-25.* http://www.esha.be/publications/publications.html, (accessed April 1, 2013).

Federal Energy Regulatory Commission (FERC). "Issued HydroKinetic Preliminary Permits." http://www.ferc.gov/industries/hydropower/gen-info/licensing/hydrokinetics/licences.asp (accessed March 21, 2013).

———. "FERC Issues First Pilot License for Tidal Power Project in New York." News release, January 23, 2012. http://www.ferc.gov/media/news-releases/2012/2012-1/01-23-12.asp.

———. "Present Development of Conventional Hydroelectric Projects." June 13, 2013. http://www.ferc.gov/industries/hydropower/gen-info/regulation/present-dev.asp.

Frankfurt School of Finance and Management (FS) and United Nations Environment Programme (UNEP). *Global Trends in Renewable Energy Investment 2012*. Frankfurt, Germany: FS, 2012.

GE Energy. "GE Energy Industrial Solutions: GE Serves Hydro." Marketing brochure, 2011.

Geothermal Energy Association (GEA). *Annual U.S. Geothermal Power Production and Development Report,* April 2012. http://geoenergy.org/pdf/reports/2013AnnualUS GeothermalPowerProductionandDevelopmentReport Final .pdf.

———. *Green Jobs through Geothermal Energy,* 2010. http://geo-energy.org/pdf/reports /GreenJobs Through Geothermal Energy Final Oct2010.pdf ?

———. "New Geothermal Terms and Definitions: A Guide to Reporting Resource Development Progress and Results to the Geothermal Energy Association," November 2010. http://geoenergy.org/pdf/NewGeothermalTermsandDefinitions January2011.pdf.

Hopwood, David. "Renewable Power Generation—2010 Figures." *Renewable Energy Focus,* January 10, 2012. http://www.renewableenergyfocus.com/view/23029/renewable-power-generation-2010- figures/.

HydroWorld.com. "Brookfield Renewable Completes Acquisition of New England Hydropower Projects," March 5, 2013. http://www.hydroworld.com /articles/2013/03 /brookfield-renewablecompletes-acquisition-of-new-england-hydrop.html.

———. "Hydroelectric Power Industry Leaders, Policy Makers React to Bill Enactment," August 12, 2013. http://www.hydroworld.com/articles/2013/08/hydroelectric-power-industry-leaders-policyleaders-react-to-bill-enactment.html.

———. "Obama Signs Hydroelectric Power Bills," August 9, 2013. http://www.hydroworld.com/articles/2013/08/obama-signs-hydroelectric-power-bills-h-r-267- and-678-now-federal-law.html.

———. "The Road to Sustainable Development: Small Hydro in the EU," February 1, 2013. http://www.hydroworld.com/articles/print/volume-21/issue-01/articles/small-hydro/the-road-tosustainable-development-small.html.

———. "Sri Lanka Micro Hydro Rehabilitation Projects Receive Boost from New ADB Loan," January 18, 2013. http://www.hydroworld.com/articles/2013/january/sri-lanka-micro-hydro-rehabilitationprojects-receive-boost-from.html.

———. "Voith Signs Deals for Two South American Small Hydroelectric Projects," January 28, 2013. http://www.hydroworld.com/articles/2013/january/voith-signs-deals-for-two-south-americansmall-hydroelectric-pro.html.

———. "World Bank Announces Renewed Support for Large Hydropower," May 30, 2013. http://www.hydroworld.com/articles/2013/05/world-bank-announces-renewed-support-for-largehydropower.html.

IBISWorld. "Geothermal Electricity Plant Operation in the U.S.," 2012.

Ingram, Elizabeth. "Banks Provide Critical Support for Hydro Project Development." *HydroWorld.com.,* November 1, 2011. http://www.hydroworld.com/articles/2011/11 /banks-provide-critical-supportfor.html.

Institute for Energy Research (IER). "The Status of Renewable Electricity Mandates," http://www.instituteforenergyresearch.org/renewable-mandates/ (accessed July 9, 2013).

International Energy Agency (IEA). 2012 *Hydropower Technology Roadmap,* 2012. http://www.iea.org/publications/freepublications/publication/name,32864,en.html.

———. *Renewable Energy: Markets and Prospects by Technology,* 2011. http://www.iea.org/publications/freepublications/publication/Renew Tech.pdf.

———. *Renewable Energy Essentials: Geothermal*, 2010. http://www.iea.org/publications/freepublications/publication/name,3914,en.html.

International Renewable Energy Agency (IRENA). "Hydropower." In *Renewable Energy Technologies: Cost Analysis Series.* Vol. 1: *Power Sector,* issue 3/5, June 2012.

Islandbanki Geothermal Research. *U.S. Geothermal Industry Overview,* February 2011. http://skjol.islandsbanki.is/servlet/file/store156/item102416/US%20Geothermal%20Energy%20 Market%20Report%20web.pdf.

Kleinschmidt. "ITC/PTC—What's in It for Hydro and Wind Developers?" www.kleinschmidtUSA.com (accessed March 7, 2013).

KPMG. *World Geothermal Market and Outlook,* n.d. http://www.kpmg.com/IS/is /utgefidefni/greinar-ogutgefid/Documents/KPMG WorldGeothermalMarketandOutlook oct2010.pdf, July 10, 2013.

Lobo, Marcello. "Global Renewable Energy Developments: Trends, Innovations and Developments in Brazil." Webinar presentation sponsored by American Council on Renewable Energy (ACORE), January 16, 2013.

Lowder, Travis, Ryan Hubbell, Michael Mendelsohn, and Karlynn Cory. *Renewable Energy Finance Tracking Initiative (REFTI): Snapshot of Recent Geothermal Financing Terms.* National Renewable Energy Labor (NREL) Technical Report, NREL/TP-6A20-54438, September 2012.

National Hydropower Association. "Hydro Works for America," n.d. http://www.hydro.org/wpcontent/uploads/2010/12/Pumped-storage5.pdf (accessed June 26, 2013).

New Times (Rwanda). "Geothermal Drilling Starts Next Month," March 27, 2013. http://www.newtimes.co.rw/news/index.php?i=15309&a=65341.

Oregon State University. "Dam Construction to Reduce Greenhouse Gases Causes Ecosystem Disruption." Press release, June 18, 2013. http://oregonstate.edu/ua/ncs/archives /2013/jun/dam-constructionreduce-greenhouse-gases-causes-ecosystem-disruption.

Ormat. "Ormat and JFE Engineering Enter into a Cooperation Agreement for Implementing Geothermal Projects in Japan." News release, June 28, 2010. http://www.ormat.com /news/general/ormat-andjfe-engineering-enter-cooperation-agreement-implementing-geothermal-projects-.

PWC. *Power in Indonesia: Investment and Taxation Guide,* 2nd Ed. April, 2013.

Renewable Energy Policy Network for the 21st Century (REN21). *Renewables 2012: Global Status Report,* 2012.

———. *Renewables: Global Futures Report 2013*, 2013. www.ren21.net/Portals/0/REN21 GFR 2013 print.pdf.

Reuters. "China Urges Hydropower Developers to Heed Environment," January 17, 2012.

http://www.reuters.com/article/2012/01/17/china-hydropower-idUSL3E8CH2AC20120117 (accessed June 12, 2013).

———. "Rwanda to Spend $935 Mln on Geothermal Power," March 22, 2011. http://www.reuters.com/article/2011/03/22/ozabs-rwanda-geothermal-idAFJOE72L0CX20110322.

Roca, Marc. "Spain Limits Payments to Renewables to Reduce Tariff Deficit," February 1, 2013.

Smith, Deonta. IBISWorld. "Hydroelectric Power in the U.S." IBISWorld Industry Report 22111c, July 2012.

South China Morning Post. "World Bank U-turn Brings Hydropower in From Cold," May 11, 2013. http://www.scmp.com/news/world/article/1234897/world-bank-u-turn-brings-hydropower-cold.

Taylor, Mark. "East Africa in Global Geothermal." Presentation at ARGEO-C4, Nairobi, November 23, 2012.

U.S. Department of Energy. Energy Information Administration (EIA). *Annual Energy Review 2011.* DOE/EIA-0384 (2011), September 2012. http://www.eia.gov/totalenergy /data/annual/index.cfm.

———. Electric Power Monthly. Data for February 2013, released April 22, 2013. http://www.eia.gov/electricity/monthly.

———. "Existing Nameplate and Net Summer Capacity by Energy Source, Producer Type and State (EIA-860)." October 2, 2012. http://www.eia.gov/electricity/data/state/.

———. "Geothermal Explained." http://www.eia.gov/energyexplained/index.cfm?page=geothermal_use (accessed November 7, 2012).

———. International Energy Statistics. http://www.eia.gov/cfapps/ipdbproject/IEDIndex3.cfm?tid=2&pid=2&aid=7 (accessed March 20, 2013).

———. *Monthly Energy Review,* October 2012.

———. Renewable Energy Consumption and Electricity Preliminary Statistics 2010. http://www.eia.gov/renewable/annual/preliminary/ (accessed January 1, 2013).

———. "Table 1.19.B. Net Generation from Geothermal by State, by Sector, Year-to-Date through October 2012 and 2011." In Electric Power Monthly, December 2012. http://www.eia.gov/electricity/monthly/pdf/epm.pdf.

———. "Today in Energy." http://www.eia.gov/todayinenergy/detail.cfm?id=3970 (accessed May 31, 2013).

U.S. Department of Energy. Geothermal Technologies Office. "American Recovery and Reinvestment Act." http://www1.eere.energy.gov/geothermal/recovery.html (accessed June 27, 2013).

U.S. Geological Survey. "Fact Sheet 2008—3082," 2008. http://pubs.usgs.gov/fs/2008/3082/ (accessed July 9, 2013).

Verdant Power. Written submission to the U.S. International Trade Commission in connection with inv. no. 332-534, *Renewable Energy and related Services: Recent Developments,* March 1, 2013.

Watanabe, Chisaki. "Japan Considering 21 Geothermal Plants after Fukushima." Bloomberg, April 9, 2013. http://www.bloomberg.com/news/2013-04-09/japan-considering-21-geothermal-plansafter-fukushima-meti-says.html.

———. "JFE Engineering Unit Wins Geothermal Plant Order." Bloomberg, February 20, 2013. http://www.bloomberg.com/news/2013-02-20/jfe-engineering-unit-wins-order-to-constructgeothermal-plant.html.

World Bank. Energy Sector Management Assistance Program (ESMAP). *Geothermal Handbook: Planning and Financing Power Generation.* Technical Report 002/12, June 2012. http://www.esmap.org/sites/esmap.org/files/FINAL Geothermal%20Handbook TR002-12.pdf.

———. "Full Steam Ahead: World Bank Seeks 'Global Push' for Geothermal Energy Revolution," March 6, 2013. http://web.worldbank.org/WBSITE/EXTERNAL/NEWS/0,,contentMDK:23367555~pagePK:343 70~piPK:34424~theSitePK:4607,00.html.

"Renewable Energy." n.d. http://web.worldbank.org/WBSITE/EXTERNAL/TOPICS/EXTENERGY2/0,,contentMDK:22505284~menuPK:4140690~pagePK:148956~piPK:216618~theSitePK:4114200,00.html (accessed June 4, 2013).

World Bank website. http://water.worldbank.org/node/84059 (accessed March 19, 2013).

SECTION 6. MARKET EFFECTS OF CLEAN ENERGY INCENTIVE PROGRAMS

Introduction

This section examines incentives that promote investment in renewable energy. It identifies the goals of such incentives, and examines the structure and implementation of feed-in tariffs (FITs), renewable portfolio standards (RPSs), and production tax credits. It then analyzes the effect of incentives on profits and renewable energy deployment.

The overarching purpose of these incentives is to "internalize" the external benefits of using renewable energy instead of other energy sources. These spillover benefits exist because electricity generated from nonrenewable sources, particularly fossil fuels, often has negative externalities (i.e., social costs imposed at the regional and global levels) in the form of pollution, health costs, and carbon emissions. For example, fossils fuels yield large quantities of carbon dioxide (CO_2) compared to renewable energy.[1] The emission of carbon dioxide may play a role in trapping heat in the earth's atmosphere, potentially leading to an acceleration in climate change. Often, firms do not face the full social costs of producing electricity using CO_2-emitting power plants.[2] Policies intended to reduce negative externalities include carbon taxes or a cap-and-trade system that would set the private cost of energy at a price closer to its social cost. However, these policies can be politically controversial and technically challenging. Providing incentives to develop and use renewable energy can thus help reduce CO_2-related negative externalities insofar as renewable energy replaces fossil fuel-based energy.

Renewable energy policies are also used to meet an array of additional goals. They can be industrial policies for governments that anticipate growing demand for renewable energy in the future (in the same way that governments promote investment in high-tech manufacturing and other industries). Policymakers often talk about the need to improve domestic competitiveness and increase their country's share of the global renewable energy market. [3]

Renewable energy incentives are also used to achieve economic development, job creation, and environmental justice (for example, reducing emissions of particulates such as soot and ash near low-income communities). Renewable energy incentives that are sufficiently powerful can accomplish these goals, but such goals are often in tension with each other. For example, a policy that aims to deploy renewable energy and create jobs simultaneously may cost more per job and per kilowatt-hour (kWh) than single-focus policies.[4]

Renewable energy incentives can be either quantity-based or price-based. Quantity-based incentives, such as RPSs, mandate a particular goal for electricity production (for example, that a particular percentage of electricity must come from renewable sources within 10 years), regardless of the cost. FITs and other fiscal incentives are price-based, meaning the cost of the policy is certain at least in a relative sense (for example, that producers will receive a particular dollar amount per kWh of electricity generated from renewable energy sources), but the total amount of energy that will be created is unknown.[5] In practice, governments often use a mix of quantity-based and price-based incentives simultaneously. For example, Texas has a statewide RPS aiming to achieve 5,880 megawatts (MW) of capacity by 2015 (and 10,000 MW by 2025), while San Antonio and Austin have municipal RPSs of 20 percent and 35 percent respectively by 2020. Texas also incentivizes renewable energy with tax benefits (corporations can deduct up to 10 percent of the amortized cost of a solar energy system from the state franchise tax, and the cost of installing solar or wind-powered systems can be deducted from property tax) and with loans and grants (the Texas Department of Rural Affairs offers grants via the Renewable Energy Demonstration Pilot Program). Additionally, governments can establish regulations and standards that facilitate renewable energy deployment without directly incentivizing it. For example, the Public Utility Commission of Texas established interconnection standards in 1999 that allow on-site distributed generation.[6]

FITs and RPSs are the most widely used renewable energy incentives. However, the variety and complexity of FITs and RPSs make it difficult to draw broad conclusions about their relative merits. It is nonetheless possible to measure the effect of the existence of FITs or RPSs. By some accounts, developers may prefer FITs, as RPSs can prompt overly aggressive bidding processes.[7] In addition, one study found that FITs may outperform RPSs in promoting wind-energy capacity installation.[8]

The Need for Incentives

The renewable energy sector is affected by technology, regulations, institutions, financial markets, human capital, politics, and physical infrastructure. All of these factors can pose barriers to current deployment and be sources of future risk, both in their absolute levels and in their fluctuations (for example, both the price of natural gas and the volatility of that price shape investment decisions about renewable energy). To drive investment in renewable energy, incentives must increase the risk-adjusted net present value of renewable energy projects enough to attract investors. However, even large renewable energy incentives may not be enough to overcome high political risk, excessive permitting requirements, limited access to capital, or other constraints.[9]

Diverse financial challenges confront efforts in this sector. Some technologically advanced renewable energy projects with high up-front costs can fall into a situation in which no one wants to finance a project without a successful demonstration, yet successful demonstrations require financing (the so-called "Valley of Death"). [10] Additionally, projects like rooftop solar panels often face a financing mismatch, such as when landlords incur the upfront costs of installing renewable energy while tenants reap the benefits of lower heating costs. These gaps can theoretically be addressed through contracts, but in practice, doing so can be difficult. Broader events can affect the financial landscape as well; for example, California's electricity crisis during 2000–01 dried up renewable energy financing in the state.[11] Greater financial challenges require larger incentives to achieve a given amount of renewable energy deployment.

Along with financial needs, an International Energy Agency (IEA) survey found that risk of policy changes was among the largest barriers to renewable energy deployment.[12] Credible commitments to long-term policies are important for investor certainty. An example is the U.S. wind production tax credit, which subsidizes wind power by 2.2 cents per kWh. The tax credit was established in 1992 but has had a checkered history: it was allowed to expire in June 1999 before being restored in December 1999, expired again in December 2001 and was restored in March 2002, expired again in December 2003 and was restored in October 2004, and expired yet again at the end of 2012 before being extended a few days later (applying only to wind projects that are under construction by the end of 2013). By some accounts, the uncertainty regarding the tax credit's post-2012 status made investors reluctant to finance wind projects (see figure 6.1).[13] Stable policy frameworks allow companies to take a long-term perspective, enabling them to take the plunge and optimize their supply chains, while "stop and go" policies can prompt firms and banks to take a wait-and-see approach.[14]

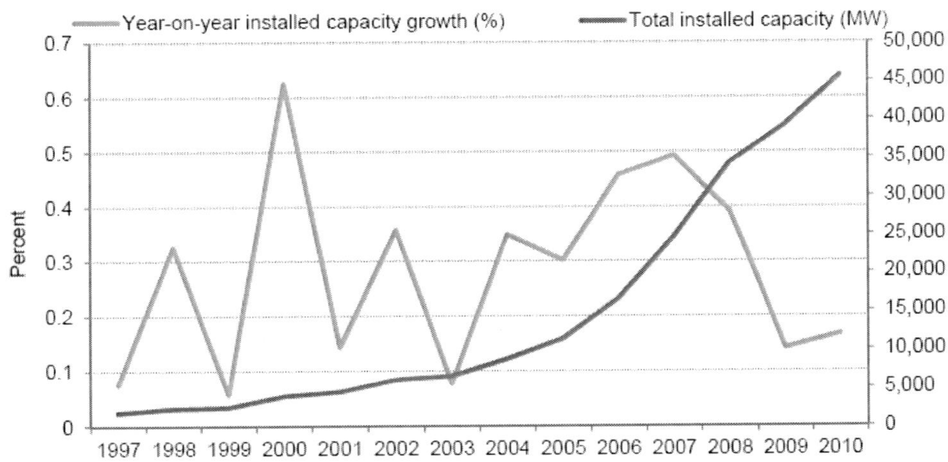

Source: U.S. Energy Information Administration database (accessed May 30, 2013).

Figure 6.1. The expiration of Production Tax Credits in 1999, 2001, and 2003 impacted wind power growth in the United States.

Regions with a good supply of relevant resources, such as wind, solar, and geothermal endowments, reliable electrical grid infrastructure, and advanced supporting industries (such as silicon wafer manufacturing and engineering services), are well positioned to deploy renewable energy, and can meet ambitious renewable energy targets at lower cost. For example, renewable energy producers in San Diego benefit from complementary defense and biotech industrial clusters; research institutions at the University of California, San Diego, and the Scripps Research Institute; proximity to the Imperial Valley (which has a lot of sunshine) and to Baja California (which has low-cost manufacturing); and the availability of venture capital.[15] Similarly, Hamburg, Germany, evolved naturally into a renewable energy industry cluster due to its robust wind resources, its well-developed financial sector, and the fact that its shipbuilders were able to build turbine blades (which are similar to propellers) and large steel wind towers in their shipyards.[16]

Other factors that influence regional adoption of renewable energy include energy security (net fossil fuel importers are more likely to deploy renewables),[17] expertise in high-tech manufacturing (which is correlated with how long countries have been using renewable energy technology),[18] and income (which may help countries afford high energy prices and/or high regulatory costs related to the use of renewable energy).[19] One study found that U.S. states' use of renewable energy was negatively correlated with electricity use per capita, the percentage of petroleum and coal manufacturing in the state product, and deregulation.[20]

Incentives are often tailored to specific phases of renewable energy deployment.[21] Initially, incentives try to provide the industry with a "jump start," aiming to create consumer confidence, insulate investors from some financial risks, and overcome the lack of supply chains and other supportive infrastructure. Then, once the industry is established, policies become more focused on reducing costs, often through scheduled reductions in incentives based on fixed time periods or on installed-capacity benchmarks. Finally, in mature energy markets, policies try to accommodate daily fluctuations in supply and demand, facilitate the use of least-cost solutions for the system as a whole, and optimize energy flows through space (by incentivizing grid extension and integration) and time (by incentivizing storage).[22]

Feed-in Tariffs

Feed-in tariffs (FITs) are contracts that guarantee a payment rate (e.g., dollars per kWh) to renewable energy producers for a period of time, often 15 to 20 years. These policies encourage investment in renewable energy generation by providing a certain degree of price certainty: the greater security of future cash flows lowers the risk of investing in renewable energy ventures, which are often capital-intensive projects with high up-front costs and high ratios of fixed to variable costs.[23] FITs can be either fixed-price, in which the tariff is absolute, or premium-price, in which the FIT provides a premium payment above spot-market prices, and is therefore tied to changes in market prices for electricity.[24] FITs often impose few eligibility restrictions, so both small and large renewable energy generators can participate.

FITs can be designed in elaborate ways. They can include regular degression (in which reimbursements are lowered over time),[25] growth caps (in which reimbursements are lowered once annual installed capacity exceeds a target), stepped tariffs (in which tariffs are set relative to the expected power output of individual plants so that locations with especially good wind or sun resources do not benefit disproportionately), and net metering (in which reimbursement depends on the grid load so that providing excess electricity is not compensated).[26] Some fixed-price FITs have full or partial inflation adjustment; for example, Ontario, Canada, offers inflation adjustment on 20 percent of the base price of electricity for wind, biomass, and hydroelectric sources over the course of a 20-year contract.[27] On average, premium-price FITs are more costly than fixed-price FITs, and result in more uncertainty, without eliminating the danger that energy producers will be overcompensated. But relying on markets allows a more efficient assignment of grid costs: when supply is scarce and demand is high, prices rise, which induces suppliers to provide more electricity at peak usage times.[28] For that reason, premium-price FITs are especially attractive in regions with large gaps between peak and off-peak demand.

FITs are widely used, particularly in Europe: they are the main support instrument for renewable energy in 20 EU countries and are used as secondary instruments in another 4 EU countries.[29] In the EU, total remuneration of onshore wind power generation in 2012 ranged from €36.6 per megawatt-hour (MWh) in Denmark, which used a fixed-price FIT system, to €101.3 per MWh in Germany, which used a sliding FIT system based on monthly average electricity prices.[30] By some measures, FITs have been the primary policy driver of renewable energy deployment in Germany, Spain, Portugal, and Denmark.[31]

Setting FITs at the right level is difficult.[32] FITs aim to provide a steady revenue stream that stimulates long-term large-scale investment, but if they are set too high they will generate windfall profits for developers, leading to unnecessary public spending or customer dissatisfaction at overpriced electricity.[33] In many cases, FITs are adjusted downwards after being established. Germany recently reduced its FIT by 15 percent for open-field solar installations and 16 percent for rooftop solar installations,[34] while Italy reduced its solar FIT by 22–30 percent in 2011.[35] In this way FITs can be victims of their success: they are reduced or eliminated when they drive growth faster than projected (see box 6.1).

Renewable Portfolio Standards (RPSs)

RPSs mandate specific amounts of renewable energy deployment, either in terms of absolute capacity or as a percentage of total electricity use. Unlike FITs, which support renewable energy generation at an established rate regardless of quantity, RPSs encourage generators to bid against each other to supply the targeted amount of electricity at the lowest cost. [36] RPS targets are intended to be cost-effective and technically feasible, but the precise target tends to be arbitrary, as RPSs are typically established via political processes (for example, the mandatory RPS targets in 13 U.S. states are all multiples of 10).

The design of each RPS specifies a target, an end date, an application process, eligibility requirements, and administration and enforcement details. In some cases there are annual targets set, with penalties for not reaching targets, though longer compliance periods help accommodate the fact that renewable energy development can be "lumpy" (involving large and irreversible up-front costs).

Thirty U.S. states currently have mandatory RPSs (up from three in 1998). These exhibit a wide diversity in type (absolute capacity versus share in total electricity mix), as well as in target chosen, time frame, and coverage (for example, some states exempt or give lower targets to municipal utilities or rural electric cooperatives) (table 6.1). The stringency of state-level RPSs varies: in Texas, RPS policies were ambitious enough to have contributed to substantial growth in renewable energy use, while Maine's RPS has largely endorsed existing levels of renewable energy generation.[37] Some states have changed their targets over time. For example, California's original RPS target called for 20 percent of statewide electricity to come from renewable sources by 2017; now it mandates 33 percent by 2020. While changing targets can create policy uncertainty, this tightening RPS program has helped build a mature renewable energy market in California.

BOX 6.1. Feed-in Tariffs in Spain

Spain's history with FITs illustrates both the effectiveness of FITs and the difficulty of calibrating them. Spain has promoted renewable energy deployment through FITs since 1998, motivated by the desire for energy security, industrial development, and climate change mitigation.[a] In 2004, renewable energy generation had increased to the point where policymakers were concerned about grid stability and voltage drops, among other issues, and generators felt that the annual FIT revisions were non-transparent and arbitrary, so the FIT program was modified.[b] It was further modified in 2007, adopting a cap-and-floor price to help control system costs and improve the security of the electricity supply.[c]

However, Spain's solar installations in 2008 exceeded expectation by a wide margin. The government forecast achieving 400 MW of solar capacity by 2010, but under the FIT system 2.6 gigawatts of capacity was installed in 2008 alone, which accounted for 40 percent of total global installations that year and resulted in Spain having the largest installed concentrated solar power (CSP) capacity in the world.[d] This rapid deployment was largely due to the fact that the costs of solar panels collapsed while FITs remained unchanged.[e] The country's solar energy market crashed in 2009; there had been excessive construction and payments,[f] and demand for electricity, which was closely tied to real estate development, had collapsed. The Spanish government temporarily closed its FIT system starting in 2012 (with an exception for plants already receiving payments and plants already registered for pre-allocations), citing the country's economic crisis.[g] The country also implemented several retroactive changes to renewable energy contracts, including a lifetime cap on FITs for plants connected to the grid before 2008, which prompted some global investors to seek international arbitration.[h] Firms, and banks that loaned money to them, had made their investing and financing decisions based on anticipated FITs.[i]

Spain's frequent FIT modifications created an unpredictable policy environment, and resulted in price uncertainty and high volatility in the renewable energy market.[j] Yet Spain was able to transform its energy mix by deploying an enormous amount of renewable energy in a short span of time. Although Spain's FITs did not provide the renewable energy industry with stable and competitive economic underpinnings, the frequent amendments and policy shifts did not discourage renewable energy investment in the country.[k]

[a] Industry representatives, interview by USITC staff, Madrid, Spain, May 8, 2013.
[b] Del Río González, "Ten Years of Renewable Electricity Policies in Spain," 2008.
[c] Del Río González, "Ten Years of Renewable Electricity Policies in Spain," 2008.
[d] SolarServer, "Spanish Government Halts PV, CSP Feed-in Tariffs," January 30, 2012.
[e] Industry representatives, interview by USITC staff, Madrid, Spain, May 8, 2013.
[f] Voosen, "Spain's Solar Market Crash Offers a Cautionary Tale," August 18, 2009
[g] SolarServer, "Spanish Government Halts PV, CSP Feed-in Tariffs," January 30, 2012.
[h] ICIS, "Cloud of Legal Action," November 30, 2011.

[i] Industry representatives, interview by USITC staff, Madrid, Spain, May 7, 2013.
[j] Industry representatives, interview by USITC staff, Madrid, Spain, May 8, 2013.
[k] Ibid.

Table 6.1. Mandatory renewable portfolio standards by U.S. state

State	Details	Target	Year
Arizona		15%	2025
California		33%	2020
Colorado	Investor-owned utilities	30%	2020
	Electric cooperatives and municipal utilities serving more than 40,000 customers	10%	2020
Connecticut		27%	2020
Delaware		25%	2027
District of Columbia		20%	2020
Hawaii		40%	2030
Illinois		25%	2025
Iowa		105 MW capacity	2000
Kansas		20%	2020
Maine		40%	2017
Maryland		20%	2022
Massachusetts	New resources	15%	2020
	Existing resources	7.1%	2009
Michigan		10%	2015
Minnesota	Xcel Energy	30%	2020
	Other utilities	25%	2025
Missouri		15%	2021
Montana		15%	2015
Nevada		25%	2025
New Hampshire		24.8%	2025
New Jersey		20.38%	2021
New Mexico	Investor-owned utilities	20%	2020
	Rural electric cooperatives	10%	2020
New York		29%	2015
North Carolina	Investor-owned utilities	12.5%	2021
	Electric cooperatives, municipal utilities	10%	2018
Ohio		12.5%	2024
Oregon	Large utilities	25%	2025
	Small utilities	10%	2025
	Very small utilities	5%	2025
Pennsylvania		18%	2021
Rhode Island		16%	2019
Texas		5,000 MW capacity	2014
Washington		15%	2020
Wisconsin		9.55%	2015

Sources: USDOE; DSIRE database (accessed November 29, 2012).

One industry representative estimates that the RPS motivated approximately $50 billion in investment in the state's renewable energy sector, including substantial investment from overseas.[38]

Most RPSs also set up a market for tradable certificates, allowing renewable energy generators to sell certificates representing one MWh of renewable electricity generation. If a large consumer or retailer of electricity is required to meet renewable energy targets under a RPS, it can purchase a certificate in lieu of deploying one MWh of its own renewable energy. Renewable energy generators can receive revenue both from selling electricity on the open market and from selling certificates to entities covered by the RPS. With a robust certificate market, regions can meet their RPSs by deploying renewable energy wherever within the area of coverage it is cheapest to do so.[39] However, the prices of tradable certificates can be volatile. In Connecticut, prices fell from more than $30 per MWh in June 2005 to less than $10 in October 2005, due to changes in the eligibility of biomass-based generators under the state's RPS. Maine's certificate prices hovered near zero through 2006, reflecting the fact that Maine's target could be met entirely with existing renewable energy generation.[40]

Most RPSs have been implemented since 2000—for example, the EU imposed mandatory renewable energy targets for member countries in 2001[41]—so there are limited data on their impact. One study on U.S. states found that RPSs do not predict the percent of renewable energy generation in a state's total electricity mix, but each additional year that a state has a RPS increases total renewable energy generation.[42]

Tax Incentives and Public Financing

Many countries and regions use tax codes to provide incentives for renewable energy investment. Such incentives effectively replicate direct government spending while being more popular and easier to implement politically. In China, advanced and new technology enterprises in the solar, wind, and geothermal energy sectors are subject to a reduced corporate income tax rate, and 50 percent of the value-added tax on sales of wind power is refunded.[43] Romania exempts renewable electricity from excise duties,[44] while the Republic of Korea reduces import duties by 50 percent for components and equipment used in renewable energy plants.[45]

Another type of tax incentive is accelerated depreciation. Many countries allow renewable energy generators to adopt accelerated depreciation schedules, in which equipment depreciates at a higher rate early during the life of the asset compared to later years. This accounting technique reduces the taxable profits, especially in the first years of operation, of firms who claim depreciation as an expense (even though their actual assets depreciate more slowly than the schedule). For example, wind farms in India built before April 1, 2012, could claim accelerated depreciation on 80 percent of the cost of equipment, and this tax incentive was estimated to be a factor in 70 percent of all India's wind installations in 2011.[46] Renewable energy projects often have large up-front costs, and accelerated depreciation attempts to soften this particular barrier to investment by deferring tax payments.

There are multiple U.S. tax incentives for investing in renewable energy, including the investment tax credit for renewable energy (Internal Revenue Code section 48) and the renewable electricity production credit (IRC section 45A). [47] Under the former, investments in solar energy property, fuel cells, and small wind systems may receive a 30 percent tax credit, while investments in geothermal systems, microturbines, and combined heat and power technologies (which simultaneously produce electricity and heat from a single fuel source) may qualify for a 10 percent tax credit.[48] The latter credit includes the most effective federal incentive for wind power, a per-kWh credit applied to the output of qualifying facilities during the first 10 years of their operation. Total 2011 tax expenditures for the investment tax credit were an estimated $300 million, while the production tax credit resulted in an estimated $1.4 billion of foregone revenue in 2011.[49] There are also tax incentives at the state level; for example, North Carolina has tax incentives for manufacturing renewable energy equipment and for investing in renewable energy property.[50] (Federal and state tax incentives apply to fossil fuel industries as well, so tax codes may not provide a net benefit to renewable energy relative to other energy sources.)[51]

Governments also provide various forms of direct public financing that facilitates the use of renewable energy. As part of the U.S. American Recovery and Reinvestment Act, Congress established a temporary grant program (the "1603 Program") that entitled project developers to receive a cash payment worth 30 percent of a project's capital cost. The 1603 Program awarded $11.6 billion to 38,000 projects before expiring at the end of 2011.[52] The Department of Energy also plays a major role in providing loan guarantees for projects that use new (or significantly improved) clean energy technologies, and as of September 2011 had made $15.1 billion in loan guarantees. [53] For example, SolarReserve's 110-MW Crescent Dunes CSP project in Nevada is backed by a $737 million loan guarantee. The guarantee, combined with a 25-year power purchase agreement with NV Energy, provides a long-term financial and policy framework that has helped facilitate a large investment in a new technology (molten salt storage).[54] However, the Department of Energy generated some controversy when it provided over $500 million in loan guarantees to the solar company Solyndra, which later went bankrupt in part due to the falling price of silicon panels. It has been argued that some defaults are to be expected to the extent that such loan programs invest in unproven technologies (by one estimate, 30–40 percent of high-potential U.S. start-ups fail).[55]

However, the Solyndra bankruptcy has made renewable energy firms less certain about the availability of public loan guarantees in the future.[56]

Incentives and Return on Equity

The Commission examined the ability of the renewable energy industry to generate profits from shareholders' equity (i.e., from assets minus liabilities). It used a global company database to assemble information about firms in 54 countries from 2003 to 2012, focusing on firms with a primary North American Industry Classification System (NAICS) code of 221119 ("other electric power generation facilities").

This code covers firms that convert solar, wind, and tidal energy (but not hydroelectric, fossil fuel, or nuclear energy) into electricity and provide such electricity to power transmission systems. The return on equity (ROE) was measured by summing the net income and summing the shareholders' equity for all available firms in each country-year, and then dividing the former by the latter, in an attempt to capture the ROE for the entire "other electricity" sector. When a particular country in a particular year had fewer than three firms, that observation was dropped.

The Commission then estimated the ROE for all industries other than NAICS 221119 in each country-year (using each country's top 30,000 firms, excluding renewable energy firms), to compare the performance of renewable energy to the performance of the rest of the economy. The Commission derived a "ROE differential" measure for each country in each year, equal to the difference between ROE in renewable energy and ROE in all non-renewable energy sectors.

The average ROE differentials are presented in figure 6.2. No clear patterns emerged, though the graph illustrates the high variance in the performance of renewable energy sectors relative to the rest of the economy in different countries. In econometric analysis, the Commission did not find a significant relationship between ROE differentials and the presence of FITs, or between total installed wind energy capacity and ROE differentials, holding other variables constant.

This suggests that the deployment of FITs may not improve the ability of the renewable energy industry to turn equity into profits. It also suggests that the quantity of installed wind energy capacity in a country may primarily be driven by factors other than the profitability of its renewable energy sector.

One possibility is that the introduction of a FIT may draw resources into the renewable energy sector rapidly, leading to high levels of competition that limit firms' profits. Anecdotally, industry representatives have suggested that profits in the renewable energy industry are limited by the competitive and fragmented nature of the market, and by underlying market conditions (for example, the stagnation or decrease in energy demand in many countries following the 2008 global financial crisis). [57] Additionally, overcapacity in the solar market, as well as low maintenance costs for wind farms (i.e., they have high up-front costs and low marginal costs), may put negative pressure on prices in competitive markets.[58]

Finally, some industry representatives note that small investors, particularly in the EU, will often put money into small-scale renewable energy plants as a "savings account," and are content with relatively low rates of return (e.g., 5 or 6 percent).[59] FITs may not be powerful enough to consistently overcome these factors and significantly increase the profitability of the renewable energy sector.

Incentives and Installed Wind Capacity

The Commission examined the relationship between FITs and installed wind capacity in 54 countries from 2006 to 2010. Data on FITs were drawn from the Renewable Energy Policy Network for the 21st Century's annual reports and supplemented with additional research, while data on installed wind capacity were drawn from the World Wind Energy Association's World Wind Energy Reports. The analysis controlled for gross domestic product (GDP) per capita, electric power consumption per capita, oil rents as a percentage of GDP,[60] and net energy imports (all from the World Bank's World Data Indicators). Summary statistics are presented in table 6.2.

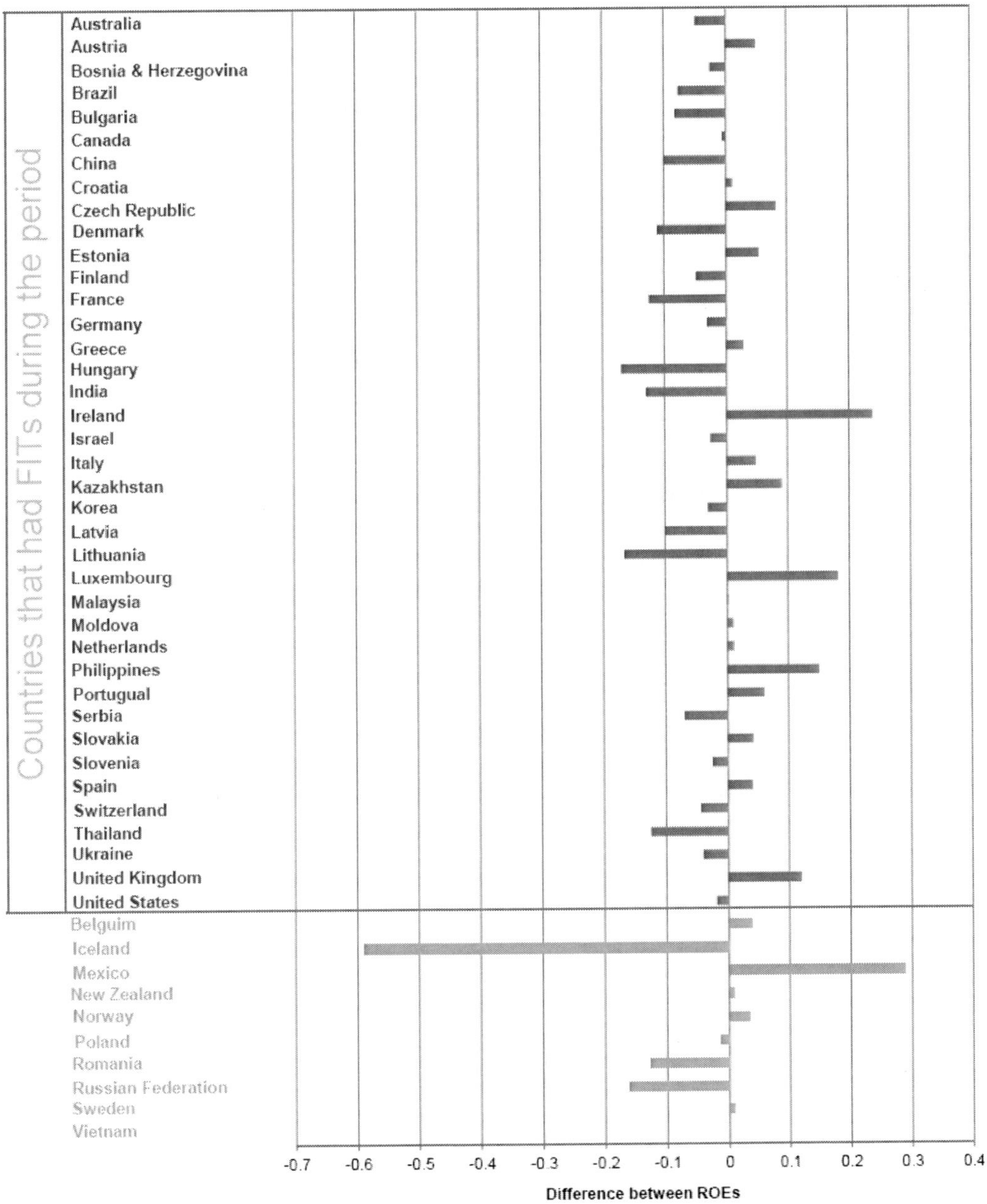

Source: Compiled by Commission staff from Bureau van Dijk, 'ORBIS database,' accessed various dates.

Figure 6.2. The average 2002–12 difference between 'Other Electricity' return-on-equity (ROE) and total country returnon-equity was not significantly different for countries that had feed-in tariffs during the period.

Table 6.2. Summary statistics

Variable	Mean	Std. Dev.	Min	Max
Installed wind capacity (MW)	2,572	6,378	0	44,733
GDP per capita	15,747	13,188	578	56,285
Electric power consumption (KWH per capita)	6,586	4,928	459	25,175
Net energy imports (percentage of energy use)	8.958	119.371	–736.16	97.565
Oil rents (percentage of GDP)	2.188	4.961	0	31.041
FITs	0.614	0.488	0	1

Source: REN21; WWEA; World Bank; USITC calculations.

FITs were time-varying, as several countries adopted FITs during the period, and were represented using a binary variable: "1" if the country in question employed a FIT that year, and "0" if not. Using this dummy variable conceals the wide variety in the significance and design of FITs across countries, so the coefficient on the FIT variable identifies the effects of FITs per se—that is, the significance of simply having a policy called a FIT, regardless of its details or magnitude.

The regression used year dummy variables and controlled for country fixed effects (which includes the natural wind resource endowments of countries). Results are presented in table 6.3. For countries in this sample, having a FIT policy was correlated with an additional 1,856 MW of installed wind capacity. The correlation was significant at the 5 percent level. Other variables were not significant. (These findings were roughly consistent with other estimations; see Jenner 2012 for an overview of empirical studies of FITs.)

Table 6.3. Regression results (the dependent variable is total installed wind capacity)

Variable	Coefficient (t score)
GDP per capita	–0.490
	(–1.12)
Electric power consumption (KWh per capita)	0.955
	(1.19)
Net energy imports (percentage of energy use)	–12.586
	(–0.66)
Oil rents (percentage of GDP)	180.362
	(0.64)
FITs	1,855.871 †
	(2.05)
Number of observations	267

Note: † indicates a 5 percent level of significance.

$$Installed\ Wind\ Capacity_{it} = \beta_0 + \beta_1(GDP\ pc_{it}) + \beta_2(Power\ consumption\ pc_{it}) + \beta_3(Net\ energy\ imports_{it}) + \beta_4(Oil\ rents_{it}) + \beta_5(FITs_{it}) + \beta_6(2006) + \cdots + \beta_{10}(2010) + (Fixed\ effects_i) + \varepsilon_{it}$$

References

Alagappan, L., R. Orans, and C. K. Woo. "What Drives Renewable Energy Development?" *Energy Policy* 39, no. 9 (September 2011): 5099–5104.

Bird, Lori, Mark Bolinger, Troy Gagliano, Ryan Wiser, Matthew Brown, and Brian Parsons. "Policies and Market Factors Driving Wind Power Development in the United States." *Energy Policy* 33, no. 11 (July 2005): 1397–1407.

Bourgeois, Matt, Kevin Breaux, Michael Chiasson, and Shawn Mauldin. "Tax Incentives of Going Green." *CPA Journal* 80, no. 11 (November 2010): 19–26.

Bureau van Dijk. Orbis database. https://orbis.bvdinfo.com/version2013527/home.serv?product=orbisneo (accessed various dates).

California Energy Commission. "Renewable Energy Program: Annual Project Activity Report to the Legislature." P500-02-068, December 2002. http://www.energy.ca.gov /reports/2003-01-13 500- 02-068.pdf.

Cardwell, Diane. "Tax Credit in Doubt, Wind Power Industry Is Withering." *New York Times*, September 20, 2012. http://www.nytimes.com/2012/09/21/business/energy-environment/as-a-tax-creditwanes-jobs-vanish-in-wind-power-industry.html?pagewanted=all.

Carley, Sanya. "State Renewable Energy Electricity Policies: An Empirical Evaluation of Effectiveness." *Energy Policy* 37, no. 8 (2009): 3071–81.

CleanTechnica. "Italy to Levy Carbon Tax, Cut Solar FIT in Wake of Record-Setting Growth," April 18, 2012. http://cleantechnica.com/2012/04/18/italy-to-levy-carbon-tax-cut-solar-fit-in-wake-ofrecord-setting-growth/.

Cory, Karlynn, Toby Couture, and Claire Kreycik. "Feed-in Tariff Policy: Design, Implementation, and RPS Policy Interactions." National Renewable Energy Laboratory. Technical Report NREL/TP6A2-45549, March 2009. http://www.nrel.gov/docs/fy09osti /45549.pdf.

Cosmo Jr., Vito A. "States Provide Tax Incentives for Investing in Alternative/Renewable Energy Products." *Journal of State Taxation* 29, no. 6 (September–October 2011).

Couture, Toby, and Yves Gagnon. "An Analysis of Feed-in Tariff Remuneration Models: Implications for Renewable Energy Investment." *Energy Policy* 38 (2010): 955–65.

Del Río González, Pablo. "Ten Years of Renewable Electricity Policies in Spain: An Analysis of Successive Feed-in Tariff Reforms." *Energy Policy* 36, no. 8 (2008): 2917–29.

Dong, C. G. "Feed-in Tariff vs. Renewable Portfolio Standard: An Empirical Test of Their Relative Effectiveness in Promoting Wind Capacity Development." *Energy Policy* 42, 2012.

Environmental Law Institute. "Estimating U.S. Government Subsidies to Energy Sources: 2002–2008," September 2009. http://www.elistore.org/Data/products/d19 07.pdf.

Ernst & Young. "Renewable Energy Country Attractiveness Indices," May 2012. http://www.ey.com/Publication/vwLUAssets/Renewable energy country attractiveness indices - Issue 33/$FILE/EY RECAI issue 33.pdf.

Gage, Deborah. "The Venture Capital Secret: 3 Out of 4 Start-Ups Fail." *Wall Street Journal,* September 19, 2012. http://online.wsj.com/article/SB10000872396390443720204578004980476429190.html.

GAO. *See* U.S. Government Accountability Office.

Ghosh, Arunabha and Himani Gangania. "Governing Clean Energy Subsidies: What, Why, and How Legal?" International Centre for Trade and Sustainable Development, Global Platform on Climate Change, Trade and Sustainable Energy, August 2012. http://ictsd.org/downloads/2012/09/governing-clean-energy-subsidies-what-why-and-howlegal.pdf.

ICIS. "Cloud of Legal Action Overshadows Spain's Solar Electricity Subsidies." November 30, 2011. http://www.icis.com/heren/articles/2011/11/30/9512912/cloud+of+legal+actions+overshadows+s pains+solar+electricity.html.

Intergovernmental Panel on Climate Change (IPCC). "Climate Change 2007: Synthesis Report." 2007. http://www.ipcc.ch/pdf/assessment-report/ar4/syr/ar4 syr.pdf.

International Energy Agency (IEA). *Renewable Energy: Policy Considerations for Deploying Renewables*, November 2011. http://www.iea.org/publications/freepublications/publication/Renew Policies.pdf.

Jenkins, Jesse and Sara Mansur. "Bridging the Clean Energy Valleys of Death." Breakthrough Institute, November 2011. http://thebreakthrough.org/blog/Valleys of Death.pdf.

Jenner, Steffen. "Did Feed-in Tariffs Work? An Econometric Assessment," August 1, 2012. http://papers.ssrn.com/sol3/papers.cfm?abstract id=2121261.

Johnson, Laurie T., and Chris Hope. "The Social Cost of Carbon in U.S. Regulatory Impact Analysis: An Introduction and Critique." *Journal of Environmental Studies and Sciences* 2, no. 3 (September 2012): 205–21. http://link.springer.com/article/10.1007%2Fs13412-012-0087-7.

Kalkuhl, Matthias, Ottmar Edenhofer, and Kai Lessman. "Renewable Energy Subsidies: Second-best Policy or Fatal Aberration for Mitigation?" *Resource and Energy Economics* 35 (2013): 217–234. http://www.pik-potsdam.de/members/edenh/publications-1/Renewableenergysubsidies.pdf.

Kho, Jennifer. "What the Solyndra Bankruptcy Means for Clean Tech Investors." Forbes, September 23, 2011. http://www.forbes.com/sites/jenniferkho/2011/09/23/solyndra-bankruptcy-impactcleantech-investing/.

KPMG International. "Taxes and Incentives for Renewable Energy," June 2012. http://www.kpmg.com/Global/en/IssuesAndInsights/ArticlesPublications/Documents/taxesincentives-renewable-energy-2012.pdf.

Marques, António, José A. Fuinhas, and J. R. Pires Manso. "Motivations Driving Renewable Energy in European Countries: A Panel Data Approach." *Energy Policy* 38, no. 11 (2010): 6877–85.

Mendelsohn, Michael, and John Harper. "§1603 Treasury Grant Expiration: Industry Insight on Financing and Market Implications." National Renewable Energy Laboratory. Technical Report NREL/TP6A20-53720, June 2012. http://www.nrel.gov/docs/fy12osti/53720.pdf.

Morris, Adele, Pietro S. Nivola, and Charles L. Schultze. "Clean Energy: Revisiting the Challenges of Industrial Policy." Brookings Institute. Climate and Energy Economics Discussion Paper, June 4, 2012. http://www.sciencedirect.com/science/article/pii/S0140988312002009.

Ragwitz, Mario, Jenny Winkler, Corinna Klessmann, Malte Gephart, and Gustav Resch. "Recent Developments of Feed-in Systems in the EU: A Research Paper for the International Feed-in Cooperation." Commissioned by the Ministry for the Environment,

Nature Conservation and Nuclear Safety (Germany), January 2012. http://www.feed-in-cooperation.org/wDefault_7/download-files/research/101105_feed-in_evaluation_update-January-2012 draft final ISI.pdf.

Sherlock, Molly. "Impact of Tax Policies on the Commercial Application of Renewable Energy Technology." Testimony before the House Committee on Science, Space, and Technology, Subcommittee on Investigations and Oversight, and Subcommittee on Energy and Environment, April 19, 2012. http://science.house.gov/sites/republicans.science.house.gov/files/documents/hearings/HHRG112-SY21-WState-MSherlock-20120419.pdf.

SolarServer. "Spanish Government Halts PV, CSP Feed-in Tariffs," January 30, 2012. http://www.solarserver.com/solar-magazine/solar-news/current/2012/kw05/spanish-governmenthalts-pv-csp-feed-in-tariffs.html.

Sovacool, Benjamin K. "Valuing the Greenhouse Gas Emissions from Nuclear Power: A Critical Survey." *Energy Policy* 36, no. 8 (August 2008): 2940–53.

Stromsta, Karl-Erik. "German Politicians Stand Firm Behind 16% Solar FIT Decrease." Recharge News, April 23, 2010. http://www.rechargenews.com/solar/article1285424.ece.

Trabish, Herman. "SolarReserve's CEO Weighs in on CSP, Policy, Jobs, Election." Greentech Solar, August 27, 2012. http://www.greentechmedia.com/articles/read/what-do-solar-power-towers-sayabout-the-doe-loan-program.

University of Texas at Austin. Bureau of Economic Geology. Center for Energy Economics. *Harmonization of Renewable Energy Credit Markets Across the U.S.* Prepared for State Energy Conservation Office, Texas, November 2009. http://www.beg.utexas.edu/energyecon/transmission forum/CEE National RECs study.pdf.

U.S. Department of Energy (USDOE). DSIRE Database of State Incentives for Renewables and Efficiency. http://www.dsireusa.org/ (accessed various dates).

U.S. Government. Interagency Working Group on Social Cost of Carbon. *Technical Support Document: Technical Update of the Social Cost of Carbon for Regulatory Impact Analysis under Executive Order 12866*, May 2013. http://www.whitehouse.gov/sites/default/files/omb/inforeg/social cost of carbon for ria 2013 update.pdf.

U.S. Government Accountability Office (GAO). *DOE Loan Guarantees: Further Actions Are Needed to Improve Tracking and Review of Applications.* GAO 12-157, March 2012. http://www.gao.gov/assets/590/589210.pdf.

Voosen, Paul. "Spain's Solar Market Crash Offers a Cautionary Tale about Feed-in Tariffs." *New York Times*, August 18, 2009. http://www.nytimes.com/gwire/2009/08/18/18greenwire-spains-solarmarket-crash-offers-a-cautionary-88308.html?pagewanted=all.

Weitzman, Martin L. "Prices vs. Quantities." *Review of Economic Studies* 41, no. 4 (October 1974): 477– 49. http://qed.econ.queensu.ca/pub/faculty/garvie/eer/weitzman.pdf.

Widiantoro, Dimas Mukhlas, Daudi Pascal, Kiet Nguyen, and Muhammad Usman Sharief. "Porter's Diamond: National Competitiveness in Solar Energy Industry," November 6, 2011. http://ssrn.com/abstract=2078915.

Wiser, Ryan, Christopher Namovicz, Mark Gielecki, and Robert Smith. "Renewables Portfolio Standards: A Factual Introduction to Experience from the United States." Ernest Orlando Lawrence Berkeley National Laboratory, April 2007. http://www.escholarship.org/uc/item/9wr9062r.

APPENDIX A. REQUEST LETTER

EXECUTIVE OFFICE OF THE PRESIDENT
THE UNITED STATES TRADE REPRESENTATIVE
WASHINGTON, D.C. 20508

DOCKET NUMBER
2906
Office of the Secretary
Int'l Trade Commission

JUL 3 0 2012

RECEIVED
JUL 3 1 2012
OFFICE OF THE SECRETARY
U.S. INTL TRADE COMMISSION

The Honorable Irving A. Williamson
Chairman
U.S. International Trade Commission
500 E Street, S.W.
Washington, DC 20436

Dear Chairman Williamson,

I am writing to request that the U.S. International Trade Commission (Commission) conduct two investigations under section 332(g) of the Tariff Act of 1930 (19 U.S.C. 1332(g)) regarding trade and market trends in the environmental services and renewable energy services sectors.

Since the publication of the Commission's investigations on environmental and renewable energy services in 2004 and 2005, the U.S. and global markets for such services have undergone significant change. In recent years, overall demand in the environmental services market has continued to rise due to new regulations, population and industry growth, and aging infrastructure. However, factors such as new technologies, tightening government budgets, and growing interest in environmental sustainability have altered the means through which such services are supplied. In the renewable energy services sector, technological improvements and decreasing prices have led to rapid growth in demand, particularly in the industry's wind and solar power segments. At the same time, changes in government incentive programs have created uncertainty regarding the future of the renewable energy market.

To assist us in better understanding recent developments in the environmental services and renewable energy services sectors, I request that the Commission conduct two investigations and prepare reports, as described below. I understand that the Commission will shape its approach to these investigations by the extent to which it can develop appropriate analytical frameworks and collect the requisite data.

Investigation 1: Based on available information, I request that the Commission provide a first report on environmental and related services that, to the extent practicable:

- Estimates the size of the U.S. and global markets for certain environmental and related services—including water and wastewater services, solid and hazardous waste services, and remediation services—identifies top suppliers and key country markets for such services, investigates factors affecting supply and demand in these market segments, and highlights market developments that have occurred within the last five years;
- Estimates the value of trade and investment in the subject environmental services segments, identifies key export and import markets for such services, and discusses recent trends in environmental services trade and investment; and
- Identifies barriers to trade and investment in the subject environmental services segments, discusses recent efforts to liberalize trade and investment in environmental services, and investigates the potential impact of further liberalization in environmental services.

I request that this report be delivered eight months from the date of receipt of this letter.

Investigation 2: Based on available information, I request that the Commission provide a second report on renewable energy and related services that, to the extent practicable:

- Defines types of renewable energy and related services, identifies leading suppliers, and generally describes the relationship of renewable energy services to the development of renewable energy projects worldwide;
- Estimates the size of the U.S. and global markets for certain renewable energy services, identifies key export and import markets for such services, and describes factors affecting supply and demand;
- Examines U.S. and global renewable energy services trade during 2007-11, and highlights recent trends in investment in renewable energy projects and firms, including new business strategies or practices;
- Identifies barriers to U.S. trade and investment in renewable energy services, and examines recent efforts to liberalize trade in leading markets for such services; and
- Examines the role of clean energy incentive programs in encouraging investment in and creating markets for renewable energy goods and services.

The report should focus on services incidental to the development, generation, and distribution of renewable energy, with particular emphasis on wind energy (onshore and offshore) and solar energy, and other technologies that the Commission's research shows to be of significance. Such services include scientific and technical consulting, services incidental to energy distribution, professional services, construction and engineering services, management consulting and related services, and maintenance and repair of equipment, among others.

I request that the second report be delivered eleven months from the date of receipt of this letter.

As we intend to make the Commission's reports available to the public, these reports should not include confidential business or national security classified information.

I appreciate the Commission's continued assistance and cooperation on this matter.

Sincerely,

Ron Kirk

Ambassador Ron Kirk

EXECUTIVE OFFICE OF THE PRESIDENT
THE UNITED STATES TRADE REPRESENTATIVE
WASHINGTON, D.C. 20508

The Honorable Irving Williamson
Chairman
U.S. International Trade Commission
Washington, DC 20436

332-534

Dear Chairman Williamson:

In a letter dated July 30, 2012, former U.S. Trade Representative Ron Kirk requested, pursuant to section 332(g) of the Tariff Act of 1930, that the U.S. International Trade Commission (Commission) institute an investigation and provide a report on recent developments in the renewable energy services sector. Ambassador Kirk asked the Commission to provide its report to the Office of the U.S. Trade Representative (USTR) by eleven months after receipt of the request, or by June 28, 2013.

I am hereby amending the request, and now ask that the Commission provide its completed report on the renewable energy sector to USTR no later than August 30, 2013.

Sincerely,

Ambassador Demetrios Marantis
Acting United States Trade Representative

APPENDIX B. *FEDERAL REGISTER* NOTICE

Federal Register / Vol. 77, No. 170 / Friday, August 31, 2012 / Notices

INTERNATIONAL TRADE COMMISSION

[Investigation No. 332–534]

Renewable Energy and Related Services: Recent Developments

AGENCY: United States International Trade Commission.

ACTION: Institution of investigation and scheduling of public hearing.

SUMMARY: Following receipt of a request on July 30, 2012 from the U.S. Trade Representative (USTR) under section 332(g) of the Tariff Act of 1930 (19 U.S.C. 1332(g)), the U.S. International Trade Commission (Commission) instituted investigation No. 332–534, Renewable Energy and Related Services: Recent Developments.

DATES: November 15, 2012: Deadline for filing requests to appear at the public hearing. November 19, 2012: Deadline for filing pre-hearing briefs and statements. November 29, 2012: Public hearing. December 17, 2012: Deadline for filing post-hearing briefs and statements. March 1, 2013: Deadline for filing all other written submissions. June 28, 2013: Transmittal of Commission report to USTR.

ADDRESSES: All Commission offices, including the Commission's hearing rooms, are located in the United States International Trade Commission Building, 500 E Street SW., Washington, DC. All written submissions should be addressed to the Secretary, United States International Trade Commission, 500 E Street SW., Washington, DC 20436. The public record for this investigation may be viewed on the Commission's electronic docket (EDIS) at https://edis.usitc.gov/edis3-internal/ app.

FOR FURTHER INFORMATION CONTACT: Project Leader Lisa Alejandro (202–205– 3486 or Lisa.Alejandro@usitc.gov) or Deputy Project Leader Samantha Brady Pham (202–205–3459 or Samantha.Pham@usitc.gov) for information specific to this investigation. For information on the legal aspects of this investigation, contact William Gearhart of the Commission's Office of the General Counsel (202–205–3091 or william.gearhart@usitc.gov). The media should contact Margaret O'Laughlin, Office of External Relations (202–205– 1819 or margaret.olaughlin@usitc.gov). Hearing-impaired individuals may obtain information on this matter by contacting the Commission's TDD terminal at 202–205–1810. General information concerning the Commission may also be obtained by accessing its Internet server (http://www.usitc.gov). Persons with mobility impairments who will need special assistance in gaining access to the Commission should contact the Office of the Secretary at 202–205–2000. Background: In his letter the USTR requested that the Commission prepare two reports, one on environmental and related services, and a second on renewable energy and related services, and deliver the reports in 8 and 11 months, respectively, after receipt of the letter. This notice announces the institution of an investigation and schedule, including the date for a public hearing, relating to the preparation of the second report, on renewable energy and related services; the Commission published notice of the institution of the first investigation, No. 332–533, Environmental and

Related Services, in the Federal Register of August 21, 2012. As requested by the USTR, the Commission will provide a report on renewable energy and related services that, to the extent practicable:

- Defines types of renewable energy and related services, identifies leading suppliers, and generally describes the relationship of renewable energy services to the development of renewable energy projects worldwide;
- Estimates the size of the U.S. and global markets for certain renewable energy services, identifies key export and import markets for such services, and describes factors affecting supply and demand;
- Examines U.S. and global renewable energy services trade during 2007–11, and highlights recent trends in investment in renewable energy projects and firms, including new business strategies or practices;
- Identifies barriers to U.S. trade and investment in renewable energy services, and examines recent efforts to liberalize trade in leading markets for such services; and
- Examines the role of clean energy incentive programs in encouraging investment in and creating markets for renewable energy goods and services.

As requested by the USTR, the report will focus on services incidental to the development, generation, and distribution of renewable energy, with particular emphasis on wind energy onshore and offshore) and solar energy, and other technologies that the Commission's research shows to be of significance. The USTR defined such services to include scientific and technical consulting, services incidental to energy distribution, professional services, construction and engineering services, management consulting and related services, and maintenance and repair of equipment, among others.

As requested, the Commission expects to deliver this second report to the USTR no later than June 28, 2013.

Public Hearing: A public hearing in connection with this investigation will be held at the U.S. International Trade Commission Building, 500 E Street SW., Washington, DC, beginning at 9:30 a.m. on November 29, 2012. Requests to appear at the public hearing should be filed with the Secretary no later than 5:15 p.m., November 19, 2012. All pre-hearing briefs and statements should be filed no later than 5:15 p.m. November 6, 2012 and all post-hearing briefs and statements should be filed no later than 5:15 p.m., December 17, 2012. All pre- and post-hearing briefs and statements must be filed in accordance with the requirements in the ''Written Submissions'' section below. In the event that no witnesses are scheduled to appear at the hearing as of the close of business on November 15, 2012, the hearing will be canceled. Any person interested in attending the hearing as an observer or nonparticipant should contact the Office of the Secretary at 202–205–2000 after November 1, 2012, for information concerning whether the hearing will be held.

Written Submissions: In lieu of or in addition to participating in the hearing, interested parties are invited to file written submissions concerning this investigation. All written submissions other than those related to the hearing) should be addressed to the Secretary, and should be received no later than 5:15 p.m., March 1, 2013. All written submissions must conform with the provisions of section 201.8 of the Commission's Rules of Practice and Procedure (19 CFR 201.8). Section 201.8 and the Commission's Handbook on Filing Procedures require that interested parties file documents electronically on or before the filing deadline and submit eight (8) true paper copies by 12:00 p.m. eastern time on the next business day. In the event that confidential treatment of a document is requested, interested parties must file, at the same time as the eight paper copies, at least four (4) additional true paper copies in which the confidential information must be deleted (see the following paragraph for further information regarding confidential business information). Persons with questions regarding electronic filing should contact the Secretary (202–205–2000).

Any submissions that contain confidential business information (CBI) must also conform with the requirements in section 201.6 of the Commission's Rules of Practice and Procedure (19 CFR 201.6). Section 201.6 of the rules requires that the cover of the document and the individual pages be clearly marked as to whether they are the ''confidential'' or ''non-confidential'' version, and that the confidential business information be clearly identified by means of brackets. All written submissions, except for confidential business information, will be made available for inspection by interested parties.

In the request letter, the USTR stated that his office intends to make the Commission's report available to the public in its entirety, and asked that the Commission not include any confidential business information or national security classified information in the report. Any confidential business information received by the Commission in this investigation and used in preparing this report will not be published in a manner that would reveal the operations of the firm supplying the information.

By order of the Commission.
Issued: August 27, 2012.
Lisa R. Barton, Acting Secretary to the Commission.
[FR Doc. 2012–21492 Filed 8–30–12; 8:45 am] BILLING CODE 7020–02–P

Federal Register / Vol. 78, No. 87 / Monday, May 6, 2013 / Notices

INTERNATIONAL TRADE COMMISSION
[Investigation No. 332–534]
Renewable Energy and Related Services: Recent Developments
AGENCY: United States International Trade Commission.
ACTION: Extension of date for transmitting report.

SUMMARY: Following the receipt of a letter on April 15, 2013, from the United States Trade Representative (USTR), the Commission has extended to August 30, 2013, the date for transmitting its report to USTR in investigation No. 332–534, Renewable Energy and Related Services: Recent Developments.

DATES: April 15, 2013: Receipt of the letter from USTR. August 30, 2013: New date for transmitting the Commission's report to USTR.

Backround

The Commission published notice of institution of the investigation in the Federal Register on August 31, 2012 (77 FR 53233). In its original notice of investigation, the Commission indicated that it would transmit its report to USTR on June 28, 2013. The notice is also available on the Commission Web site at http://www.usitc.gov. All other information about the investigation, including a description of the subject matter to be addressed, contact information, and Commission addresses, remains the same as in the original notice. The public record for this investigation may be viewed on the Commission's electronic docket (EDIS) at http://www.usitc.gov/secretary/ edis.htm.

Issued: April 30, 2013.
By order of the Commission.
Lisa R. Barton, Acting Secretary to the Commission.
[FR Doc. 2013–10535 Filed 5–3–13; 8:45 am]
BILLING CODE 7020–02–P

APPENDIX C. POSITIONS OF INTERESTED PARTIES

Summary of Views of Interested Parties

The Commission invited interested parties to file written submissions for its investigation on U.S. and global markets for renewable energy services. This appendix summarizes the views expressed to the Commission via written submissions and reflects only the principal points made by each party. The views summarized are those of the submitting parties and not the Commission. In preparing this summary, the Commission did not confirm the accuracy of, or otherwise correct, the information summarized. For the full text of all written submissions, see entries associated with investigation no. 332-534 at the Commission's Electronic Docket Information System (https://edis.usitc.gov).

Semiconductor Equipment and Materials International (SEMI)[1]

In a written submission, the Semiconductor Equipment and Materials International (SEMI) stated that it is a global trade association representing the equipment makers and material suppliers in the microelectronics industries. SEMI said that it has 2,000 members globally, of which approximately 500 are involved in the production of solar photovoltaics (PV). SEMI stated that there are more than 5,000 companies in the U.S. solar value chain, with at least 39 active facilities manufacturing PV components (polysilicon, wafers, cells, modules, inverters) spread across 17 states. According to SEMI, while some solar manufacturing operations have closed due to obsolete equipment, uncompetitive technology, or other reasons, a strong supply chain does still exist in the United States. SEMI stated that U.S. companies are represented in every step of the value chain, including manufacturing equipment, materials, solar cells and modules, and balance of system components.

SEMI said that the U.S. PV industry accounted for over 119,000 jobs in 2012, with only 24 percent of these jobs in manufacturing, since a majority of PV cells and modules are made overseas. According to SEMI, GTM Research found that the U.S. solar supply chain had a $2 billion trade surplus in 2010, but that this changed to a $1.5 billion deficit in 2011 and that, as a result, some U.S. material manufacturing plants and Research and Development centers are likely to move overseas. SEMI indicated that after years of continuous growth, the solar PV industry now finds itself in a sustained period of rapidly falling prices for PV cells and modules, mostly due to overcapacity of production, even as global installation numbers continue to grow. The global industry is worth billions of dollars and employs hundreds of thousands of people, according to SEMI, but competition is fierce, and this has resulted in trade friction.

SEMI indicated that there are a number of trade complaints between the United States, China, the European Union, The Republic of Korea (Korea), India, and Canada, and that other trade and market barriers have arisen in areas related to investment, government procurement, local-content requirements, and conflicting standards and certification requirements. SEMI stated that there is a need to level the global playing field for PV products in order to support U.S. competitiveness, and that what is needed is a high-standard, comprehensive, and global accord to control for improprieties that have contributed to the current market surplus. It recommended the establishment of a global free trade agreement for renewable energy products, and indicated that the World Trade Organization's Information Technology Agreement (ITA) is a model for such an agreement.

Verdant Power (Verdant)[2]

In a written submission to the Commission, Mr. William H. Taylor, Co-founder and President of Verdant Power, said that marine and hydrokinetic (MHK) energy is derived from ocean tides, waves, and currents; temperature gradients in the ocean; and free-flowing rivers and streams. Mr. Taylor stated that an important MHK-related service involves resource assessment, wherein potential MHK energy resources and sites are analyzed to determine the technical and economic feasibility of deploying of MHK devices and long-term project development. Mr. Taylor noted that MHK resource assessment services are not fully developed globally and represent an opportunity for U.S. firms. According to Mr. Taylor, the global market for MHK resources is an estimated 400,000 GW, but there are few firms focused on providing such detailed resource assessment services. Mr. Taylor noted that most investment in the MHK sector has been related to MHK device developers, but in recent years investment has declined due to the Great Recession and cheap natural gas in North America. Mr. Taylor stated that MHK resource assessment services have been secondary to MHK device development and limited mostly to macro assessments of MHK resources.

According to Mr. Taylor, a primary barrier to the provision of site-specific resource assessment is the experience required to provide these services effectively. Mr. Taylor noted that the service provider must understand resource assessment/project feasibility from a scientific perspective and from the perspective of a device/project developer. Mr. Taylor stated that Verdant, based in New York City, is uniquely equipped with this combined expertise. He noted that the firm installed and operated the world's first grid-connected array of MHK tidal devices through its Roosevelt Island Tidal Energy Project located in the East River of New York. Mr. Taylor stated that this project has increased the firm's resource assessment techniques as well as its device/project developer capabilities. Mr. Taylor said that Verdant has used its resource assessment methodology to assess sites in Canada and the United States. According to Mr. Taylor, an additional barrier to site-specific resource assessments is a lack of public funding. Mr. Taylor noted that funding mechanisms should be developed for MHK resource assessments in order to increase domestic expertise in this area and to expand U.S. exports of these services.

End Notes for Executive Summary

[1] Excludes hydropower sources with generating capacities greater than 50 megawatts (MW). For more information on definitional issues related to renewable hydropower, see section 5 of this report.

[2] Feed-in tariffs are government incentives that offer a set price for renewable energy to investors (which may include producers of all sizes).

End Notes for Section 1

[1] As wind turbine and solar photovoltaic (PV) panel prices have fallen, manufacturers have elevated the importance of services in their business lines, since services may be more lucrative and, in the case of operations and maintenance, provide long-term revenue streams.

[2] The USTR made this request under section 332(g) of the Tariff Act of 1930. In its request letter, the USTR also asked that the Commission conduct an investigation on environmental services. That report, *Environmental and Related Services*, USITC Publication 4389, was delivered to the USTR on March 29, 2013.

[3] Coverage of services related to biomass energy and biofuels are not included in this report.

[4] This channel uses three of the four modes of supply defined in the General Agreement on Trade in Services (GATS): cross-border supply (mode 1), consumption abroad (mode 2), and movement of natural persons (mode 4).

[5] This channel uses the remaining GATS mode of supply—commercial presence (mode 3).

[6] The checklist approach makes it easier to schedule commitments without requiring significant changes to the W/120, helps WTO members develop a common agreement about the full range of applicable services, and offers a mechanism for assessing the value of countries' market access and national treatment offers.

[7] Details about countries' specific commitments on energy services included in the checklist appear in appendix C of USITC Publication 3805, *Renewable Energy Services: An Examination of U.S. and Foreign Markets* (2005), available at http://www.usitc.gov/publications/332/pub3805.pdf.

[8] WTO, General Agreement on Trade in Services.

End Notes for Section 2

[1] BNEF database (accessed March 11, 2013).

[2] EIA, "What Is Renewable Energy?" (accessed January 30, 2013).

[3] REN-21, *Renewables 2013,* 2013, 21.

[4] Ibid.

[5] EIA, *International Energy Outlook 2011* (interactive table viewer, accessed March 25, 2013); REN-21, *Renewables 2012,* 2012, 21.

[6] REN-21, *Renewables 2012,* 2012, 21.

[7] EIA, *International Energy Outlook 2011*, 2011, 1, 4.

[8] REN-21, *Renewables 2012,* 2012, 13.

[9] REN-21, *Renewables 2012,* 2012, 23. Includes hydropower sources with capacities of 50 megawatts (MW) or more.

[10] Global renewable energy capacity as reported by Bloomberg New Energy Finance (BNEF) excludes hydropower sources with generating capacities of more than 50 MW. For more information on definitional issues related to renewable hydropower, see section 5 of this report.

[11] BNEF database (accessed March 11, 2013).

[12] A BTU is a common measure of energy consumption that refers to the quantity of heat required to raise the temperature of one pound of liquid water by one degree Fahrenheit at the temperature at which water has its greatest density (approximately 39 degrees Fahrenheit).

[13] EIA, "What Is Renewable Energy?" (accessed January 30, 2013); EIA, *Monthly Energy Review*, March 2013, table 1.3. Total energy consumption refers to the consumption of primary energy sources (e.g., fossil fuels like crude petroleum or natural gas) before they are transformed into other forms of energy (i.e., electricity).

[14] Increased vehicle fuel economy standards coupled with the recent U.S. recession have helped to lower demand for transportation fuels like gasoline. Stricter environmental regulations for coal and record low prices for natural gas have helped to cut coal consumption, particularly in coal-fired power plants that can use natural gas as a substitute fuel. EIA, *Monthly Energy Review*, March 2013, table 1.3; Bloomberg, *Sustainable Energy in America 2013*, 2013, 7.

[15] Biomass, including wood and waste, is used to produce heat and steam in industrial applications and for residential heating. Biofuels, another form of biomass, are used as transportation fuels. EIA, "What is Renewable Energy?" (accessed January 30, 2013).

[16] EIA, *Energy Perspectives 2011*, September 2012, table 32. EIA, *Monthly Energy Review*, March 2013, table 7.2a. Includes hydropower sources with capacities of 50 MW or more.

[17] Ibid.

[18] Bloomberg, *Sustainable Energy in America 2012*, 2013, 9.

[19] EIA, *International Energy Outlook 2011*, September 2011, 4.

[20] REN-21, *Renewables 2013,* 2013, 69. For more information on the market effects of clean energy incentive programs, see section 6 of this report.

[21] REN-21, *Renewables 2013*, 2013, 80–82.

[22] SEIA, "Solar Investment Tax Credit (ITC)," n.d. (accessed February 5, 2013); North Carolina Solar Center, DSIRE database, "Business Energy Investment Tax Credit (ITC)," January 3, 2013.

[23] For more information on the wind sector, see section 4 of this report.

[24] Public Law 111-5.

[25] An electricity generator may have the option of purchasing a tradable renewable energy certificate (REC) or credit in order to comply with an RPS. An REC or similar credit represents a claim to have purchased electricity generated from an eligible renewable source. NREL, "Renewable Portfolio Standards" (accessed February 1, 2013); North Carolina Solar Center, DSIRE database, "Glossary" (accessed February 5, 2013).

[26] NREL, "Renewable Portfolio Standards" (accessed February 1, 2013).

[27] REN-21, *Renewables 2013,* 2013, 72.

[28] REN-21, *Renewables 2012,* 2012, 67, 119; North Carolina Solar Center, DSIRE database (accessed February 5, 2013). The following 29 states have RPS policies mandating the use of renewable energy: Arizona, California, Colorado, Connecticut, Delaware, Hawaii, Illinois, Iowa, Kansas, Maine, Maryland, Massachusetts, Michigan, Minnesota, Missouri, Montana, Nevada, New Hampshire, New Jersey, New Mexico, New York, North Carolina, Ohio, Oregon, Pennsylvania, Rhode Island, Texas, Washington, and Wisconsin. The following 8 states have voluntary goals to increase the use of renewable energy: Indiana, North Dakota, Oklahoma, South Dakota, Utah, Vermont, Virginia, and West Virginia.

[29] NREL, "Feed-In Tariffs" (accessed February 2, 2013).

[30] Payments can vary, and are affected by technology type and cost, resource availability, and installation size. IEA, *Deploying Renewables*, 2011, 79; NREL, "Feed-In Tariffs" (accessed February 2, 2013); REN-21, *Renewables 2012,* 2012, 74.

[31] REN-21, *Renewables 2013,* 2013, 72.

[32] REN-21, *Renewables 2012,* 2012, 66, 118.

[33] Clover, "Spain Suspends Feed-in-Tariffs, Receives EU Criticism," February 3, 2012.

[34] For more information on FITs, see section 6 of this report.

[35] An exception is large hydropower, which has historically been price-competitive with fossil fuels to generate electricity. However, in many instances, the development of large hydropower projects raises environmental and social concerns. For more information on recent developments in large hydropower, see section 5 of this report.

[36] REN-21, *Renewables 2012,* 2012, 65–67.

[37] USITC, *Renewable Energy Services*, 2005, 1-2.

[38] Ibid.

[39] Steenblick and Geloso Grosso, "Trade in Services Related to Climate Change," 2011, 10.

[40] BOS refers to all components of a renewable energy power station other than the electricity-generating apparatus, such as a wind turbine or solar panel. These can include supporting infrastructure, including land, as well as wiring, switches, support racks, inverters, and batteries.

[41] McBee, "Solar Panel Installation in the U.S.," 2012, 9; REN-21, *Renewables 2012,* 2012, 59.

[42] Deloitte, "European Wind Services Study," 2012, 4; Broad, "Post-Warranty Wind O&M Costs Proven to Increase by 25%," January 18, 2013.

[43] A gearbox increases the rotational speed of a shaft that connects the wind turbine's rotor and blades to a generator. A low-speed shaft feeds into the gearbox, and a high-speed shaft feeds from the gearbox into the generator. Gearboxes are a main cause of breakdowns and require regular repairs. Bartholl and Oleownik, "Wind Services: New Growth Opportunities," 2012, 8; David, "Wind Turbines," 2009, 2.

[44] Bartholl and Oleownik, "Wind Services: New Growth Opportunities," 2012, 8.

[45] Deloitte, "European Wind Services Study," 2012, 12.

[46] Ibid.

[47] A dearth of qualified crane technicians in Brazil reportedly hinders the ability of project developers to erect wind towers. Industry representative, interview with USITC staff, São Paulo, Brazil, May 8, 2013.

[48] Deloitte, "European Wind Services Study," 2012, 16.

[49] Ibid.

[50] Deloitte, "European Wind Services Study," 2012, 16; Bartholl and Oleownik, "Wind Services: New Growth Opportunities," 2012, 12.

[51] Estimates of the size of the global and U.S. market for renewable energy services presented in this section were developed by the Commission by obtaining industry estimates of the share of a project cost that each service constitutes and then applying the estimates to installed capacity for a given market.

[52] For more information on the global market for solar PV services, see section 3 of this report.

[53] Lucintel, "Growth Opportunities in Solar Photovoltaic," 2012.

[54] For more information on the global market for renewable energy services in the solar sector, see section 3 of this report.

[55] For more information on the global market for wind-energy services, see section 4 of this report.

[56] Deloitte, "European Wind Services Study," 2012, 4; USITC estimates. Deloitte estimated the size of the global market at €4.6 billion, which was converted to U.S. dollars by multiplying by an average exchange rate of 1.35. For more information on the global market for renewable energy services in the wind sector, see section 4 of this report.

[57] Deloitte, "European Wind Services Study," 2012, 5.

[58] For more information on the global market for hydropower and geothermal services, see section 5 of this report.

[59] For more information on the U.S. market for wind services, see section 4 of this report.

[60] For more information on the U.S. market for hydropower and geothermal services, see section 5 of this report.

[61] Refers to new investment in renewable energy in the following asset classes: asset finance (new development and installation of renewable energy generating projects); small distributed capacity (residential-scale projects of less than 1 MW); funds raised by firms both privately (i.e., as private equity or venture capital) or through capital markets; reinvested equity; and corporate and government research and development. Asset finance accounts for the majority of new investment in renewable energy (60 percent of the total in 2012), followed by small distributed capacity (32 percent). BNEF, *Market Sizing*, November 2012.

[62] FS and UNEP, *Global Trends in Renewable Energy Investment 2012*, 2012, 12.

[63] Ibid., 12, 23, 26, 33–34.

[64] Investment data cited by the REN-21 is from FS and UNEP and the BNEF database. The BNEF database is frequently updated and revised with new investment data, and as a result, more recent investment data collected by Commission staff in figure 2.7 may differ slightly from those shown in figure 2.8. In addition, Commission staff is unable to collect data on new investment for the same regions as those reported by REN-21 and FS and UNEP due to subscription limitations on BNEF's database.

[65] FS and UNEP, *Global Trends in Renewable Energy Investment 2012*, 2012, 12, 26, 23.

[66] Ibid., 24.

[67] Ibid., 13.

[68] See also Monkelbaan, "Sustainable Energy Services in a SETA," 2013, 29.

[69] Kim, "Facilitating Trade in Services," 2011, 19.

[70] Monkelbaan, "Sustainable Energy Services in a SETA," 2013, 12. Services affiliated with renewable energy technologies that are produced in one country and exported to another do not necessarily imply that those services are exported from that country to another.

[71] Pankratz, "Vestas' Colorado Factories to Build Components," April 9, 2013. Vestas is a wholly owned U.S. subsidiary of Denmark-based Vestas Wind Systems. Services supplied by its U.S. subsidiary to the project located in Alberta, Canada, are considered a services export from Denmark to Canada.
[72] Monkelbaan, "Sustainable Energy Services in a SETA," 2013, 22.
[73] Tulacz, *The Global Sourcebook*, 2012, 18.
[74] Most renewable energy technologies are classified in basket categories that contain other products within the Harmonized Commodity Description and Coding System (HS) at the 6-digit level.
[75] GTIS, Global Trade Atlas database for HS 8502.31 (accessed April 16, 2013). Wind-powered generating sets include the wind turbine, nacelle (which houses the components of the wind turbine, such as the controller, gearbox, and generator), and when imported with the nacelle, other components such as blades.
[76] GTIS, Global Trade Atlas database for HS 8502.31 (accessed April 16, 2013). Includes intra-EU trade.
[77] Steenblick and Geloso Grosso, "Trade in Services Related to Climate Change," 2011, 37; Monkelbaan, "Sustainable Energy Services in a SETA," 2013, 12.
[78] Steenblick and Geloso Grosso, "Trade in Services Related to Climate Change," 2011, 4.
[79] Ibid., 4.
[80] Ibid., 37.
[81] For a list of selected country-specific schedules of commitments on professional, technical, and business services and on construction services, see USITC, *Renewable Energy Services*, 2005.
[82] Major importers of architectural, engineering, and other technical services (a subset of other professional, technical, and business services) are the EU, India, Canada, Brazil, and Russia. Major importers of construction services are the EU, Japan, Russia, Kazakhstan, and China. Kim, "Facilitating Trade in Services," 2011, 15, 24.
[83] Kim, "Facilitating Trade in Services," 2011, 33; Steenblick and Geloso Grosso, "Trade in Services Related to Climate Change," 2011, 38.
[84] Kim, "Facilitating Trade in Services," 2011, 33–34; Steenblick and Geloso Grosso, "Trade in Services Related to Climate Change," 2011, 38.
[85] Industry representative, interview by USITC staff, São Paolo, Brazil, May 8, 2013.
[86] Fessenden, written testimony before the U.S. House Ways and Means Subcommittee on Trade, September 20, 2012, 4.
[87] Localization barriers to trade can broadly be defined as measures designed to protect, favor, or stimulate domestic industries, services providers, and/or intellectual property (IP) at the expense of goods, services, or IP from other countries. Marantis, Statement to the Senate Committee on Finance, March 19, 2013, 4.
[88] Peszko and Ketterer, "Local Content Requirements for Renewable Energy," November 23, 2012.
[89] Fessenden, written testimony before the U.S. House Ways and Means Subcommittee on Trade, September 20, 2012, 4–5.
[90] Ibid., 4–5.
[91] *Inside U.S. Trade*, "USTR Explores Smaller Environmental Goods Tariff," November 27, 2009; USITC, *Certain Environmental Goods*, 75 Fed. Reg. 28652 (May 21, 2010).
[92] WTO, "Energy Services: Overview of Classification Issues," March 5, 2012.
[93] Executive Office of the President, "The President's Climate Action Plan," June 2013, 19.
[94] Ibid., 19–20.
[95] APEC, "Annex C—APEC List of Environmental Goods," September 8–9, 2012.
[96] Marantis, Statement to the Senate Committee on Finance, March 19, 2013, 4.
[97] *Inside U.S. Trade*, "Danish Minister Says Time Is Ripe," May 3, 2012.
[98] Ibid.
[99] USTR, "Outlines of the Trans-Pacific Partnership Agreements," November 12, 2011.

End Notes for Section 3

[1] This section covers PV technologies, which convert sunlight directly to electricity and account for the largest share of the market among solar technologies. Concentrated solar power (CSP) technologies, which use sunlight to heat a fluid to generate electricity in large power plants, and other solar technologies, such as solar water heating, will not be covered.

[2] Crystalline silicon (c-Si) modules account for the majority of the global market and have been in production for the longest period of time. Thin-film technologies use a thin layer of a photosensitive material such as cadmium telluride (CdTe), amorphous silicon (a-Si), and copper indium gallium (di)selenide (CIGS). One of the largest U.S. services firms, First Solar, produces and installs CdTe modules exclusively.

[3] All references to system sizes in this section, whether in watts (W), kilowatts (kW), megawatts (MW), or gigawatts (GW), are in direct current (DC) unless specifically noted as an alternating current (AC) system. Equipment other than the PV modules—specifically, inverters and other equipment such as the racking and wiring—is referred to as the balance of system (BOS).

[4] For the purposes of this section, residential and nonresidential firms will be characterized as installers, while utility firms will be characterized as project developers and engineering, procurement, and construction (EPC) firms. In actuality, there is not a clear dividing line between industry sectors, and the installation of some large nonresidential projects will look more like the utility project development process than the residential installation process.

[5] Firms generate their sales leads through marketing, customer referrals, and other sales channels, such as stores and roofing contractors. Marketing and other services not tied to specific installations, such as the distribution of modules, inverters, and other equipment, are not discussed in this section. Real Goods Solar, "Form 10-K," March 15, 2012, 8; Sunvalley Solar, "Form 10-K," April 16, 2012, 8; SolarCity, "Form S-1," October 5, 2012, 27, 35, 96, 98, 103–104.

[6] A PPA is a long-term agreement to purchase electricity. In the wholesale market, this is generally between a utility and an independent power producer (IPP). There are also PPAs between PV firms and on-site users such as homeowners and businesses.

[7] Some forms of financing, such as loans, may be handled by the customer rather than the installer or third-party owner.

[8] EPRI, "Addressing Solar Photovoltaic," July 30, 2010.

[9] "Installations" are PV modules that are installed in a given year, measured in watts. They are also correlated with demand for services, as every module installed requires related installation services and, once installed, O&M services.

[10] GTM does not include concentrating photovoltaic (CPV) systems in its PV installation data; this segment accounted for only about 1 percent of installations in 2012. GTM Research and SEIA, *U.S. Solar Market Insight Report: 2012 Year-in-Review*, 2013, 5–6.

[11] GTM Research and SEIA, *U.S. Solar Market Insight Report: 2011 Year-in-Review*, 2012, 3; GTM Research and SEIA, *U.S. Solar Market Insight Report: 2010 Year-in-Review*, 2011, 2; GTM Research and SEIA, *U.S. Solar Market Insight Report: 2012 Year-in-Review*, 2013, 2.

[12] U.S. utility installations increased from 6 percent of the market in 2007 to 54 percent in 2012. GTM Research and SEIA, *U.S. Solar Market Insight Report: 2011 Year-in-Review*, 2012, 9, 11; GTM Research and SEIA, *U.S. Solar Market Insight Report: 2010 Year-in-Review*, 2011, 6, 10; Barbose, Darghouth, and Wiser, *Tracking the Sun V*, November 2012, 12.

[13] Data to calculate the value of the U.S. PV solar services market in 2012 are not available.

[14] Services market values are based on installations completed in that year and may differ from company revenues, since firms may recognize these revenues based on the percent of an installation that is completed. This estimate is based on the assumption that services contributed 47 percent of the value of this market, as reported by Ardani. There are varying estimates of the share of the market represented by services. Using the lowest estimate of this share would put the value of the services market at $0.5 billion; using the highest estimate would put it at about $1.1 billion in 2011. The Ardani estimate was used over data from the New York State Energy Research and Development Authority (NYSERDA) and *Photon* magazine because the latter may incorporate some installer markups on equipment costs, and therefore may overstate the share of the value of installations accounted for by equipment. The Ardani estimate was chosen over data from Woodlawn Associates because half of the data in the Woodlawn estimate is from 2012, when module prices were lower, thus likely reflecting a lower share for equipment costs than was the case in much of 2011. Finally, NYSERDA data only separate out modules and inverters, so the remaining cost may contain some balance of system components. Woodlawn Associates, "Solar Installation Effectiveness," September 10, 2012, 8; NYSERDA PowerClerk database (accessed February 1, 2013); GTM Research and SEIA, *U.S. Solar Market Insight Report: 2011 Year-in-Review*, 2012, 9–11; Ardani, "Benchmarking Soft Costs," May 17, 2012, 7; Bosworth and Hirsch, "One Size Fits All," April 2011, 74–77.

[15] The value of the nonresidential market, like that of the residential market, was estimated by using a share of installed system costs accounted for by services. However, estimates for nonresidential systems were

complicated by the widely varying system size and prices associated with those systems. In order to calculate a market share, systems were first stratified according to size and then an estimated value in each market segment was calculated using data from Lawrence Berkeley National Laboratory (LBNL). A share of these values was then calculated using estimated shares accounted for by system sizes in each range based on values from Ardani and *Photon* (excluding an estimated equipment markup). While the Ardani data were from 2010, these were the best data available to include as an input in this calculation. GTM Research and SEIA, *U.S. Solar Market Insight Report: 2011 Year-in-Review*, 2012, 9–11; Bosworth and Hirsch, "One Size Fits All," April 2011, 74–77; Ardani, "Soft Costs in the U.S. Solar Markets," November 8, 2012, 11; Barbose, Darghouth, and Wiser, *Tracking the Sun V*, November 2012, data file.

[16] The value of the services share of the utility market was estimated by calculating the goods and services value of installations using an average cost per MW and subtracting the estimated value of equipment associated with these installations. The value of the modules associated with these installations was calculated using Energy Information Administration (EIA) shipment and price data for the utility sector of the market, and the inverter and balance of system equipment costs were calculated using an average cost per watt. GTM Research and SEIA, *U.S. Solar Market Insight Report: 2011 Year-in-Review*, 2012, 9–11; EIA, *Solar Photovoltaic*, September 2012, 7, 16; GTM Research, "PV BOS Cost Analysis," July 25, 2011.

[17] O&M costs are difficult to estimate due to a lack of available data, variability in O&M costs and pricing depending on the company and system characteristics, changing costs as the system ages, and differences as to whether customers perform O&M in-house or contract it out. In addition, any estimate of the market value is more theoretical than actual, given that a large number of problems at a single project could significantly change the value of the market in a given year. This estimate is based on a compilation of the estimated O&M costs in each market segment for cumulative installed capacity at the end of 2011. Since some of these systems were installed during 2011 and not operational for the full year, the O&M value could be lower. For residential systems, O&M costs were estimated at 2.5 cents per installed watt based on NREL data reported by Cameron and Goodrich (though this excludes inverter repair, which could increase costs). In the residential market, there may essentially be no O&M costs for many years, as installed systems generally do not require O&M unless there is a problem. Some companies do sell additional O&M contracts, but the focus of this estimate is on actual expenditures and not the contracted value. For the nonresidential segment of the market, O&M costs were estimated at 2.1 cents per installed watt, based on a small sample of nonresidential systems for which O&M contracts were available. The prices used here for distributed installations are within the range of 0.6 to 2.7 cents per watt reported by the Electric Power Research Institute (EPRI). In the utility sector, EPRI reports O&M costs from 4.7 to 6 cents per watt per year, depending on the type of system. Actual project O&M costs were identified up to almost 11 cents per installed watt, though some newly installed systems were also significantly lower than the range estimated by EPRI. This paper uses 5 cents per installed watt. Real Goods Solar, "Form 10-K," March 15, 2012, 16; SolarCity, "Form S-1," October 5, 2012, 105; EPRI, "Addressing Solar Photovoltaic," July 2012, 8; Cameron and Goodrich, "The Levelized Cost of Energy," n.d. (accessed March 5, 2013), 2; industry representative, interview by USITC staff, January 14, 2013; individual project level cost data.

[18] NPD Solarbuzz, "PV Installed Base," September 11, 2012.

[19] Woody, "Solar Industry Anxious over Defective Panels," May 28, 2013.

[20] Greentech Media, "US Residential Solar Financing," February 11, 2013; Kann, "The U.S. PV Market," October 18, 2011.

[21] This section will include national data on residential and nonresidential installations where possible. In some cases comprehensive national data were not available, so individual state datasets were used. In particular, the CSI dataset is used extensively, since California has large residential and nonresidential markets; moreover, this dataset covers most of the period of this report and includes a significant amount of data about each installation. CSI working dataset, January 2, 2013 (accessed January 9, 2013).

[22] Greentech Media projects that this market will grow to $5.7 billion in 2016. The value of this market is not additive to the value of other service sector markets above. Greentech Media, "US Residential Solar Financing," February 11, 2013.

[23] CSI working dataset, January 2, 2013 (accessed January 9, 2013).

[24] Distributed installations are residential and nonresidential installations, as opposed to utility installations. Industry representatives, interviews by USITC staff, October 22, 2012, October 23, 2012, and February 1, 2013; Kann, "The U.S. PV Market," October 18, 2011; Trabish, "Emerging Solar Strategies," November 23, 2012; SolarCity, "Form S-1," October 5, 2012, 95; Barbose, Darghouth, and Wiser, *Tracking the Sun V*, November 2012, data file.

[25] Distributed energy sources are typically systems installed close to energy demand, such as those installed at homes and businesses; as noted earlier, they generally include residential and nonresidential PV systems. North Carolina Solar Center, DSIRE, "Renewable Portfolio Standard Policies," February 2013; North Carolina Solar Center, DSIRE, "Renewable Portfolio Standard Policies with Solar/Distributed Generation Provisions," February 2013; Real Goods Solar, "Form 10-K," March 15, 2012, 5–6; Sherwood, "Market Prospects for Solar in North America," October 18, 2011; Kann, "The U.S. PV Market," October 18, 2011; Gibson, "Electric Utilities and Solar: Threat or Opportunity?" April 26, 2011; SolarCity, "Form S-1," October 5, 2012, 12; First Solar, "Form 10-K," February 27, 2013, 9.

[26] Sherwood, "Market Prospects for Solar in North America," October 18, 2011; Sherwood, "Big Time for Solar," July/August 2013, 40-41; SolarCity, "Form S-1," October 5, 2012, 14, 108.

[27] Go Solar California website, http://www.gosolarcalifornia.ca.gov/about/csi.php (accessed March 4, 2013); Yonan, "State Unveils New Rules," November 9, 2012.

[28] SolarCity, "Form S-1," October 5, 2012, 13–14; First Solar, "Form 10-K," February 27, 2013, 9.

[29] If a system (1) was placed in service during 2009–11, or (2) the developer started construction during 2009–11, submitted the application by October 1, 2012, and will complete the system before 2017, the developer could elect to receive a payment equal to the amount of the tax credit, rather than taking the ITC. This payment was commonly known as the ITC grant or grant in lieu of the ITC. Sherwood, "Utility-Scale Installations Lead Growth," July/August 2011, 31; industry representative, interview by USITC staff, October 23, 2012; Sherwood, "Market Prospects for Solar in North America," October 18, 2011; Sherwood, "Big Time for Solar," July/August 2012, 40–41; SEIA website, http://www.seia.org/policy/finance-tax/solarinvestment-tax-credit (accessed March 4, 2013); U.S. Treasury, "Payments for Specified Energy Property," April 2011; SolarCity, "Form S-1," October 5, 2012, 14; First Solar, "Form 10-K," February 27, 2013, 9.

[30] SolarCity, "Form S-1," October 5, 2012, 95, 108; Ardani and Margolis, *2010 Solar Technologies Market Report*, November 2011, 83–84; REC website, http://www.recsolar.com/bonus-depreciation (accessed March 5, 2013); IRS, *How to Depreciate Property*, February 15, 2013.

[31] Industry representatives, interviews by USITC staff, October 22, 2012, October 23, 2012, December 19, 2012, and February 6, 2013; Bosworth, "Split Ends," April 2012, 57.

[32] SEPA, *Photovoltaic Incentive Programs Survey*, November 2009, 9; Wesoff, "What Really Motivates Consumers to Install Residential Solar?" March 23, 2011; Itron and Kema, *CPUC California Solar Initiative*, June 2010, 11-3, 11-8, 11-21 to 11-22; Shelton Group survey cited in Crume, Crume, and Koshmrl, "Selling Solar," April 2011, 32.

[33] Sills and Martin, "U.S. Utilities," November 30, 2010; Hering, "A New Day," 16, 18; Hamm, "Why Are Utilities?" September 2, 2008; EER, "New Study," December 3, 2009; Murphy, "Private Sector Taking Up," April 26, 2011; EER, "US Utility Solar PV Markets," November 2009; *PV News*, "Strategies for the U.S. Utility-Scale PV Market," October 2009, 1; Sherwood, *U.S. Solar Market Trends 2010,* June 2011, 7; industry representative, interview by USITC staff, February 6, 2013.

[34] Real Goods Solar, "Form 10-K," March 15, 2012, 16; Bosworth, "Split Ends," April 2012, 55.

[35] Installers, as discussed here, are the firms that are responsible for the PV system installation, though they may subcontract some parts of the installation to other firms such as electrical contractors. They may sell the systems themselves or be contracted by other system sellers—such as third-party owners—to install the system. The discussion of residential installers in this study focuses on national trends, but there may be substantial regional differences. Market shares discussed in this section are based on publicly available data compiled by USITC from various sources, including state/utility datasets from Arizona, California, Connecticut, Massachusetts, Maryland, and New Jersey; data from the Energy Information Administration (EIA), the Open PV Project, SEIA, and Bloomberg New Energy Finance database; and media articles, company websites, news releases, and other sources. USITC collected data on 2.7 GW of PV installations in the residential, nonresidential, and utility market segments combined, representing about 82 percent of all installations in 2012, including 83 percent of residential installations, 71 percent of nonresidential installations, and 88 percent of utility installations. These data, therefore, likely provide a fairly representative snapshot of the solar industry, though percentages may vary slightly if an installer has a large market share in a state for which only limited data are available. Data on the EPC contractor and O&M provider are not available for all utility projects. Company market shares exclude self-installed systems. Subsequently this data will be referenced as "data compiled by USITC."

[36] Real Goods Solar, "Form 10-K," March 15, 2012, 16; data compiled by USITC.

[37] Data compiled by USITC.

[38] Ibid.

[39] Ibid.

[40] Based on data compiled by USITC.

[41] CSI working dataset, January 2, 2013 (accessed January 9, 2013).

[42] Based on data compiled by USITC.

[43] Information on firms' approaches to expanding their presence in markets is based on industry representative, interview by USITC staff, October 23, 2012; Real Goods Solar, "Form 10-K," March 15, 2012, 3, 25–26; Shimogawa, "Haleakala, Solar Wave Set to Merge," February 3, 2012; *Maui News*, "Hawaii Solar Companies," February 4, 2012; Matz, "Going National," April 2011, 42–43; Cocke, "RevoluSun Plans Mainland Expansion," February 11, 2011; B-GC, "Interview with Jared Haines," October 12, 2009; Sunvalley Solar, "Form 10-K," April 16, 2012, 13; Lighthouse Solar website, http://www.lighthousesolarfranchising.com /franchise-information/ (accessed January 23, 2012); SolarCity, "Form S-1," October 5, 2012; Real Goods Solar website, http://realgoodssolar.com/solar-near-you/ (accessed January 23, 2013); Lighthouse Solar, "The Lighthouse Solar Franchising Company," February 17, 2010.

[44] Industry representative, telephone interview by USITC staff, February 6, 2013; Trabish, "Emerging Solar Strategies," November 23, 2012.

[45] Industry representatives, interviews by USITC staff, October 23, 2012 and December 12, 2012; Krulewitz, "Who Reigns Supreme," February 2012, 11; Shimogawa, "Haleakala, Solar Wave Set to Merge," February 3, 2012; Matz, "Going National," April 2011, 43–44; Hardcastle, "Power Player," August 31, 2012; Sunvalley Solar, "Form 10-K," April 16, 2012, 8.

[46] Commonwealth of Massachusetts, Executive Office of Energy and Environmental Affairs, RPS Solar Carve-Out data (accessed January 16, 2013).

[47] In this paragraph, the analysis of California data only includes installers with more than one installation during 2008–12. CSI working dataset, January 2, 2013 (accessed January 9, 2013).

[48] Industry representative, telephone interview by USITC staff, January 14, 2013; Matz, "Going National," April 2011, 44.

[49] Real Goods Solar, "Form 10-K," March 15, 2012, 16.

[50] CSI working dataset, January 2, 2013 (accessed January 9, 2013).

[51] This discussion will focus on the residential sector, for which data for multiple states are available. Wesoff, "Borrego Joins," July 31, 2012; Greentech Media, "US Residential Solar Financing," February 11, 2013; industry representative, telephone interview by USITC staff, February 25, 2013; Bosworth, "Split Ends," April 2012, 57–58; Ross, "Expanding Residential Solar Markets," October 20, 2011.

[52] Wesoff, "Borrego Joins," July 31, 2012.

[53] Maryland Residential Clean Energy Grant Program, Historic Award Data Report Excel worksheet (accessed January 17, 2013).

[54] Greentech Media, "US Residential Solar Financing," February 11, 2013.

[55] These firms may also develop projects for other energy sectors and/or install residential and commercial PV systems.

[56] Both crystalline silicon and thin-film manufacturers have vertically integrated into project development, primarily by acquiring project development companies. Through these acquisitions, solar manufacturers gained access to preexisting project pipelines and sought to improve their access to large-scale projects over the long term. In addition, entry into project development gave manufacturers a more reliable market in the event of a downturn in demand and let them minimize costs and optimize systems using the company's modules. As discussed in this paragraph, their perception of the relationship between project development and production has changed over time. Industry representative, telephone interview by USITC staff, December 11, 2009; industry representatives, interviews by USITC staff, October 15, 2009, and May 30, 2013; First Solar, "Form 10-K," February 28, 2011, 1–2; Mehta and Krulewitz, "The Direction of the Vertically Integrated," March 2011, 10–11; Lacey, "SunPower to Acquire PowerLight," November 15, 2006.

[57] Firms that have operations that are regulated as utilities often also have businesses that develop projects and/or act as independent power producers and are not regulated as utilities.

[58] An IPP is an entity that primarily produces electricity for sale on the wholesale market. It is not a utility, does not own electricity transmission, and does not have a designated service area. EIA, "Electric Industry Overview 2007" (accessed July 1, 2013).

[59] Based on a review of firms completing projects in 2012.

[60] SunEdison is included in this calculation for the entire time period, though it was not acquired by equipment manufacturer MEMC until 2009. MEMC, "MEMC Completes Acquisition of SunEdison," November 23, 2009.

[61] Bloomberg New Energy Finance (BNEF) data were used for the calculations of installations from 2007 to 2011. This database does not separate information on large installations by sector, so these data may also include some large commercial projects. Data for 2012 were compiled by USITC and only include utility projects. BNEF database, http://www.bnef.com (accessed January 22, 2013); data compiled by USITC.

[62] First Solar, "Form 10-K," February 29, 2012, 3, 124; First Solar, "Form 10-Q," November 2, 2012, 42–43; MEMC, "Form 10-K," February 29, 2012.

[63] Based on data compiled by USITC staff.

[64] Market share of these firms includes those where more than one developer was listed. BNEF database (accessed January 22, 2013).

[65] Data compiled by USITC; BNEF database, http://www.bnef.com (accessed January 22, 2013).

[66] Industry representatives, interviews by USITC staff, October 23, 2012, February 1, 2013, February 25, 2013, and February 26, 2013; data compiled by USITC; First Solar, "Form 10-K," February 27, 2013, 3.

[67] Data compiled by USITC.

[68] Industry representatives, interviews by USITC staff, October 23, 2012, and February 1, 2013.

[69] Mortenson, "Solar Energy Industry," January 2012, 8; industry representatives, interviews by USITC staff, February 1, 2013, and February 26, 2013; *Climate Change Business Journal*, "Bechtel Applies," Fall 2012, 1–2.

[70] Based on data compiled by USITC.

[71] It is not clear whether all of these distributed facilities are PV. True South Renewables website, http://truesouthrenewables.com/press/?p=46 (accessed March 6, 2013).

[72] Other services include "research and development, finance and accounting, legal work, or other ancillary services that support the solar industry." Project development was added as a category in 2012, so it is not possible to determine the exact employment increase during 2011–12. The Solar Foundation, *National Solar Jobs Census 2012*, November 2012, 17, 39, 42.

[73] Estimates of the size of the global PV market in 2012 range from 29 GW to 32 GW. The data included here are based primarily on data from European Photovoltaic Industry Association (EPIA). For 2007–11, data for Europe are estimated PV installations. For 2012, data are grid-connected capacity. In addition, for 2012 USITC updated data for several non-European countries, including Australia, China, Japan, and the United States. EPIA, *Global Outlook for Photovoltaics until 2016*, May 2012, 19, 66–67; EPIA, *Global Market Outlook for Photovoltaics 2013–2017*, 2013, 16, 18, 31; GTM Research and SEIA, *U.S. Solar Market Insight Report: 2012 Year-in-Review*, 2013, 5; JPEA, *Statistics on Shipments*, February 26, 2013, 5; Government of Japan, METI, "Announcement Regarding the Present Status," March 13, 2012; IHS, "Photovoltaic Industry to Enjoy," January 25, 2013; NPD Solarbuzz, "Solar Photovoltaic Demand," February 21, 2013; Energy Foundation and CREIA, *China Solar*, April 2013, 5–6; Watt and Passey, "PV in Australia 2012," May 2013, viii.

[74] The different ranking between Germany and Italy in 2012 installations and the 2011 services market reflects a significant decline in Italian installations in 2012 and lower installed costs in Germany in 2011.

[75] These values are based on estimates by USITC using publicly available data, but the actual market value could be higher or lower. The calculation of the market value by USITC is based on the value of the U.S. services market discussed above, and a separate calculation of the value of the rest of the global goods and services market and the share of the market accounted for by services. This involved calculating the value of installed systems in each market, based on multiplying the price per watt in each market by the number of MW installed. Data on the price per watt of systems installed in each country generally came from IEA's Photovoltaic Power Systems Programme and BNEF database. The countries for which specific price per watt data were available (including the United States) accounted for about 91 percent of 2011 installations. For those countries for which system prices were not available, the average for all other countries was used. For each market, the price per watt for modules, inverters, and balance of system components was multiplied by the MW installed to derive an equipment cost. Where available, costs specific to the individual market were used. Where those were not available, global average prices were used. This was then subtracted from the total value of the installations to derive the share of the market accounted for by services.

The total value of new installations of $91.9 billion is within the range of estimates published by consulting firms such as Clean Edge, which estimated global PV revenues in 2011 at $91.6 billion, and Solarbuzz, which estimated global revenues at $93 billion. Pernick, Wilder, and Winnie, *Clean Energy Trends 2012*, March 2012, 4; Solarbuzz, "World Photovoltaic Market Grew," March 19, 2012.

[76] Pernick, Wilder, and Winnie, *Clean Energy Trends 2012*, March 2012, 3–4; Mints, "U.S. Supply/Demand," December 12, 2012, 16; SMA website, http://www.sma.de/en/investor-relations.html (accessed February–June 2013).
[77] BSW-Solar, "Statistic Data," February 2013; industry representatives, interviews by USITC staff, Santiago, Chile, May 13, 2013 and May 15, 2013.
[78] REN21, *Renewables 2012*, 2012, 66, 70–72; First Solar, "Form 10-K," February 27, 2013, 8–10; industry representative, interview by USITC staff, Tokyo, May 10, 2013; MEMC, "Form 10-K," March 1, 2013, 14; SunPower, "Form 10-K," February 25, 2013, 14.
[79] See box 3.1 for a discussion of net metering. REN21, *Renewables 2012*, 2012, 66, 70–72; First Solar, "Form 10-K," February 27, 2013, 9; MEMC, "Form 10-K," March 1, 2013, 14; SunPower, "Form 10-K," February 25, 2013, 14.
[80] Industry representative, interview by USITC staff, Santiago, Chile, May 13, 2013.
[81] Data sources for the leading firms do not separate large nonresidential projects from utility projects; therefore, this section groups these sectors together as large projects.
[82] BNEF database (accessed July 17, 2013); data compiled by USITC.
[83] IMS Research, "First Solar Ranked," March 27, 2013.
[84] BNEF database (accessed February 28, 2012); data compiled by USITC.
[85] Data compiled by USITC.
[86] Official data on PV services trade are not available; therefore, this section will use installation data to measure U.S. services trade.
[87] Based on data compiled by USITC.
[88] SunPower is publicly traded on NASDAQ and is based in the United States, but since France-based Total acquired 66 percent of its shares in 2011 and 2012, it is considered a non-U.S. firm for the purposes of this report, and the services that it provided in the United States are classified as imports. Data compiled by USITC; SunPower Corp., "Form 10-K," February 25, 2013, 4.
[89] This is based on the company that provided the financing. It does not look at the companies investing in the funding sources for these companies.
[90] Wesoff, "Borrego Joins," July 31, 2012.
[91] Maryland Residential Clean Energy Grant Program, Historic Award Data Report Excel worksheet (accessed January 17, 2013); industry representative, telephone interview by USITC staff, February 25, 2013.
[92] Greentech Media, "US Residential Solar Financing," February 11, 2013; industry representative, interview by USITC staff, February 25, 2012; Conergy website, http://www.conergy.us/Portaldata/1/Resources/usa/pdf-downloads/Solar_Energy_Commercial_Project_Services_Brochure.pdf (accessed March 8, 2013).
[93] CSI working dataset, January 2, 2013 (accessed January 9, 2013); data compiled by USITC.
[94] Based on data compiled by USITC.
[95] Kennedy, "The Changing Marketplace ," September 13, 2012.
[96] SolarCity, "Form S-1," October 5, 2012, 31.
[97] For the purposes of determining exports, project development data will also include large nonresidential projects, since data sources do not necessarily separate out these projects.
[98] MEMC, "Fourth Quarter 2012," 11–12; MEMC, "SunEdison and CAP," January 31, 2013; MEMC, "SunEdison Closes," November 28, 2012; First Solar, "First Solar Acquires," January 9, 2013; First Solar, "Form 10-K," February 27, 2013, 3; industry representatives, interviews by USITC staff, Santiago, Chile, May 13, 2013, and May 15, 2013.
[99] MEMC, "Fourth Quarter 2012," February 13, 2013, 12.
[100] First Solar, "Form 10-K," February 27, 2013, 3.
[101] Penn Energy Renewables website, http://www.pennenergyrenewables.com/ (accessed May 31, 2013).
[102] This is based on projects tracked by BNEF. It is not known whether there were other projects completed that are not included in their database. BNEF database (accessed March 6, 2012).
[103] This paragraph focuses on these firms and does not include project developers that also provide EPC services.
[104] White Construction website, http://whiteconstruction.com (accessed March 6, 2013).
[105] Signal Energy website, http://www.signalenergy.com/experience/projects (accessed March 7, 2013); Swinerton website, http://www.swinerton.com/web/do/content?oid=renewable-energy (accessed May 29, 2013).
[106] Industry representatives, interviews by USITC staff, February 1, 2013, and February 26, 2013.
[107] IMS Research, "BELECTRIC Tops," May 10, 2012.
[108] Stephens, "Solar Plant's Production Costs Are Now Competitive," 2011.

[109] Black & Veatch website, http://bv.com/home/projects/project?pid=2e9de66c-5fb6-4381-8b1c40be48748355 (accessed March 6, 2013).

[110] Industry representative, interview by USITC staff, March 4, 2013; CH2M Hill website, http://www.ch2m.com/corporate/markets/energy/solar.asp (accessed March 6, 2013).

[111] 3TIER website, http://www.3tier.com/en/about/ (accessed March 6, 2013); industry representative, interview by USITC staff, March 4, 2013.

[112] True South Renewables website, http://www.truesouthrenewables.com/3rd_party_commissioning.html (accessed March 6, 2013); True South Renewables, "True South Renewables to Commission," December 13, 2011.

[113] Stuart, "SunPower Completes," February 1, 2012; Herndon, Martin, and Goossens, "Total to Buy," April 29, 2011.

[114] Recurrent Energy, "Sharp Corporation," September 21, 2010.

[115] Herndon, "OCI of Korea," January 18, 2011; LDK Solar, "LDK Solar Agrees to Acquire," January 6, 2011.

[116] First Solar, "First Solar Acquires," January 9, 2013; Woody, "California's Sungevity," November 17, 2011.

[117] Berrill, "SunEdison Bumps up European Expansion," January 22, 2009.

[118] Ahearne, "South African PV," June 1, 2012; Ontario Power Authority, "Feed-in Tariff Contract (FIT Contract)," December 14, 2012, exhibit C, 7–8; Ontario Power Authority, "Feed-in Tariff Program: FIT Rules," December 14, 2012, 30; industry representative, telephone interview by USITC staff, May 30, 2013.

[119] USTR, *2013 National Trade Estimate Report on Foreign Trade Barriers*, March 2013, 187; industry representative, telephone interview by USITC staff, March 4, 2013; NRDC and CEEW, *Laying the Foundation*, April 2012, 20.

[120] Greece's law specified that the local-content requirement would apply after the issuance of a ministerial decision with additional details on implementation. It is not clear if such a declaration has been issued. Eleuteri, "The Wait Is Over," June 2011, 69; USTR, *2013 National Trade Estimate Report on Foreign Trade Barriers*, March 2013, 153; RES Legal website, updated March 4, 2013, http://www.reslegal.eu/search-by-country/turkey/single/s/res-e/t/promotion/aid/feed-in-tariff-7/lastp/207/ (accessed June 28, 2013); Government of France, Ministry of Ecology, Sustainable Development and Energy, "Mesures d'urgence pour la relance de la filière" [Emergency measures for relaunching], January 7, 2013; Mourgelas and Associates, "Mourgelas Greek Law Update," April 2012, 3; Watson, Farley, and Williams, "Greece: Energy Briefing," September 2012, 10–11.

[121] A feed-in tariff (FIT) guarantees the purchase of renewable energy at a set price for a period of time.

[122] RES Legal website, updated March 4, 2013, http://www.res-legal.eu/search-bycountry/turkey/single/s/res-e/t/promotion/aid/feed-in-tariff-7/lastp/207/.

[123] Nielsen, "Local Content Rule," August 9, 2012

[124] The complaint in this matter was filed by Japan in September 2010. WTO website, Dispute DS412: Canada—Certain Measures Affecting the Renewable Energy Generation Sector, http://www.wto.org/english/tratop_e/dispu_e/cases_e/ds412_e.htm (accessed July 2, 2013).

[125] There are separate FIT programs for large and small projects. Ontario Ministry of Energy, FIT Program Directive, June 12, 2013, 3

[126] WTO website, Dispute DS456: India—Certain Measures Relating to Solar Cells and Solar Modules, http://www.wto.org/english/tratop_e/dispu_e/cases_e/ds456_e.htm (accessed July 2, 2013).

[127] WTO website, Dispute DS452: European Union and certain Member States—Certain Measures Affecting the Renewable Energy Generation Sector, http://www.wto.org/english/tratop_e/dispu_e/cases_e/ds452_e.htm (accessed July 2, 2013).

[128] Ontario Power Authority, "Feed-in Tariff Contract (FIT Contract)," Version 2.1, December 14, 2012, exhibit C, 7–8; Ontario Power Authority, "Feed-in Tariff Program: FIT Rules," Version 2.1, December 14, 2012, 30.

[129] First Solar, "Form 10-K," February 29, 2012, 6, 9.

[130] MEMC, "MEMC and its SunEdison Subsidiary," July 18, 2011; MEMC, "SunEdison First to Announce," April 15, 2011.

[131] The information in this paragraph is based on USTR, *2013 Report on Technical Barriers to Trade*, April 2013, 46–47; industry representative, telephone interview by USITC staff, May 30, 2013; WTO, Committee on Technical Barriers to Trade, "Minutes of the Meeting Held on March 24–25, 2011," May 26, 2011, 48–49; WTO, Committee on Technical Barriers to Trade, "Minutes of the Meeting Held on June 13–15, 2012," September 18, 2012, 26–27.

[132] USTR, *2013 Report on Technical Barriers to Trade*, April 2013, 46–47; industry representative, telephone interview by USITC staff, May 30, 2013; WTO, Committee on Technical Barriers to Trade, "Minutes of the

Meeting Held on March 24–25, 2011," May 26, 2011, 48–49; WTO, Committee on Technical Barriers to Trade, "Minutes of the Meeting Held on June 13–15, 2012," September 18, 2012, 26– 27.

133 NREL, "Technical Study," July 2012; Matz, "Ten Things Consumers," February 2012, 70–71; Steenblik, Matsuoka, and Hight, "Facilitating Trade," 2009, 21–22.

134 Steenblik, Matsuoka, and Hight, "Facilitating Trade," 2009, 21–22; industry representative, interview by USITC staff, Tokyo, May 14, 2013.

135 Poissant et al., "National Survey Report of PV Power Applications in Canada 2011," June 2012, 7–8, 15–16; IEA PVPS, *PVPS Report: A Snapshot of Global PV 1992–2012,* 2013, 11.

136 Ayoub, Martel, and Dignard-Bailey, "National Survey Report of PV Power Applications," May 2008, 7.

137 First Solar, "Form 10-K," February 27, 2013, 8.

138 Poissant et al., "National Survey Report of PV Power Applications in Canada 2011," June 2012, 8.

139 Industry representative, telephone interview by USITC staff, May 30, 2013.

140 Solarbuzz, "China's Feed-in Tariff," October 10, 2011; Energy Foundation and CREIA, *China Solar*, April 2013, 5–6; EPIA, *Global Outlook for Photovoltaics until 2016*, May 2012, 67.

141 USITC calculation based on data from Energy Foundation and CREIA, *China Solar*, April 2013, 5– 6, 66–68, and BNEF database (accessed May 28, 2013).

142 Industry representative, interview by USITC staff, Beijing, May 17, 2013; Bloomberg New Energy Finance database (accessed May 28, 2013); *Solar Industry*, "Jinko Solar Planning 200 MW PV Project," December 20, 2012.

143 Ali-Oettinger, "China: Wirsol to Develop Projects with Suntech," March 29, 2012.

144 SunPower Corp., "SunPower Signs," December 3, 2012.

145 Wesoff, "First Solar's Q1," May 6, 2013.

146 Alexopoulou, "Solaria Corporation to Construct," May 13, 2013.

147 EPIA, *Global Outlook for Photovoltaics until 2016*, May 2012, 68–69.

148 EPIA, *Global Outlook for Photovoltaics until 2016*, May 2012, 66–67; BSW-Solar, "Statistic Data," February 2013, 1.

149 Based on calculation by USITC. See the earlier discussion of the global services market value for the methodology.

150 Wissing, "National Survey Report of PV Power Applications in Germany," July 2012, 7–8; JägerWaldau, *PV Status Report 2012*, 2012, 19–20; Podewils, "The Air is Getting Thinner," November 2012, 54– 57.

151 IMS Research, "First Solar Ranked," March 27, 2013.

152 Industry representative, interview by USITC staff, Hamburg, Germany, May 15, 2013; data compiled by USITC.

153 IMS Research data cited in GP Joule, "One Stop Solar Solutions," n.d., 4 (accessed July 9, 2013).

154 IEA, PVPS, *Trends in Photovoltaic Applications*, 2012, 29.

155 *Solar Industry*, "Forecast: Many German PV Installers," February 27, 2013.

156 Q-Cells was subsequently acquired by Korean firm Hanwha. Radowitz, "Gehrlicher Confirms Insolvency Move," July 9, 2013; Radowitz, "Conergy to File for Insolvency," July 5, 2013; Stromsta, "Hanwha SolarOne Says Q-Cells," November 25, 2012.

157 Industry representatives, interviews by USITC staff, Tokyo, May 10, 2013, May 13, 2013, and May 14, 2013; EPIA, *Global Outlook for Photovoltaics until 2016*, May 2012, 67; Government of Japan, METI, "Announcement Regarding the Present Status," May 17, 2013; Bloomberg New Energy Finance database (accessed June 2, 2013).

158 Industry representative, interview by USITC staff, Tokyo, May 10, 2013; industry representative, telephone interview by USITC staff, May 28, 2013; Nakayama and Sato, "Sekisui Expects," July 10, 2009.

159 See, for example, Canadian Solar, "Canadian Solar Announces," January 14, 2010; SunPower Corp., "SunPower Corporation and Toshiba Extend," December 4, 2012.

160 Industry representatives, interviews by USITC staff, Tokyo, May 10, 2013, May 13, 2013, and May 14, 2013; industry representative, telephone interview by USITC staff, May 28, 2013.

161 Ibid.

162 Watanabe, "Toshiba, SunEdison," April 19, 2013; Solar Frontier, "Solar Frontier and Yano Industry," March 21, 2013; juwi Shizen website, http://ww.juwi-shizenenergy.com (accessed June 2, 2013); Gestamp, "Gestamp Solar and Kankyo-Keiei," October 4, 2012.

163 Industry representative, telephone interview by USITC staff, May 28, 2013.

164 Industry representative, telephone interview by USITC staff, May 28, 2013; industry representatives, interviews by USITC staff, Tokyo, May 10, 2013, and May 14, 2013; Gestamp, "Gestamp Solar and KankyoKeiei,"

October 4, 2012; juwi Shizen website, http://ww.juwi-shizenenergy.com (accessed June 10, 2013); Kyocera, "KYOCERA Establishes," July 31, 2012.

165 SolarPlaza, "Mexico More Attractive," February 18, 2013; Krulewitz, "Si Se Puede!" January 24, 2013; Sunsong, "Latin America," January 2, 2013; Sheppard, "Latin America," January 21, 2013; industry representatives, interviews by USITC staff, Santiago, Chile, May 13, 2013, May 14, 2013, and May 15, 2013.

166 EPIA, *Global Outlook for Photovoltaics until 2016*, May 2012, 67.

167 SolarPlaza, "Mexico More Attractive," February 18, 2013; Krulewitz, "Si Se Puede!," January 24, 2013; Sunsong, "Latin America" January 2, 2013; Sheppard, "Latin America," January 21, 2013.

168 BNEF database (accessed May 29, 2013).

169 Industry representatives, interviews by USITC staff, Santiago, May 13, 2013, May 14, 2013, and May 15, 2013; First Solar, "First Solar Acquires," January 9, 2013; MEMC, "SunEdison and CAP Sign Agreement," January 31, 2013; Sciaudone, "AES Gener," November 26, 2012.

170 Industry representatives, interviews by USITC staff, Santiago, May 13, 2013 and May 15, 2013.

171 BNEF database (accessed June 10, 2013); Rosell, "Varying Rates," December 2012, 29–30.

172 Industry representative, interview by USITC staff, Santiago, May 13, 2013; industry representatives, interviews by USITC staff, Brazil, May 7, 2013, May 9, 2013, and May 10, 2013; MEMC, "SunEdison and Petrobras," April 18, 2013; Nielsen, "Brazil Promoting Rooftop Solar," April 19, 2012; PROINSO, "PROINSO Supplies 100kW to Several Net Metering PV Projects in Brazil," January 29, 2013.

End Notes for Section 4

1 Wind energy services are those services involved in the development, construction, and ongoing operation of large, utility-scale power plants, or wind farms. Wind energy entails the capture of the wind's natural kinetic energy by wind turbines, which then generate electricity. IEA, "Topic: Wind Power," n.d. (accessed September 13, 2012).

2 Wind energy services may be contracted for installation and maintenance of either small wind energy systems or utility-scale projects. Small wind systems account for only a small share of the wind energy services market and are less service intensive. Utility-scale projects can require hundreds of wind turbines generating hundreds of megawatts (MW) of capacity and are most often owned and operated by either a utility or an independent power provider (IPP), which sells the power it generates to a utility. AWEA, *Wind Energy Siting Handbook*, February 2008, 2-1.

3 A project developer typically is responsible for identifying a potential site; performing the necessary resource assessments; acquiring the necessary permits; negotiating land agreements; negotiating the power purchase agreement; securing financing; and overseeing site preparation and the installation and erection of the wind turbines, as well as the necessary supporting infrastructure.

4 A turnkey project is one in which one party (usually a developer) manages all stages from design through construction and commission, then sells the project to an owner/operator. Owner/operators retain ownership of a wind farm and are responsible for the day-to-day operation and maintenance of the facility.

5 Industry representative, interview by USITC staff, Washington, DC, January 25, 2013.

6 Industry representative, interview by USITC staff, Washington, DC, January 23, 2013.

7 In some cases, the degree of service provision varies across markets, with developers offering turnkey projects in one market but only limited services in another. Developers may solicit bids from local and international firms or may have established relationships with particular providers. Either way, developers award these subcontracts primarily based on price, as well as a number of other factors such as technical quality, experience, availability, and safety. Industry representative, telephone interview by USITC staff, January 18, 2013; industry representative, telephone interview by USITC staff, January 28, 2013; industry representative, telephone interview by USITC staff, January 30, 2013.

8 During wind analysis, wind assessment specialists use data captured and transmitted from a potential site. Data typically begin to illustrate a trend after six months that indicates whether a project can move forward. Industry representative, interview by USITC staff, Washington, DC, January 23, 2013.

9 Financing will be difficult to secure for a project unless it generates enough wind to produce a threshold internal rate of return. Industry representative, interview by USITC staff, January 18, 2013; AWEA, *Wind Energy Siting Handbook*, February 2008, 2-3; industry representative, interview by USITC staff, March 12, 2013; industry representative, interview by USITC staff, Washington, DC, January 23, 2013.

[10] Industry representative, interview by USITC staff, January 23, 2013; industry representative, interview by USITC staff, October 23, 2012.

[11] Specialized engineering firms, project developers, or large engineering, procurement, construction (EPC) firms.

[12] Industry representative, interview by USITC staff, October 23, 2012.

[13] As an owner's engineer or independent consultant for the bank, the firm is independent of the EPC contractor and acts as an advocate for its client, overseeing the engineering, construction, and design process. Connel and Grennan, "Who Needs an Owner's Engineer?" March 1, 2011; industry, interview by USITC staff, March 12, 2013.

[14] An interconnection agreement allows a project to be connected to the existing electrical grid. Applying for the interconnection agreement is the first real benchmark in the development process, as it requires a down payment or earnest money. Industry representative, interview by USITC staff, Washington, DC, January 23, 2013.

[15] An environmental impact statement for public lands can add multiple years to the project timeline as well as additional expense. As a result, this can create a preference among developers for projects only on private land—in fact, one industry representative estimated that over 90 percent of U.S. wind farms are sited on private land, a figure the representative attributed largely to the greater ease of doing business with private landowners. Industry representative, interview by USITC staff, Washington, DC, January 23, 2013.

[16] A number of contracts must be drawn up, typically beginning with a land agreement once an appropriate site has been found. Other contracts include EPC and construction agreements, power service contracts, and ownership agreements related to turnkey projects. Industry representative, interview by USITC staff, January 23, 2013.

[17] Stoel Rives, LLP website, http://www.stoel.com/showindustry.aspx?show=1980 (accessed March 20, 2013); Foley & Lardner LLP website. http://www.foley.com/windenergy/ (accessed March 20, 2013); and Alston & Bird LLP website. http://www.alston.com/services/industries/energy/ (accessed March 20, 2013).

[18] Wind farm owners and developers typically do not buy land for projects, but instead take out a 25–30 year lease. Industry representative, interview by USITC staff, January 23, 2013.

[19] Industry representative, interview by USITC staff, January 23, 2013.

[20] A power purchase agreement (PPA) is a long-term contract wherein the buyer agrees to purchase the electricity generated by the wind farm. PPA's can range from 15 to 20 years and are critical to the financial success of a wind energy project because they guarantee a steady revenue stream for the project and are typically a condition of financing, which is necessary for the construction and installation phase. Early project development is highly speculative and financed by the project developer, but once a site has been identified and the permitting process begun, the project developer typically begins looking for a buyer or power purchaser. The PPA is frequently described as a significant turning point in the lifecycle of a wind energy project; once a PPA is secured a project's probability of failure is significantly reduced. Industry representative, telephone interview with USITC staff, January 30, 2013; industry representative, interview with USITC staff, October 25, 2012; World Bank. "Power Purchase Agreements (PPAs)," June 18, 2008.

[21] Jordan and Steger, *American Wind Farms*, September 2012, 8–29.

[22] DLA Piper, "EPC Contracts in the Power Sector," 2011, 1, 4; Massy, "Banks Opting for Lazy Route Could Increase Risk," October 17, 2010.

[23] Industry representative, telephone interview by USITC staff, January 30, 2013.

[24] The selection of the turbine manufacturer for a wind project can influence the choice of a construction firm, as companies must be certified (requiring education and training) in the turbine equipment they are installing. In fact, OEM participation in the installation process can vary from sending an individual representative to oversee the turbine installation to overseeing and executing the entire construction and installation process itself. Industry representative, interview by USITC staff, Washington, DC, January 25, 2013.

[25] Industry representative, interview by USITC staff, January 18, 2013; industry representative, interview by USITC staff, February 8, 2013; Vestas website, http://www.vestas.com/en/wind-powerplants/construction.aspx#/vestas-univers (accessed March 20, 2013).

[26] Industry representative, interview by USITC staff, Washington, DC, January 25, 2013; industry representative, telephone interview by USITC staff, January 28, 2013.

[27] The components and services covered by warranties vary widely depending on the needs of the customer and services offered by the OEM. Ireland, "Wind Service," November 2011, 23–24; Canada Wind Energy Association, "The Many Steps to Planning and Building a Wind Farm," n.d. (accessed January 16, 2013).

[28] OEMs have increased their presence in the O&M sector, attracted by larger profit margins, particularly as margins on equipment have fallen. OEMs have the resources, such as engineering expertise, modeling capabilities, and ability to foresee potential problems, to allow them to offer profitable services contracts; in

turn, they are able to continually upgrade and improve their machines. However, they face competition, both from in-house O&M departments and from ISPs. Industry representative, telephone interview by USITC staff, December 12, 2012; industry representative, telephone interview by USITC staff, January 22, 2013.

[29] Industry representative, telephone interview by USITC staff, January 22, 2013.

[30] Industry representatives, interview by USITC staff, Santiago, Chile, May 14, 2013.

[31] *Wind Power Monthly*, "GE and Remote Turbine Monitoring," March 10, 2010; industry representative, interview by USITC staff, Washington, DC, January 23, 2013.

[32] Installed wind energy capacity measures utility-scale wind energy capacity, defined by AWEA as installations of wind turbines larger than 100 kW. USITC staff calculations based on AWEA, *AWEA U.S. Wind Industry Annual Market Report 2011*, 2012, 9.

[33] GWEC website, http://www.gwec.net/publications/global-wind-report-2/global-wind-report-2012/ (accessed July 9, 2013).

[34] EIA, *Electric Power Annual 2011*, January 20, 2013, table 2.4, "Average Retail Price of Electricity to Ultimate Customers"; industry representative, interview by USITC staff, January 30, 2013.

[35] One source estimates that new capacity additions could be as low as 1,000 MW in 2013. Brown, "U.S. Renewable Electricity," June 20, 2012; Cusick, "With Congress at Its Back," January 3, 2013; industry representative, interview by USITC staff, Washington, DC, January 28, 2013; industry representative, interview by USITC staff, October 23, 2012; industry representative, interview by USITC staff, January 30, 2013.

[36] Estimates are based on the breakdown of capital costs associated with the development and installation of a new wind farm. Capital costs include the cost of the wind turbine, distribution, balance of plant/erection, engineering, procurement, and construction management services, along with the owner's cost. It does not include any O&M activities. USITC staff calculations using data from AWEA, *AWEA U.S. Wind Industry Annual Market Report 2011*, 2012; Wiser and Bolinger, *2011 Wind Technologies Market Report*, August 2012, 3; and Black & Veatch, *Cost and Performance Data for Power Generation Technologies*, February 2012, 49.

[37] These estimates are based on average national costs. In reality, costs can vary substantially across projects and regions due to a number of factors including siting, and project and turbine size. Likewise, O&M costs also vary significantly across projects, due in large part to the variation in service offerings and providers. Values presented are USITC staff calculations estimated using data from Wiser and Bolinger, *2011 Wind Technologies Market Report*, August 2012. BNEF, "Operations and Maintenance Price Index," October 23, 2012; Tegen, et. al, *2010 Cost of Wind Energy Review*, April 2012, 26; and *Wind Power Monthly*, "US O&M spending to reach $6 billion by 2025," July 11, 2012.

[38] Industry representative, interview by USITC staff, October 25, 2012; industry representative, interview by USITC staff, October 23, 2012.

[39] On average, this share is around a fifth of a provider's energy portfolio, although there is considerable variation among states. For example, Florida's RPS program requires a 7.5 percent share from renewable sources, while Hawaii requires 40 percent. Boyland, "Gone with the Wind," December 2012, 30.

[40] Between 2012 and 2020, existing RPS programs are projected to drive new annual installations of energy from all renewable sources (not just wind) by 4–5GW a year. These totals are well below the capacity of new wind installations in recent years, suggesting that RPS programs do not themselves drive wind development. Data further illustrate that in many U.S. regions, RPS levels have either been met or are close to being met, suggesting some degree of oversupply of renewable energy unless standards are raised from their current levels. Wiser and Bolinger, *2011 Wind Technologies Market Report*, August 2012, 58; industry representative, email to USITC staff, December 12, 2012.

[41] Industry representative, telephone interview by USITC staff, March 25, 2013; industry representative, telephone interview by USITC staff, January 25, 2013; industry representative, telephone interview by USITC staff, January 30, 2013; Brown, "U.S. Renewable Electricity," June 20, 2012, 7.

[42] Database of State Incentives for Renewables and Efficiency website, "Renewable Portfolio Standard Policies," March 2013; Wiser and Bolinger, *2011 Wind Technologies Market Report*, August 2012, 58.

[43] States vary in their permitting requirements; some, such as Kansas, require only county permits, while others, such as Oregon and Minnesota, have state siting councils. However, industry representatives note that it is very rare for permitting difficulties to permanently halt a project. Industry representative, interview by USITC staff, March 25, 2013; Anderson, Corbin, and McMahan, "The Law of Wind," n.d. (accessed March 28, 2013).

[44] Industry representative, telephone interview by USITC staff, March 25, 2013.

[45] Brown, "U.S. Renewable Electricity," June 20, 2012, 8.

[46] 42 U.S.C. § 13211–13219.

[47] Wiser and Bolinger, *2011 Wind Technologies Market Report*, August 2012, 57; Boyland, "Gone with the Wind," December 2012, 31.

[48] Industry sources note that in areas with low energy prices or marginal wind resources, the credits such as the PTC are not sufficient to make wind power generation competitive with traditional fossil fuels. Industry representative, email to USITC staff, March 8, 2013.

[49] Periods during which the PTC lapsed resulted in lulls in new installations; conversely, years leading up to the scheduled expiration of the PTC, such as 2012, typically see strong growth in wind development and construction activities. The PTC lapsed in 2000, 2002, and 2004. Wiser and Bolinger, *2011 Wind Technologies Market Report*, August 2012, 57; industry representative, email to USITC staff, March 3, 2013; industry representative, telephone interview by USITC staff, March 25, 2013.

[50] As an incentive for investment, owners of wind energy projects are typically able to write off the value of their equipment on an accelerated time frame (5 years rather than 20). The Tax Relief, Unemployment Insurance Reauthorization, and Job Creation Act of 2010 (P.L. 111-312, 124 Stat. 3296) offered an additional incentive, allowing owners a first-year bonus depreciation of 100 percent for projects commissioned by the end of 2011, and 50 percent for projects commissioned during 2012. Wiser and Bolinger, *2011 Wind Technologies Market Report*, August 2012, 57; Windustry.org, "Section 10: Tax Incentives," n.d. (accessed January 30, 2013).

[51] Both the investment tax credit and the cash grant were provisions of The American Recovery and Reinvestment Act and were available for projects commissioned by the end of 2012. Wiser and Bolinger, *2011 Wind Technologies Market Report*, August 2012, 57.

[52] Industry representative, email to USITC staff, March 8, 2013.

[53] GWEC, *Global Wind Statistics 2011*, July 2, 2012, 65.

[54] Rising consumption of electricity increases demand for energy from renewable sources, such as wind energy, which then drives demand for wind services to develop, provide, and maintain such resources. Boyland, "Gone with the Wind," December 2012, 5; industry representative, interview by USITC staff, Washington, DC, January 28, 2013.

[55] Wiser and Bolinger, *2011 Wind Technologies Market Report*, August 2012, 33; *Wind Power Monthly*, "Balancing Act for Hard Pressed Supply Chain," August 25, 2011.

[56] In addition to the warranty, turbine sales increasingly include an additional service agreement, wherein the OEM typically provides scheduled maintenance for a period of 2 to 5 years, although longer service contracts for up to 10 years are increasingly common. Scheduled maintenance refers to routine procedures performed on a regular basis. By comparison, unscheduled maintenance refers to unexpected repairs that are typically more complex, requiring troubleshooting and diagnostic strategies to get a turbine back online. *Wind Power Monthly*, "Balancing Act for Hard Pressed Supply Chain," August 25, 2011; industry representative, telephone interview by USITC staff, January 22, 2013.

[57] Longer contracts allow the provider better knowledge of the machine and the project, so they can better optimize performance. Industry representative, telephone interview by USITC staff, January 18, 2013; industry representative, telephone interview by USITC staff, January 22, 2013; *Wind Power Monthly*, "Balancing Act for Hard Pressed Supply Chain," August 25, 2011.

[58] Broehl, "Operations & Maintenance—Industry Prepares," April 1, 2010.

[59] Industry representative, telephone interview by USITC staff, December 12, 2012.

[60] Industry representative, interview by USITC staff, Washington, DC, January 23, 2013.

[61] Wiser and Bolinger, *2011 Wind Technologies Market Report*, August 2012, 3; industry representative, interview by USITC staff, Washington, DC, January 23, 2013; industry representative, telephone interview by USITC staff, December 12, 2012.

[62] Industry representative, interview by USITC staff, Washington, DC, January 23, 2013.

[63] *Wind Power Monthly,* "Operations and Maintenance," September 1, 2009.

[64] OEMs traditionally focused on warranty and maintenance services because these add value to turbine sales as well as provide manufacturers with valuable knowledge from the opportunity to monitor their turbines in the field over time. Industry representative, interview by USITC staff, October 24, 2012; industry representative, interview by USITC staff, October 23, 2012; industry representative, interview by USITC staff, Santiago, Chile, May 12, 2013.

[65] Competitiveness in the wind energy equipment market has increased due to declining turbine prices, overcapacity, and uncertainty regarding government incentives and other favorable policies. Wiser and Bolinger, *2011 Wind Technologies Market Report*, August 2012, 33; Boyland, "Gone with the Wind," 20;

industry representative, interview by USITC staff, October 24, 2012; industry representative, telephone interview by USITC staff, January 22, 2013.

66 Industry representative, telephone interview by USITC staff, January 22, 2013. U.S. project developers not only design and build wind projects, but many also maintain ownership and operate these facilities. Industry sources estimate that 90 percent of wind farms built are owned by project developers who sell the energy under a PPA. Statistics suggest that in 2011, independent power producers (such as project developers) owned 73 percent of all new capacity additions. Industry representative, interview by USITC staff, March 25, 2013; Wiser and Bolinger, *2011 Wind Technologies Market Report*, 2012, 29.

67 BTM Consult, *World Market Update 2011*, March 2012, 32.

68 The leading project developers vary from year to year depending on the projects commissioned. Table 4.2 presents the five largest developers from 2012.

69 Industry representative, email message to USITC staff, March 8, 2013.

70 10 largest project developers in each year based on annual installed capacity. USITC staff calculations using data from AWEA, *AWEA U.S. Wind Industry Fourth Quarter 2011 Market Report*, January 2012; AWEA, *AWEA Year End 2009 Market Report,* January 2010.

71 One industry representative estimated that there are roughly 38 firms that provide construction or EPC services for the wind industry, of which 22 are actively competing in the U.S. market. Industry representative, telephone interview by USITC staff, February 8, 2013.

72 One representative noted that seven to eight years ago, there were few firms capable of providing comprehensive EPC services. Industry representative, telephone interview by USITC staff, March 25, 2013.

73 One industry representative estimates that for a $100 million project, turbines and other equipment account for $75 million, while $23 million is EPC services and $2 million is other services. Industry representative, telephone interview by USITC staff, March 25, 2013.

74 Industry representative, telephone interview by USITC staff, February 8, 2012.

75 Industry representative, telephone interview by USITC staff, February 8, 2012; industry representative, telephone interview by USITC staff, March 25, 2013.

76 Industry representative, telephone interview by USITC staff, February 8, 2012.

77 By one estimate, asset owners and OEMs each account for roughly 40 percent of the O&M market, with third-party providers making up the remainder. Industry representative, interview by USITC staff, January 23, 2013; industry representative, telephone interview by USITC staff, January 22, 2013.

78 EDF Renewable Services is the O&M services branch of EDF Renewable Energy, a wind farm developer and owner. Industry representative, interview by USITC staff, January 23, 2013; industry representative, interview by USITC staff, October 23, 2012.

79 Industry representative, telephone interview by USITC staff, January 18, 2013; industry representative, telephone interview by USITC staff, January 22, 2013.

80 One industry representative estimated 95 percent of O&M services are for scheduled maintenance. Industry representative, telephone interview by USITC staff, January 22, 2012.

81 Industry representative, interview by USITC staff, January 22, 2012.

82 Industry representative, interview by USITC staff, December 12, 2012.

83 Industry representative, interview by USITC staff, December 12, 2012; industry representative, interview by USITC staff, January 22, 2012.

84 Industry representative, interview by USITC staff, January 18, 2013.

85 GWEC website, http://www.gwec.net/publications/global-wind-report-2/global-wind-report-2012/ (accessed July 9, 2013).

86 GWEC, *Global Wind Statistics 2012,* February 11, 2013; GWEC, *Global Wind 2007 Report*, 2008.

87 BTM Consult, *World Market Update 2011*, March 2012, 35.

88 2011 is the most recent year for which data are available. Pernick, Wilder, and Winnie, *Clean Energy Trends 2012*, March 2012, 3.

89 These values are estimated by USITC using publicly available data. The value of the development and installation services market was calculated multiple ways. First, available cost share estimates from Black & Veatch and IRENA were applied to Pernick, Wilder, and Winnie's estimates of the overall wind energy market. Estimates were also made by calculating and summing the cost of installation for the top 10 markets (by total installed capacity), as well as the rest of the world, then subtracting the estimated turbine cost. Estimated turbine cost is the sum of average wind turbine prices for each of the 10 leading markets multiplied by annual installations (in MW) in those markets plus the global average turbine price multiplied by annual installations (in MW) in the rest of the world. Global or regional averages were applied to countries for which annual turbine prices or

installation costs were not available. The resulting estimates of the services markets were not exactly the same, but were comparable. USITC staff calculations using data from Pernick, Wilder, and Winnie, *Clean Energy Trends 2012*, March 2012, 4; GWEC, *Global Wind 2011 Report,* 2012, 12; IEA, *IEA Wind 2011 Annual Report*, July 2012; IRENA, *Renewable Power Generation Costs in 2012: An Overview*, 2013, 32.

[90] Data on O&M costs in 2011 were available for only 14 countries, mostly European, representing 78 percent of the market by installed capacity. Data used for the UK are for full-service O&M contracts, and thus may overestimate the market. The global average (21 percent) was used for countries that did not report average capacity factors. Costs for the rest of the world were calculated using the global average capacity factor and the U.S. O&M cost ($0.01/kWh). O&M costs in the United States are reportedly among the lowest in the world, suggesting that the actual value of the 2011 O&M services market is higher than estimated. BNEF reports that prices for full-service O&M contracts are highest in Eastern Europe; they also tend to be higher than average in emerging markets such as Brazil, Mexico, and South Africa. IRENA, *Cost Analysis of Wind Power*, June 2012; GWEC, *Global Wind 2011 Report*, 2012; BNEF, "Operations and Maintenance Price Index," October 23, 2012.

[91] Data on O&M costs in 2011 are available for Austria, China, Denmark, Finland, Germany, Italy, Japan, the Netherlands, Norway, Spain, Sweden, Switzerland, the UK, and the United States. In the absence of a country-specific average capacity factor, the world average was used for Austria, China, the Netherlands, Spain, and Sweden.

[92] Global average capacity factor was used to calculate the value of China's O&M market. USITC calculations using data from IRENA, *Cost Analysis of Wind Power*, June 2012; GWEC, *Global Wind 2011 Report*, 2012; and *Wind Power Monthly*, "Service Market Grows as Installations Fall," July 2012.

[93] Electricity generation using fossil fuels increased in many countries in Latin America during 2007– 11, including in countries like Brazil and Venezuela that traditionally relied on hydroelectric power to meet much of their energy demand. Further, many countries in the region use oil for electricity generation, and changes in oil prices, therefore, can have a significant impact on the cost of electricity generation. Industry representative, telephone interview by USITC staff, January 18, 2013; industry representative, interview by USITC staff, Beijing, May 16, 2013; industry representative, interview by USITC staff, São Paulo, Brazil, May 8, 2013; EIA, International Energy Statistics database (accessed August 19, 2013); Yepez-García, Johnson, and Andrés, *Meeting the Electricity Supply/Demand*, September 2010, 12.

[94] Industry representative, interview by USITC staff, Santiago, Chile, May 14, 2013.

[95] Industry representative, interview by USITC staff, Beijing, May 16, 2013.

[96] FITs guarantee a certain rate per kilowatt-hour for wind-generated energy. In 2011, China, Germany, India, Spain, and the UK all had FIT programs. Ontario Power Authority, "General Information about the FIT and microFIT Programs" (accessed January 31, 2013); IEA, *IEA Wind 2011 Annual Report*, July 2012, 10; GWEC, *India Wind Energy Outlook 2012,* November 2012, 11.

[97] Launched in June 2010, the European Wind Initiative is a plan to support development of the wind power industry and the integration of growing amounts of wind electricity into the grid. One of its stated goals is for wind energy to supply 20 percent of EU electricity consumption by 2020. IEA, *IEA Wind 2011 Annual Report*, July 2012, 96.

[98] Industry representative, telephone interview by USITC staff, January 30, 2013.

[99] BTM Consult, *International Wind Energy Development*, March 2012, viii; Scott, "In Europe, Green Energy Takes a Hit from Debt Crisis," November 13, 2012; Sills, "Spain Halts Renewable Subsidies to Curb $31 Billion of Debts," January 27, 2012.

[100] BTM Consult, "World Market Update 2011," March 2012, 55.

[101] Ibid.

[102] Ibid., 59.

[103] Industry representative, telephone interview by USITC staff, January 30, 2013.

[104] Industry representative, interview by USITC staff, October 23, 2012; industry representative, telephone interview by USITC staff, January 18, 2013.

[105] Industry representative, telephone interview by USITC staff, January 30, 2013.

[106] Ibid.

[107] Ibid.

[108] Industry representative, interview by USITC staff, Washington, DC, January 23, 2013.

[109] For example, during the first nine months of 2011, 71 percent of investment by Italian wind energy firms occurred outside Italy. O'Brien, "Uncertainty Drives Italian Firms Abroad," December 23, 2011; Morales and

Sills, "Spain Ejects Clean-Power Industry," May 30, 2012; industry representative, interview by USITC staff, São Paulo, Brazil, May 8–10, 2013.

[110] Industry representative, telephone interview by USITC staff, January 30, 2013.

[111] Industry representative, telephone interview by USITC staff, January 18, 2013; industry representative, interview by USITC staff, São Paulo, Brazil, May 8, 2013.

[112] Industry representative, telephone interview by USITC staff, January 28, 2013.

[113] Industry representative, telephone interview by USITC staff, March 12, 2013.

[114] Nielsen, "Cheapest Wind Energy Spurring Renewable Energy Deals," January 29, 2013.

[115] OEMs GE Wind, Vestas, and Siemens, among others, have manufacturing facilities in Brazil and India. Sen, "Wind Energy Majors Set Up Low-Cost Plants," February 23, 2011; *Cleantechinvestor.com*, "Brazil—Wind Manufacturing Hub," August 15, 2011.

[116] Industry representative, telephone interview by USITC staff, January 25, 2013.

[117] Industry representative, interview by USITC staff, Washington, DC, January 23, 2013.

[118] Rankings compiled by USITC using project data from BNEF database. Available data cover 84 percent of projects commissioned in 2012. These Chinese firms will be discussed further in the foreign market profiles section.

[119] Industry representative, telephone interview by USITC staff, January 28, 2013.

[120] BNEF, "Operations and Maintenance Price Index," October 23, 2012.

[121] Industry representative, telephone interview by USITC staff, January 28, 2013.

[122] Qi, "China's Service Market Grows as Installations Fall," July 1, 2012.

[123] BNEF, "Operations and Maintenance Price Index," October 23, 2012.

[124] Ibid.

[125] Ibid.

[126] Industry representative, telephone interview by USITC staff, January 22, 2013.

[127] BNEF, "Operations and Maintenance Price Index," October 23, 2012.

[128] Ibid.

[129] Qi, "China's Service Market Grows as Installations Fall," July 1, 2012.

[130] Ibid.

[131] Industry representative, telephone interview by USITC staff, January 22, 2013.

[132] North, "Ainscough Completes Windcon Deal," October 24, 2012.

[133] Schoenberger, "Timken Buys Wazee Companies," January 2, 2013.

[134] Types of services trade are discussed in more detail in section 1.

[135] Industry representative, telephone interview by USITC staff, March 12, 2013.

[136] Industry representative, telephone interview by USITC staff, January 18, 2013; industry representative, telephone interview by USITC staff, January 30, 2013; industry representative, telephone interview by USITC staff, January 28, 2013; industry representative, interview by USITC staff, São Paulo, Brazil, May 10, 2013.

[137] GL Garrad Hassan website, http://www.gl-garradhassan.com (accessed April 3, 2013).

[138] The only other U.S.-based firm in the top 10 was Clipper, which accounted for 1.5 percent of installations. The top 5 firms, other than GE, are Siemens (Germany, 20 percent of installations), Vestas (Denmark, 14 percent), Gamesa (Spain, 10 percent), and REpower/Suzlon (India, 6 percent). Siemens, Vestas, and Gamesa all manufacture nacelles and blades in the United States; Vestas also produces towers in the United States. REpower is Suzlon's German-based subsidiary. AWEA, "AWEA U.S. Wind Industry Annual," May 2013, 2. For a list of nacelle and blade plants in the United States as of March 2012, see David and Fravel, "U.S. Wind Turbine Export Opportunities," July 2012, 7.

[139] USITC calculations using data from BNEF database (accessed April 4, 2013).

[140] Industry representative, interview by USITC staff, Washington, DC, January 23, 2013.

[141] Ibid.

[142] USITC calculations using data from BNEF database (accessed April 4, 2013) and ENR *Global Sourcebook* 2012, December 10, 2012, 18, 47. Mortenson Construction website, http://www.mortenson.com/INdustry_RenewableEnergy.aspx (accessed April 4, 2013).

[143] *Wind Power Monthly*, "Balancing Act for Hard Pressed Supply Chain," August 2011; industry representative, telephone interview by USITC staff, February 8, 2013.

[144] Xiao, "Goldwind, Change is in the Air," December 2012, 46.

[145] Wind-powered generating sets include nacelles and any items imported with the nacelle, such as the blades or hub. If these components are imported or exported separately from the nacelle, they are included in different

HS subheadings. Wind-powered generating sets are included in 8502.31 in the international Harmonized Commodity Description and Coding System (HS).

146 China accounted for 6.9 percent of global exports. Chinese export data may include some exports by non-Chinese firms from their plants in China. GTIS, Global Trade Atlas database (accessed July 9, 2013).

147 McGovern, "China Puts Bolivia on Wind Map," March 11, 2013; industry representative, interview by USITC staff, Beijing, May 17, 2013; David and Fravel, "U.S. Wind Turbine Export Opportunities," June 2012; Qi, "Sany Wins 153MW Ethiopian Contract," May 20, 2013.

148 Industry representative, interview by USITC staff, Beijing, May 16 and May 17, 2013; Nielsen, "China Grabs Share," November 20, 2012; Qi, "Sany Wins," May 20, 2013.

149 Qi, "Sany Wins 153MW Ethiopian Contract," May 20, 2013; McGovern, "China Puts Bolivia on Wind Map," March 11, 2013; Qi, "Hydrochina Wins 30MW Deal in Pakistan," June 4, 2013.

150 Weyndling, "Frustrated Spanish Wind Sector Looks Abroad," June 28, 2011; Scott, "A Spanish Blowout," October 29, 2008.

151 Anderson and Gibson, "Despite Uncertainty, Foreign Investors Eye U.S. Market,"April 2012, 28; Lima, "EDP Buys Horizon Wind from Goldman," March 27, 2007; BNEF database, n.d. (accessed January 11, 2013).

152 Analysis by USITC using data from BNEF.

153 Quilter, "AMP Capital to Invest $100 Million," February 12, 2013.

154 Goosens, "AES Says U.S. Wind Market Is Weak," November 4, 2010.

155 Industry representative, email to USITC staff, March 8, 2013.

156 Industry representative, telephone interview by USITC staff, January 25, 2013.

157 Countries with local-content requirements include Turkey, Ukraine, Croatia, and South Africa. EWEA, "WTO Rules Against Ontario," December 20, 2012; Tardieu, "EWEA Trade Policy Views," March 29, 2012; Creed and Kordvani, "The WTO Report on Local Content Requirements," February 13, 2013; industry representative, telephone interview by USITC staff, March 8, 2013; industry representative, telephone interview by USITC staff, January 18, 2013; industry representative, telephone interview by USITC staff, February 8, 2013.

158 Industry representative, telephone interview by USITC staff, January 30, 2013.

159 Industry representative, telephone interview by USITC staff, January 18, 2013.

160 The complaint in this matter was filed by Japan in September 2010. WTO Dispute DS 412: Canada, (accessed July 2, 2013).

161 There are separate FIT programs for large and small projects. Ontario Ministry of Energy, FIT Program Directive, June 12, 2013, 3.

162 Industry representative, interview by USITC staff, São Paulo, Brazil, May 8, 2013; Nielsen, "BNDES Raises Local-Content Requirement," September 13, 2012.

163 Industry representative, interview by USITC staff, São Paulo, Brazil, May 8, 2013.

164 Industry representative, email to USITC staff, March 8, 2013; industry representative, telephone interview by USITC staff, January 25, 2013.

165 One representative estimated that only 20 percent of wind power projects are awarded to foreigners, stating that most often, requests for proposals (RFPs) are awarded to local firms. Industry representative, telephone interview by USITC staff, January 28, 2013.

166 Industry representative, telephone interview by USITC staff, January 28, 2013; industry representative, telephone interview by USITC staff, March 12, 2013.

167 GWEC website, http://www.gwec.net/wp-content/uploads/2013/04/8Top-10-New-InstalledCapacity-Jan-Dec-2012.jpg (accessed July 9, 2013).

168 Industry representatives, interviews by USITC staff, Rio de Janeiro and São Paulo, Brazil, May 7– 10, 2013. Reportedly, the northeastern portion of Brazil benefits from strong, steady winds that tend to rarely change direction—optimal conditions for wind turbines.

169 Brazil accounted for 2.4 percent of global new installed capacity in 2012. GWEC, *Global Wind Statistics 2012*, February 11, 2013; GWEC, *Global Wind 2012 Report*, April 2013.

170 Through the PROINFA program, which ended in 2011, the government actively encouraged development of renewable energy in an effort to diversify its power sources. Currently, low water levels are hindering the country's reliance on hydropower, and the high price of diesel makes thermal plants very expensive. In contrast, wind energy is cost competitive, due to the strong, consistent equatorial winds already mentioned, and projects are typically able to be constructed and commissioned with a relatively short turnaround time. Industry representative, interview by USITC staff, São Paulo, Brazil, May 9, 2013; industry representative,

interview by USITC staff, São Paulo, Brazil, May 8, 2013; industry representative, interview by USITC staff, Rio de Janeiro, Brazil, May 7, 2013.

171 Developers include subsidiaries of Iberdrola (Spain), EDP Renováveis (Portugal/Spain), and GDF Suez (France). Industry representative, telephone interview by USITC staff, April 11, 2013.

172 BTM Consult, *World Market Update 2012*, 78.

173 Currently, OEMs provide nearly all O&M services in the Brazilian market. However, industry representatives believe this may change, either through the development of O&M ISPs or the development of in-house capabilities by wind farm owners. Industry representative, interview by USITC staff, São Paulo, Brazil, May 8, 2013; industry representative, interview by USITC staff, São Paulo, Brazil, May 9, 2013.

174 600 MW of wind power installed in 2012 lacked grid connections. Spatuzza, "Brazil—Lack of Grid Connections Keeps Wind Farms Offline," March 1, 2013.

175 Industry representative, interview by USITC staff, São Paulo, Brazil, May 8, 2013; industry representative, interview by USITC staff, Rio de Janeiro, Brazil, May 7, 2013.

176 Industry representative, telephone interview by USITC staff, April 11, 2013.

177 Previously Brazil had a 60 percent local-content requirement. The government recently introduced new regulations that require that certain stages in the manufacturing process or certain components must be manufactured in Brazil in order to qualify for financing from the national development bank, BNDES. Financing from BNDES is considered essential, as it offers very low rates compared to commercial banks, and accounts for the majority of project finance in the market. Industry representative, interview by USITC staff, São Paulo, Brazil, May 8, 2013; industry representative, interview by USITC staff, Rio de Janeiro, Brazil, May 7, 2013.

178 Industry representatives, interviews by USITC staff, São Paulo, Brazil, May 8–10, 2013.

179 Statistics on trade in wind energy services are not available. Industry representative, telephone interview by USITC staff, February 8, 2013; Bailey, "Building Boom Draws in Global Players," October 1, 2012.

180 USITC calculations using data from GWEC, *Global Wind Statistics 2012*, February 11, 2013, and GWEC, *Annual Market Update 2011*, March 2012, 27.

181 GWEC, *Global Wind Statistics 2012*, February 11, 2013. The Canadian Wind Energy Association (CanWEA) reports 936 MW of new wind energy capacity installed in 2012, for a total of 6,500 MW. CanWEA, "Wind by the Numbers," January 2013.

182 Hydro-Québec held wind tenders in 2003, 2005, and 2009, and Quebec has the second highest installed wind capacity among Canadian provinces. Industry representative, telephone interview by USITC staff, February 8, 2013; David and Fravel, "U.S. Wind Turbine Export Opportunities," July 2012, 12; Canadian Wind Energy Association (CANWEA) website, http://www.canwea.ca/farms/index_e.php (accessed August 20, 2013).

183 Industry representative, telephone interview by USITC staff, February 8, 2013.

184 Ibid.

185 Industry representative, telephone interview by USITC staff, January 30, 2013.

186 Bailey, "Building Boom Draws in Global Players," October 1, 2012.

187 Ibid.

188 *Xinhua.net*, "China's Largest Wind Power Developer," July 13, 2011; Bailey, "Building Boom Draws In Global Players," October 1, 2012.

189 From 2008 through 2012, installed capacity grew at a compound annual rate of 58 percent. USITC calculations using data from GWEC, *Global Wind 2012 Report*, April 2013, 30.

190 From 2008 through 2011, installed capacity grew at a compound annual rate of 73 percent. USITC calculations using data from GWEC, *Global Wind 2012 Report*, April 2013, 30.

191 At the end of 2012, only 50 GW of wind energy capacity was connected to the grid. Johnston, "China Reaches 50 GW," February 21, 2013.

192 Annual installations in 2012 totaled 12,960 MW, compared to 17,631 in 2011. USITC calculations using data from GWEC, *Global Wind Report 2012*, April 2013, 31.

193 Curtailment reportedly cost the Chinese wind energy industry $1.6 billion (10 billion yuan) in 2012, as wind energy expansion was reined in. This curtailment was largely due to insufficient grid infrastructure, but also in order to allow thermal and nuclear power to meet quotas. Qi, "China's Wind Sector Lost $1.6 Billion," January 25, 2013; Qi, "Debt Hampers Chinese Wind," May 2013, 29.

194 Although China's overall consumption of energy rose nearly 4 percent in 2012, it fell by 3.6 percent per unit of GDP. Associated Free Press, "China Energy Consumption Rises 3.9% in 2012," February 23, 2013; GWEC, *Global Wind Report 2012*, April 2013, 31; Qi, "Debt Hampers Chinese Wind," May 2013, 29; *ChinaDaily.com*, "China's GDP Growth Eases to 7.8%," January 18, 2013; industry representative, interview

by USITC staff, Beijing, May 16, 2013; industry representative, interview by USITC staff, Beijing, May 17, 2013.

[195] From 2006 through 2010, nearly 90 percent of Chinese wind power projects were invested, constructed and completed by state-owned enterprises. GWEC, *2012 China Wind Energy Outlook*, 2012, 41; industry representative, interview by USITC staff, Beijing, May 16, 2013.

[196] *Xinhuanet.com*, "SOEs dominate China's Wind Power Projects," April 10, 2013; Navigant Research, *World Market Update 2012*, March 2013, 28.

[197] Industry representative, interview by USITC staff, Beijing, May 16, 2013. Turbine manufacturers provide maintenance services during a short warranty period (two years is standard, but Chinese OEMs tend to throw in additional years for free). All services connected with design and installation of turbines are provided by the customer. Industry representative, interview by USITC staff, Beijing, May 16, 2013.

[198] Industry representative, interview by USITC staff, Beijing, May 17, 2013; industry representative, interview by USITC staff, Beijing, May 16, 2013; Li et al., "China Wind Energy Outlook," n.d., 25.

[199] Industry representative, interview by USITC staff, Beijing, May 16, 2013.

[200] Industry representative, interview by USITC staff, Beijing, May 17, 2013; industry representative, interview by USITC staff, Beijing, May 16, 2013; industry representative, telephone interview by USITC staff, March 12, 2013.

End Notes for Section 5

[1] Deonta Smith, "Hydroelectric Power in the U.S.," July 2012, 7–8; industry representatives, interviews with USITC staff, March 4, 12, and 14, 2013. The environmental and social impacts of a single large hydropower project are likely to be greater than those of individual smaller projects, but the cumulative impacts of many smaller hydroplants may equal or even exceed those of a single large plant generating the same output. In other instances, the overall environmental sustainability of a project involving a few large plants may actually receive more scrutiny and control than a larger number of separate, smaller projects. Consequently, the International Energy Agency (IEA) recommends assessing particular projects on their overall sustainability performance, rather than their size. IEA, *2012 Hydropower Technology Roadmap,* 2012, 31. In another recent study, researchers concluded that the cumulative impact of small dams on China's Nu River system has caused significant environmental damage to the local ecosystem. Oregon State University, "Dam Construction to Reduce Greenhouse Gases Causes Ecosystem Disruption," June 18, 2013.

[2] For example, small-scale hydro, as referenced in individual country laws, is defined as project generation capacity below 1.5 MW in Sweden, but below 50 MW in China. IEA, *2012 Hydropower Technology Roadmap,* 2012, 15.

[3] This cutoff follows the definition used by Bloomberg New Energy Finance (BNEF), the source of much of the renewable energy project data in this report.

[4] Although the production of electricity using hydropower is a clearly defined industry, there is no widely accepted list of services related to hydroelectricity generation. In an effort to provide such a list, the tables list the services offered by individual companies, as described in submissions to the Commission, company websites, and marketing brochures.

[5] According to one U.S. industry representative on China's Three Gorges Dam project, the budget for such social services was larger than the construction budget. Industry representative, telephone interview with USITC staff, March 14, 2013.

[6] GE Energy, "GE Energy Industrial Solutions: GE Serves Hydro," 2011, 4.

[7] Upgrades often involve adding larger turbines to an existing site, and may or may not involve expanding the dam that is part of a hydropower site. Since hydropower sites may be decades old, such upgrades are not uncommon and are seen as an efficient way to maximize the capacity of an existing site. In the United States, upgrades need to be relicensed by the U.S. Federal Energy Regulatory Commission (FERC)—a process that may or may not take less time than the original licensing of an entirely new project. IRENA, "Renewable Energy Technologies," June 2012, 21; industry representative, telephone interview with USITC staff, March 4, 2013.

[8] Industry representative, telephone interview with USITC staff, March 14, 2013.

[9] Ibid.

[10] Industry representative, interview with USITC staff, Tokyo, March 10, 2013.

[11] Voith has manufacturing operations in the United States, but it is not clear whether the goods and services in these deals will be supplied from Germany or elsewhere. HydroWorld.com, "Voith Signs Deals," January 28, 2013; industry representative, interview with USITC staff, São Paulo, Brazil, May 9, 2013.

[12] Pumped storage is a type of hydroelectric power generation that stores energy in the form of water in an upper reservoir; the water is pumped from a second reservoir at a lower elevation. During periods of high electricity demand, the stored water is released through turbines in the same way as a conventional hydro station. Excess energy refills the reservoir by pumping the water back to the upper reservoir, usually during nights and weekends when electricity demand is low. National Hydropower Association, "Hydro Works for America," n.d. (accessed June 26, 2013).

[13] These statistics reflect data as of 2011. Non-federal public owners include irrigation districts, cities, and water districts. The U.S. Federal Energy Regulatory Commission (FERC) regulates privately owned and non-federal publicly owned sites but does not regulate federally owned sites. FERC, "Present Development of Conventional Hydroelectric Projects," June 10, 2013.

[14] BNEF, *Sustainable Energy in America 2013 Factbook,* January 2013, 39.

[15] This number does not include federally owned hydropower plants, which accounts for the difference from the total of 2,400 hydropower plants cited above.

[16] FERC, "Present Development of Conventional Hydroelectric Projects," June 10, 2013.

[17] U.S. Energy Information Administration, *Electric Power Monthly*, April 22, 2013.

[18] To estimate the size of the market, the Commission relied on data for global hydropower capacity from the U.S. Department of Energy (DOE) Energy Information Agency (EIA), along with estimates of the share of development and O&M services in a hydropower project from Black & Veatch (B&V). According to B&V, using estimates based on an average 500 MW project located in the U.S. Midwest region, engineering, procurement, and construction management (EPC) costs are approximately 7 percent of total project costs, and owner's costs (primarily services such as feasibility studies, environmental impact studies, resource assessments, and permitting and other legal costs) are approximately 23 percent of total project costs. Hence about 30 percent of total project costs are likely to be services. B&V notes that these estimates can vary within a fairly wide range of +/– 35 percent. This estimate does not include the costs of materials and labor for actual plant and civil works construction, which vary widely for hydropower projects based on particular site characteristics, but it does include construction management and project management costs. For O&M services, B&V estimated average costs in 2010 for a 500 MW plant in the United States to be $15/kW-yr in fixed costs and $6/MWh in variable costs. These estimates include fixed and variable costs for maintenance and repairs, and supervision costs for existing plants. Industry representative, telephone interview with USITC staff, March 14, 2013; EIA, "Annual Energy Review 2011," 2012; B&V, "Cost and Performance Data for Power Generation Technologies," February 2012, 35.

[19] Fixed costs are set for the project, and include such items as maintenance, rent, and employee salaries. Examples of variable costs include electricity and other utilities necessary to run the project, and unscheduled maintenance costs not covered by a warranty or OEM contract. Industry representative, email communication with USITC staff, March 20, 2013.

[20] Industry representatives, interviews with USITC staff, March 4, 2013.

[21] Eligible projects include adding capacity at an existing hydropower facility; adding hydropower to an existing dam; conduit energy; and ocean, tidal, or wave projects. Industry representative, email message to USITC staff, June 12, 2013.

[22] Deonta Smith, "Hydroelectric Power in the U.S.," July 2012, 7–8.

[23] Kleinschmidt, "ITC/PTC—What's in It for Hydro and Wind Developers?" (accessed March 7, 2013).

[24] "The Investment Tax Credit (ITC) allows project developers to take a tax credit equal to 30 percent of the cost of constructing their project, and is in effect through the end of 2013 for wind, biomass, geothermal, hydropower, marine and tidal energy." BNEF, *Sustainable Energy in America 2013 Factbook,* January 2013, 21-2.

[25] Kleinschmidt, "ITC/PTC—What's in It for Hydro and Wind Developers?" (accessed March 7, 2013).

[26] BNEF, *Sustainable Energy in America 2013 Factbook,* January 2013, 22, 38; IER, "The Status of Renewable Electricity Mandates," n.d. (accessed March 29, 2013) ; Center for Climate and Energy Solutions, "Renewable and Alternative Energy Portfolio Standards," March 21, 2013.

[27] Exempted projects do not require renewed applications to FERC to continue operating, as opposed to licenses, which have terms of 30–50 years and must then undergo a relicensing process.

[28] Industry representatives, interviews with USITC staff, March 4 and 12, 2013; BNEF, *Sustainable Energy in America 2013 Factbook,* January 2013, 39.

[29] Industry representatives, interviews with USITC staff, March 4 and 12, 2013.

[30] The length of the warranty period varies according to the manufacturer.

[31] Company websites; industry representative, interview with USITC staff, Washington, DC, March 14, 2013.

[32] Industry representative, interview with USITC staff, Washington, DC, March 14, 2013.

[33] Deonta Smith, "Hydroelectric Power in the U.S.," July 2012, 12.

[34] The U.S. Department of Energy, Department of Interior, and Army Corps of Engineers signed a memorandum of understanding for hydropower in March 2010, with a focus on increasing generation from federal hydropower facilities and reducing environmental impacts. IEA, *2012 Hydropower Technology Roadmap,* 22.

[35] Voith Website, http://voith.com/en/markets-industries/industries/hydro-power/large-hydroplants/main-4354_m_special_ohio_river-4354.html (accessed June 14, 2013).

[36] IEA, *2012 Hydropower Technology Roadmap,* 2012, 9.

[37] REN21, *Renewables 2012,* 2012, 97.

[38] BNEF compiles data on small hydropower (capacity less than or equal to 50 MW) on a project-byproject basis. These calculations are based on BNEF estimates of 24.8 GW of small hydropower capacity and IEA estimates of 1,067 GW of total hydropower capacity, both for 2011. The actual small hydro share may be higher, as the BNEF database likely does not include all small hydropower projects in operation around the world. IEA, *2012 Hydropower Roadmap,* 2012, 18.

[39] BNEF database, accessed March 6, 2013. Data reported by BNEF for the United States show a significantly smaller number of projects than does data from FERC, which is likely to be more much more complete. However, FERC provides only U.S. data. We report the BNEF data here to allow cross-country comparisons of small hydro projects.

[40] Industry representative, interview with USITC staff, Tokyo, May 10, 2013; BNEF database, accessed March 6, 2013. Data for China reflect projects with reported commissioning dates, which were only about half of all reported projects.

[41] Even though B&V estimates apply to a hydropower plant located in the United States, this study uses the same estimates for the global market because detailed cost breakdowns are not available for all countries. The Commission excluded construction labor costs from its estimates because for most projects, construction workers are hired locally, at local labor rates, which vary dramatically by country. These services are unlikely to be performed by non-local companies, and it is highly unlikely that foreign construction labor comprises a significant export market for U.S. service providers. It is likely that EPC and owners' costs will also vary by country, but less so than labor costs. As noted earlier, for O&M services, B&V estimated average costs in 2010 for a 500 MW plant in the United States to be $15/kW-yr in fixed costs and $6/MWh in variable costs.

[42] This varies somewhat from the B&V estimates used above. The Commission uses the B&V to estimate the market size for services in this section because B&V provides a breakdown of costs between services and goods. IEA, *Renewable Energy: Markets and Prospects by Technology*, 2011, 28.

[43] Plant load factor represents the output produced by actual operation of a given site, relative to its maximum capacity. So, for a 20 MW capacity project, a plant with a 30 percent load factor would generate an average of 6 MW of electricity over time, while a plant with a 70 percent load factor would generate an average of 14 MW over time. The actual load factor for a given plant depends on specific site conditions and the equipment chosen for the site.

[44] Industry representative, telephone interview with USITC staff, February 6, 2013. According to another analysis, compared with large hydropower, the cost per installed kW tends to be higher because in most cases, small hydropower sites have low water heads (meaning that the water drops only a small distance); costs per kW decrease rapidly as head increases. At about 15 m, the rate of decrease levels out and, eventually, the cost stabilizes. ESHA, "*State of the Art of Small Hydropower*, 9 (accessed April 1, 2013).

[45] IEA, *Renewable Energy: Markets and Prospects by Technology,* 2011, 28.

[46] IEA, *2012 Hydropower Roadmap, 2012,* 18-24; REN21, *Renewables 2013 Global Status Report,* 2013, 93.

[47] Industry representative, telephone interview with USITC staff, Beijing, May 17, 2013.

[48] IEA, *2012 Hydropower Roadmap,* 2012, 18–20; IEA, *Renewable Energy: Markets and Prospects by Technology,* 2011, 28–30.

[49] IEA, *Renewable Energy: Markets and Prospects by Technology,* 2011, 28–30.

[50] However, U.S. exports of hydroelectricity were $83.3 million in 2012, primarily to Canada, and imports were $1.8 billion, again mostly from Canada. The U.S. and Canadian electricity transmission systems are closely interconnected along the border, and Canada generates significantly more hydropower than it can use. Smith, "Hydroelectric Power in the U.S.," July 2012, 4; industry representative, telephone interview with USITC staff, March 14, 2013.

[51] *HydroWorld.com*, "Brookfield Renewable Completes Acquisition," March 5, 2013; BNEF, *Sustainable Energy in America 2013 Factbook,* January 2013, 39–40.

[52] Industry representative, interview with USITC staff, Tokyo, May 10, 2013.

[53] REN21, *Renewables 2012,* 2012, 45.

[54] BNEF, *Sustainable Energy in America 2013 Factbook,* January 2013, 39.

[55] BNEF, "Global Trends In Clean Energy Investment," January 14, 2013, 25–26

[56] BNEF, "Global Trends in Renewable Energy Investment 2012, Chart Pack," n.d., 5.

[57] As noted, this refers to projects of 50MW or less, including both greenfield and refurbishment projects. BNEF, "New Investment in Clean Energy Fell 11%," January 14, 2013.

[58] World Bank website, n.d. http://water.worldbank.org/node/84059 (accessed March 19, 2013).

[59] Industry representative, interview with USITC staff, Tokyo, March 10, 2013; industry representative, interview with USITC staff, Beijing, May 17, 2013.

[60] *Environmental Business Journal,* "Multilateral Development Banks Push Ahead," 2012, 33.

[61] *BNEF Newswatch,* "Asian Development Bank Invests in Indian Renewable Power Company," May 2, 2013.

[62] Government representative, interview with USITC staff, Tokyo, May 13, 2013.

[63] Industry representative, telephone interview with USITC staff, May 14, 2013.

[64] Industry representative, telephone interview with USITC staff, February 6, 2013.

[65] For projects up to 15 MW, local content must be a minimum of 64.20 percent for goods and 86.06 percent for services, or 70.76 percent combined. This is gradually reduced as projects get larger, with projects rated at greater than 150 MW required to have 47.82 percent local content for goods and 46.98 percent for services, with a combined rate of 47.60 percent. PWC, *Power in Indonesia,* 2013, 16.

[66] BNEF, *Climatescope 2012,* 6.

[67] Ibid., 38, 60, 68, 76, 96.

[68] Brazil relies on large hydro facilities for 67 percent of its power. BNEF, *Climatescope 2012,* 43.

[69] Industry representative, interview with USITC staff, Santiago, Chile, May 15, 2013.

[70] REN21, *Renewables 2013,* 2013, 51.

[71] Industry representatives, interviews with USITC staff, São Paulo, May 9 and 10, 2013.

[72] Industry representatives, interviews with USITC staff, São Paulo, May 8 and 9, 2013.

[73] Industry representative, interview with USITC staff, São Paulo, May 9, 2013.

[74] BNDES likely considers the services component of hydropower construction costs to count toward the local-content requirements, as construction costs are a large majority of total hydropower project development costs. Industry representative, interview with USITC staff, Rio de Janeiro, May 7, 2013.

[75] Industry representatives, interviews with USITC staff, Rio de Janeiro, May 7, 2013, and Santiago, Chile, May 14, 2013; Lobo, "Global Renewable Energy Developments," January 16, 2013; IEA, *2012 Hydropower Roadmap,* 2012, 44.

[76] Industry representative, interview with USITC staff, Tokyo, March 10, 2013.

[77] REN21, *Renewables 2013,* 2013, 48.

[78] Industry representative, telephone interview with USITC staff, February 6, 2013.

[79] Reuters, "China Urges Hydropower Developers to Heed Environment," January 17, 2012.

[80] REN21, *Renewables 2013,* 2013, 47.

[81] Industry representative, telephone interview with USITC staff, Beijing, May 17, 2013; Bloomberg News, "Alstom Building Center in China to Expand Hydro Business," July 11, 2012.

[82] Government representative, interview with USITC staff, Tokyo, May 13, 2013.

[83] Small hydro was defined as having a project capacity of less than 30 MW for the FIT because that was the dividing line for a project's ability to cover its investment costs. Industry representative, interview with USITC staff, Tokyo, March 10, 2013.

[84] Industry representative, interview with USITC staff, Tokyo, March 10, 2013; REN21, *Renewables: Global Futures Report 2013,* 2013, 46.

[85] IRENA, "Renewable Energy Technologies: Hydropower," June 2012, 5.

[86] REN21, *Renewables 2013,* 2013, 50.

[87] IEA, *2012 Hydropower Technology Roadmap,* 2012, 23.

[88] Under the new system, the following tariffs apply: for plants rated at 1–20 kW, the tariff is €57/MWh; for plants rated at 20–500 kW, it is €219/MWh; and for plants rated at 500–1,000 kW, it is €115/MWh. *HydroWorld.com*, "The Road to Sustainable Development," February 1, 2013.

[89] Roca, "Spain Limits Payments to Renewables," February 1, 2013; BNEF database, http://www.bnef.com (accessed February 12, 2013).

[90] REN21, *Renewables 2012,* 2012.
[91] EIA, "Geothermal Explained."
[92] Lowder et al., *Renewable Energy Finance Tracking Initiative,* 2012.
[93] Industry observer, interview by USITC staff, Washington, DC, December 19, 2012.
[94] USITC, *Renewable Energy Services*, 2005.
[95] Industry representative, interview by USITC staff, Jakarta, July 3, 2013.
[96] EIA, *Annual Energy Review 2011,* 2012, 224 and 258.
[97] IBISWorld, "Geothermal Electricity Plant Operation in the U.S.," 2012.
[98] EIA, *Annual Energy Review 2011,* 2012.
[99] EIA, *Annual Energy Review 2011,* 2012, 224. Note that the 2005 USITC Renewable Energy Services study reported that geothermal energy generated 40 percent more electricity than wind technology, suggesting that the wind power sector has grown significantly since then.
[100] EIA, *Annual Energy Review 2011,* 2012, 224. For more information on the solar and wind energy services sectors, see sections 3 and 4 of this report.
[101] IEA, *Renewable Energy Essentials,* 2010.
[102] According to an industry observer, O&M services in the global geothermal power industry are usually provided by utility companies or independent power producers (IPPs). Few independent service providers are active in this service area. Industry observer, interview by USITC staff, Washington, DC, December 19, 2012.
[103] The Commission estimated the size of the market for project development-related geothermal energy services based on a constant share of total development costs for new projects, multiplied by the total of new capacity developed in the most recent data year (2010). In order to estimate the size of the market, the Commission relied on data for global geothermal capacity from the EIA, and estimates of the share of services in a geothermal project, for both development and O&M services, from B&V. According to B&V estimates, EPC costs are approximately 8 percent of total project costs, and owners' costs (primarily services such as feasibility studies, environmental impact studies, resource assessments, and permitting and other legal costs) are approximately 17 percent of total project costs, so about 25 percent of total project costs are likely to be services. This estimate does not include the costs of materials and labor for actual plant and civil works construction, which vary significantly, based on geothermal resources and particular site characteristics, but does include construction management and project management costs. For O&M services, B&V estimated average costs in 2010 to be $31/MWh in variable costs, with zero fixed O&M costs. Industry representative, telephone interview with Commission staff, March 14, 2013; EIA, International Energy Statistics, http://www.eia.gov/cfapps/ipdbproject/IEDIndex3.cfm?tid=2&pid=2&aid=7 (accessed March 20, 2013); calculations by the Commission. Information on share of services is from B&V, "Cost Estimates and Performance Data," 2012, 35.
[104] EIA, "U.S. Has Large Geothermal Resources," November 18, 2011.
[105] Geothermal Technologies Office, "American Recovery and Reinvestment Act."
[106] EIA, *Annual Energy Outlook*, 2012, 11.
[107] EIA, "U.S. Has Large Geothermal Resources," November 18, 2011.
[108] Ibid.
[109] GEA, *Annual U.S. Geothermal Power Production,* April 2012.
[110] Ibid.
[111] Ibid.
[112] When accounting for unconfirmed projects, the range of PCA in development is about 1961– 2023 MW. Of this, 949–956 MW are advanced-stage (phase 3–4) geothermal projects. GEA, *Annual U.S. Geothermal Power Production*, April 2012, 9–10; IBISWorld, "Geothermal Electricity Plant Operation in the U.S.," 2012.
[113] EIA, "Existing Nameplate and Net Summer Capacity," October 2012.
[114] EIA, "Table 1.19.B. Net Generation from Geothermal by State," December 2012.
[115] IBISWorld, "Geothermal Electricity Plant Operation in the U.S.," 2012.
[116] BNEF database (accessed February 10, 2013).
[117] REN21, *Renewables 2012*, 2012, 40.
[118] Ibid., 40–41. The Ring of Fire, a seismically active belt that circles the Pacific Ocean, includes Bolivia, Canada, Chile, Indonesia, Japan, Mexico, New Zealand, the Philippines, Russia, the United States, and other countries.
[119] Ibid., 40–41.
[120] Recent trends in geothermal energy investment (discussed below) have underscored the importance of emerging markets and developing countries when it comes to growth in the geothermal energy market. Industry

observer, interview by USITC staff, Washington, DC, December 19, 2012; GEA, *Geothermal: International Market Overview Report*, May 2012.

[121] Taylor, "East Africa in Global Geothermal," November 23, 2012.

[122] Even though B&V estimates apply to a geothermal power plant located in the United States, the Commission uses the same estimates for the global market because detailed cost breakdowns are not available for all countries. The estimates exclude construction labor costs because for most projects, construction workers are hired locally, at local labor rates, which vary dramatically by country. These services are unlikely to be performed by non-local companies, and it is highly unlikely that foreign construction labor comprises a significant export market for U.S. service providers. EPC and owners' costs are also expected to vary by country, but less so than labor costs. As noted earlier, for O&M services, B&V estimates average variable O&M costs of $31/MWh for a standard hydrothermal power plant in the United States in 2010, with zero fixed O&M costs.

[123] REN21, *Renewables 2012,* 2012.

[124] Government representatives, interviews with USITC staff, Jakarta, July 3 and 4, 2013.

[125] Industry representatives, interviews with USITC staff, Jakarta, July 3 and 4, 2013.

[126] REN21, *Renewables 2012,* 2012; industry representatives, interviews with USITC staff, Jakarta, July 3 and 4, 2013.

[127] World Bank, ESMAP, *Geothermal Handbook,* June 2012, 25.

[128] Chevron website, updated April 2013, http://www.chevron.com/deliveringenergy/geothermal/.

[129] Industry representative, interview by USITC staff, Jakarta, July 4, 2013.

[130] KPMG, *World Geothermal Market and* Outlook, n.d., 10–11.

[131] World Bank, ESMAP, *Geothermal Handbook,* June 2012, 26; industry representative, interview with USITC staff, Jakarta, July 3, 2013.

[132] USITC staff analysis of BNEF data.

[133] Industry observer, interview by USITC staff, Washington, DC, December 19, 2012.

[134] FS and UNEP, *Global Trends in Renewable Energy Investment 2012,* 2012, 15.

[135] Ibid., 21.

[136] BNEF, "East Africa in Global Geothermal," 2012.

[137] World Bank, "Full Steam Ahead," March 6, 2013. This figure includes direct lending and loan guarantees from the International Finance Corporation, the International Bank for Reconstruction and Development, the International Development Association, and the Multilateral Investment Guarantee Agency. "Renewable Energy," World Bank website, n.d.

[138] World Bank, "Full Steam Ahead," March 6, 2013.

[139] For projects up to 5 MW, local content must be a minimum of 31.30 percent for goods and 89.18 percent for services, or 42.00 percent combined. This is gradually reduced for larger projects with projects rated at greater than 110 MW required to have 16.00 percent local content for goods and 58.40 percent for services, with a combined rate of 28.95 percent. PWC, *Power in Indonesia,* 2013, 17.

[140] Industry representative, interview with USITC staff, December 19, 2012.

[141] Electricity Law No. 30/2009. PWC, *Power in Indonesia,* 2013, 9, 13.

[142] Industry and government representatives, interviews with USITC staff, Jakarta, July 3 and 4, 2013.

[143] World Bank, ESMAP, *Geothermal Handbook,* June 2012, 30.

[144] World Bank, ESMAP, *Geothermal Handbook,* June 2012, 30.

[145] World Bank, "Full Steam Ahead," March 6, 2013.

[146] Reuters, "Rwanda to Spend $935 Mln on Geothermal Power," March 22, 2011; *New Times,* "Geothermal Drilling Starts Next Month," March 27, 2011.

[147] Cichon, "Rwanda Set to Start Major Geothermal Development," March 27, 2013; *New Times,* "Geothermal Drilling Starts Next Month," March 27, 2011.

[148] World Bank, ESMAP, *Geothermal Handbook,* June 2012, 30.

[149] Industry representative, interview by USITC staff, Tokyo, May 13, 2013; KPMG, *World Geothermal Market and* Outlook, n.d., 10–11.

[150] Industry representatives, interviews by USITC staff, Tokyo, May 13, 2013, and May 14, 2013; Crowe, "Japan's Geothermal Resources Get a Closer Look," May 11, 2011; Watanabe, "Japan Considering," April 9, 2013.

[151] Industry representatives, interviews by USITC staff, Tokyo, May 13, 2013, and May 14, 2013.

[152] Ibid.

[153] BNEF database (accessed June 3, 2013).

End Notes for Section 6

[1] A coal generator without scrubbing systems has estimated lifecycle emissions of 1,050 gCO2e per kWh (the equivalent of grams of carbon dioxide per kilowatt-hour), while an offshore wind generator has estimated lifecycle emissions of 9 gCO2e per kWh. Sovacool, "Valuing the Greenhouse Gas Emissions from Nuclear Power," 2008, 2950.

[2] A 2013 estimate by the U.S. government puts the social cost of carbon at about $36 per ton of CO2. U.S. Government, Interagency Working Group on Social Cost of Carbon, *Technical Update*, May 2013. Other estimates, relying on different methodologies and discount rates, put such costs as high as $266 per ton of CO2. Johnson and Hope, "The Social Cost of Carbon in U.S. Regulatory Impact Analysis," September 2012. For more information on climate change, see IPCC, "Climate Change 2007," 2007.

[3] "President Barack Obama, May 6, 2011, Remarks at Allison Transmission Headquarters, Indianapolis, Indiana. Cited in Morris et al., "Clean Energy," June 4, 2012, 1.

[4] Industry representatives, interview by USITC staff, San Francisco, California, November 1, 2012.

[5] Price and quantity instruments should theoretically have the same effect on social welfare. However, quantity instruments may be more efficient when the "benefit function" is steeply sloped (curved) and the "cost function" is flat (linear), meaning that it is better for a policy to cost more than expected than to have fewer benefits than expected; by contrast, price instruments may be more efficient when benefits are linear and costs are curved. In wind energy, costs are mostly linear, suggesting in theory that an RPS, as a quantity instrument, may be more efficient. Weitzman, "Prices vs. Quantities," October 1974; Dong, "Feed-in Tariff vs. Renewable Portfolio Standard," 2012, 477.

[6] North Carolina Solar Center DSIRE database (accessed November 30, 2012).

[7] Alagappan et al., "What Drives Renewable Energy Development?" 2011.

[8] Dong, "Feed-in Tariff vs. Renewable Portfolio Standard," 2012.

[9] Industry representatives, interview by USITC staff, Bonn, Germany, May 14, 2013.

[10] Jenkins and Mansur, "Bridging the Clean Energy Valleys of Death," November 2011.

[11] Bird et al., "Policies and Market Factors Driving Wind Power," 2005. California's utilities were paying extremely high prices for electricity on spot markets, putting them under financial pressure and eventually driving the state's largest utility (the Pacific Gas and Electric Company) into bankruptcy. As a result, new electricity generators found it difficult to secure power purchase agreements and obtain financing. California Energy Commission, "Renewable Energy Program," December 2002, 15.

[12] IEA, "Renewable Energy: Policy Considerations," November 2011, 43.

[13] Cardwell, "Tax Credit in Doubt," September 20, 2012.

[14] Industry representatives, interview by USITC staff, Madrid, Spain, May 8, 2013.

[15] Industry representatives, interview by USITC staff, San Diego, California, October 31, 2012.

[16] Industry representatives, interview by USITC staff, Hamburg, Germany, May 15, 2013.

[17] IEA, "Renewable Energy: Policy Considerations," November 2011, 27.

[18] Widiantoro et al., "Porter's Diamond," November 6, 2011.

[19] Marques et al., "Motivations Driving Renewable Energy in European Countries," 2010.

[20] Carley, "State Renewable Energy Electricity Policies," 2009.

[21] Industry representatives, interview by USITC staff, Bonn, Germany, May 14, 2013.

[22] Industry representatives, interview by USITC staff, Hamburg, Germany, May 15, 2013.

[23] Couture and Gagnon, "An Analysis of Feed-in Tariff Remuneration Models," 2010.

[24] Cory et al., "Feed-in Tariff Policy," March 2009, 4.

[25] For example, Minnesota offers higher tariffs for the first 10 years of contract terms, while in Slovenia premiums drop by 5 percent after the first 5 years of the project's life and 10 percent after 10 years. Couture and Gagnon, "An Analysis of Feed-in Tariff Remuneration Models," 2010.

[26] Ragwitz et al., "Recent Developments of Feed-in Systems in the EU," January 2012, 11. Under net metering schemes, producers can reduce their electricity bills by providing electricity to the grid. The value of that electricity (and therefore the compensation provided to producers) varies depending on overall supply and demand.

[27] Couture and Gagnon, "An Analysis of Feed-in Tariff Remuneration Models," 2010.

[28] Ibid.

[29] Ragwitz et al., "Recent Developments of Feed-in Systems in the EU," January 2012, 4.

[30] About $45 to $125. Ragwitz et al., "Recent Developments of Feed-in Systems in the EU," January 2012, 15.

[31] Couture and Gagnon, "An Analysis of Feed-in Tariff Remuneration Models," 2010.

[32] Industry representatives, interview by USITC staff, Madrid, Spain, May 10, 2013.
[33] Industry representatives, interview by USITC staff, San Francisco, California, November 1, 2012.
[34] Stromsta, "German Politicians Stand Firm," April 23, 2010.
[35] CleanTechnica, "Italy to Levy Carbon Tax," April 18, 2012.
[36] Wiser et al., "Renewables Portfolio Standards," April 2007, 2.
[37] Ibid., 6.
[38] Industry representatives, interview by USITC staff, San Francisco, California, November 1, 2012.
[39] Some proposed federal RPS programs would facilitate interstate certificate trading, which would further increase the size and liquidity of the market. Where cross-border trade is allowed and certificates are trustworthy, regions that do not have RPSs still have incentives to develop renewable energy in order to get certificates that can be sold to their neighbors. University of Texas at Austin, *Harmonization of Renewable Energy Credit Markets*, November 2009.
[40] Wiser et al., "Renewables Portfolio Standards," April 2007, 11.
[41] EU directive 2001/77/EC.
[42] Carley, "State Renewable Energy Electricity Policies," 2009.
[43] KPMG International, "Taxes and Incentives for Renewable Energy," June 2012, 17.
[44] Ibid., 36.
[45] Ibid., 37.
[46] Ernst & Young, "Renewable Energy Country Attractiveness Indices," May 2012, 27.
[47] There is also a deduction for energy-efficient commercial buildings (IRC section 179D), an energy-efficient appliance credit (IRC section 45MA), and an energy-efficient home credit (IRC section 45LA), as well as a nonbusiness energy property tax credit (IRC section 25C) and a residential energy-efficient property tax credit (IRC section 25D) for improvements that affect energy efficiency of dwellings. Bourgeois et al., "Tax Incentives of Going Green," November 2010.
[48] Sherlock, "Impact of Tax Policies," April 19, 2012, 2. Typically such credits apply to the eligible basis of energy property placed in service during the taxable years.
[49] Sherlock, "Impact of Tax Policies," April 19, 2012, 3.
[50] Cosmo, "States Provide Tax Incentives," September–October 2011.
[51] Environmental Law Institute, "Estimating U.S. Government Subsidies to Energy Sources," September 2009. Significant federal tax preferences for fossil fuel industries include the expensing of exploration and development costs, as well as the "excess of percentage over cost depletion" deduction, which allows oil, gas, and coal producers to deduct a percentage of gross income from production instead of recovering the cost of investments based on the fraction of resources extracted. This deduction is typically more favorable than the standard depreciation deduction.
[52] Mendelsohn and Harper, "§1603 Treasury Grant Expiration," June 2012, iii.
[53] GAO, *DOE Loan Guarantees,* March 2012.
[54] Industry representatives, interview by USITC staff, Madrid, Spain, May 10, 2013. The loan guarantee will result in an estimated $300 million in interest payments to the U.S. government. Trabish, "SolarReserve's CEO Weighs In on CSP," August 27, 2012.
[55] Gage, "The Venture Capital Secret," September 19, 2012.
[56] Kho, "What the Solyndra Bankruptcy Means," September 23, 2011.
[57] Industry representatives, interview by USITC staff, Bonn, Germany, May 14, 2013.
[58] Ibid.
[59] Ibid.
[60] Oil rents are the difference between the value of crude oil production at world prices and the total costs of production.

End Notes for Appendix C

[1] SEMI, written submission to the USITC,
[2] Verdant, written submission to the USITC, March 1, 2013.

In: Renewable Energy
Editor: Ebony L. Dunbar

ISBN: 978-1-62948-941-4

Chapter 2

RENEWABLE ENERGY AND ENERGY EFFICIENCY INCENTIVES: A SUMMARY OF FEDERAL PROGRAMS*

Lynn J. Cunningham and Beth A. Roberts

SUMMARY

Energy is crucial to the operation of a modern industrial and services economy. Recently, there have been growing concerns about the availability and cost of energy and about environmental impacts of fossil energy use. Those concerns have rekindled interest in energy efficiency, energy conservation, and the development and commercialization of renewable energy technologies.

Many of the existing energy efficiency and renewable energy programs have authorizations tracing back to the 1970s. Many of the programs have been reauthorized and redesigned repeatedly to meet changing economic factors. The programs apply broadly to sectors ranging from industry to academia, and from state and local governments to rural communities.

Since 2005, Congress has enacted several major energy laws: the Energy Policy Act of 2005 (EPACT 2005; P.L. 109-58); the Energy Independence and Security Act of 2007 (EISA; P.L. 110- 140); the Energy Improvement and Extension Act (EIEA), enacted as Division B of the Emergency Economic Stabilization Act (EESA; P.L. 110-343); and the American Reinvestment and Recovery Act (ARRA; P.L. 111-5). Each of those laws established, expanded, or modified energy efficiency and renewable energy research, development, demonstration, and deployment (RDD&D) programs. The Department of Energy (DOE) operates the greatest number of efficiency and renewable energy incentive programs. The Department of the Treasury and the Department of Agriculture (USDA) operate several programs. A few programs can also be found among the Departments of Interior (DOI), Labor (DOL), Housing and Urban Development (HUD), Veterans Affairs (VA), and the Small Business Administration (SBA).

---

* This is an edited, reformatted and augmented version of Congressional Research Service, Publication No. R40913, dated October 18, 2013.

This report describes federal programs that provide grants, loans, loan guarantees, and other direct or indirect incentives for energy efficiency, energy conservation, and renewable energy. For each program, the report provides the administering agency, authorizing statute(s), annual funding, and the program expiration date. The appendixes provide summary information in a tabular format and also list recently expired programs.

INTRODUCTION

The United States has an abundance of natural resources. For much of the nation's history, energy was not a concern as commerce and industry needs could be met by domestic supplies. However, industrialization and population growth, and the continuing development of a consumer-oriented society, soon led to the necessity of obtaining foreign sources of energy to supplement the demands of a growing economy.

Recognition of the implications of dependence on foreign sources of energy, coupled with concerns over the volatility of prices driven by fluctuations in supply spurred by world events, have led to efforts to increase U.S. energy independence and reduce domestic consumption. The result has been the emergence of a number of programs focused on energy efficiency and conservation of domestic resources and on research programs that target the development of renewable sources of energy. Many of these programs have roots going back almost 40 years and have been redesigned many times over that period.

Many of the current programs have been reauthorized and redesigned periodically to meet changing economic conditions and national interests. The programs apply broadly to sectors ranging from industry to academia, and from state and local governments to rural communities. Each program has been designed to meet current needs as well as future anticipated challenges.

Since 2005, Congress has enacted several major energy laws: the Energy Policy Act of 2005 (EPACT 2005; P.L. 109-58); the Energy Independence and Security Act of 2007 (EISA; P.L. 110- 140); the Energy Improvement and Extension Act (EIEA), enacted as Division B of the Emergency Economic Stabilization Act (EESA; P.L. 110-343); and the American Reinvestment and Recovery Act (ARRA; P.L. 111-5). Each of those laws established, expanded, or modified energy efficiency and renewable energy research, development, demonstration, and deployment (RDD&D) programs. The Department of Energy (DOE) operates the greatest number of efficiency and renewable energy incentive programs. The Department of the Treasury and the Department of Agriculture (USDA) operate several programs. A few programs can also be found among the Departments of Interior (DOI), Labor (DOL), Housing and Urban Development (HUD), Veterans Affairs (VA), and the Small Business Administration (SBA).

This report outlines current federal programs and provisions providing grants, loans, loan guarantees, and other direct or indirect incentives for energy efficiency, energy conservation, and renewable energy RDD&D. The programs are grouped by administering agency with information links to applicable federal agency websites. Incentives are summarized and indexed in the appendixes.

Most program descriptions were compiled from authorizing statutes, the U.S. Code, and Administration budget request documents. Other program descriptions and some funding information were compiled from The Database of State Incentives for Renewables and

Efficiency (DSIRE), the *Catalog of Federal Domestic Assistance* (CFDA) and the Energy Star website. Most budgetary figures were compiled from executive agency budget justifications and congressional committee reports.

FY2013 Appropriations Uncertainty

The final amount of FY2013 Energy and Water Development appropriations for DOE energy technologies has not yet been established.[1] These appropriations were considered in the context of the Budget Control Act of 2011 (BCA; P.L. 112-25), which established discretionary spending limits for FY2012-FY2021.

On September 28, 2012, President Obama signed into law the Continuing Appropriations Resolution, 2013 (P.L. 112-175). For Energy and Water Development programs, the act continued appropriations through March 27, 2013, at 0.612% above the FY2012-enacted levels. Pursuant to the BCA, as amended,[2] President Obama ordered that the joint committee sequester be implemented on March 1, 2013.

On March 26, 2013, the President signed the FY2013 Defense and Military Construction/VA, Full Year Continuing Resolution (P.L. 113-6). The act funded Energy and Water Development accounts for DOE energy technologies at the FY2012 enacted level for the remainder of FY2013, subject to the BCA sequestration requirements.

The sequester will ultimately be applied at the program, project, and activity (PPA) level within each account. Because the sequester was implemented at the time that a temporary continuing resolution was in force, the reductions were calculated on an annualized basis and will be apportioned throughout the remainder of the fiscal year. Although full-year FY2013 funding has been enacted, the effect of these reductions on the funding amounts that will ultimately be available for energy technology programs at either the account or PPA level remain unclear until further guidance is provided by the Office of Management and Budget (OMB) on how agencies should apply these reductions.

In addition, the House-passed FY2014 Energy and Water Development appropriations bill (H.R. 2609, Title V) proposes to rescind $157 million of unobligated prior-year balances from within the DOE Energy Efficiency and Renewable Energy account.[3]

I. Department of Energy Office of Energy Efficiency and Renewable Energy

Renewable Energy

Biomass

1. Bioenergy Technologies Program (formerly the Biomass and Biorefinery Systems R&D Program)

- Administered by: Office of Energy Efficiency and Renewable Energy (EERE)
- Authorization: Federal Nonnuclear Energy Research and Development Act of 1974 (P.L. 93-577); Energy Policy and Conservation Act of 1975 (EPCA; P.L. 94-163); Energy Conservation and Production Act of 1976 (ECPA; P.L. 94-385); Department

of Energy Organization Act of 1977 (P.L. 95-91); Energy Tax Act (P.L. 95-618); National Energy Conservation Policy Act of 1978 (NECPA; P.L. 95-619); Powerplants and Industrial Fuel Use Act of 1978 (P.L. 95-620); Energy Security Act of 1980 (P.L. 96-294); National Appliance Energy Conservation Act of 1987 (P.L. 100-12); Federal Energy Management Improvement Act of 1988 (P.L. 100-615); Renewable Energy and Energy Efficiency Technology Competitiveness Act of 1989 (P.L. 101-218); Clean Air Act Amendments of 1990 (P.L. 101-549); Solar, Wind, Waste, and Geothermal Power Production Incentives Act of 1990 (P.L. 101-575); Energy Policy Act of 1992 (EPACT; P.L. 102-486); Biomass Research and Development Act of 2000 (Title III of Agricultural Risk Protection Act of 2000; P.L. 106-224); Farm Security and Rural Investment Act of 2002 (P.L. 107-171); Healthy Forest Restoration Act of 2003 (P.L. 108-148); Energy Policy Act of 2005 (EPACT 2005; P.L. 109-58); Energy Independence and Security Act of 2007 (EISA; P.L. 110-140); The Food, Conservation, and Energy Act of 2008 (P.L. 110-234); American Recovery and Reinvestment Act of 2009 (ARRA; P.L. 111-5)
- Annual funding: $89.8 million for FY2006; $196.3 million for FY2007; $195.6 million for FY2008; $214 million for FY2009; an additional $777 million in FY2009 from ARRA; $220 million for FY2010; $180 million for FY2011; $195 million for FY2012; $200.5 million for FY2013; and $282 million requested for FY2014
- Scheduled termination: None
- Description: This program works with industrial partners, national laboratories, universities, and other stakeholders to develop the technologies and systems needed to cost-effectively transform the nation's renewable and abundant domestic biomass resources into clean, affordable, and sustainable biofuels, bioproducts, and biopower. In recent years, the program has been primarily geared toward development and deployment of ethanol from non-food feedstocks, but is now expanding its scope to additional alternative fuels, such as bio-butanol, green gasoline, jet fuel, and diesel.
- Qualified applicant: Colleges and universities; profit organizations
- Qualified technologies: Biomass
- For more information: DOE's Bioenergy Technologies Program overview; DOE's Bioenergy Technologies Office – Financial Opportunities online resource; and program number 81.087 at the *Catalog of Federal Domestic Assistance* (CFDA) website

2. Regional Biomass Energy Grant Programs

- Administered by: Bioenergy Technologies Office, EERE
- Authorization: Department of Energy Organization Act of 1977 (P.L. 95-91); Energy and Water Development Appropriations Act for FY1987 (P.L. 99-500)
- Annual funding: $395,000 for FY2007; an estimated $75,131 for FY2008; an estimated $25,705 for FY2009; an estimated $4.8 million for FY2010; $0 for FY2011; $0 for FY2012; $0 for FY2013; data for FY2014 is currently unavailable
- Scheduled termination: None
- Description: This program provides assistance to increase America's use of fuels, chemicals, materials, and power made from domestic biomass on a sustainable basis. Assistance may be used to develop and transfer any of several biomass energy technologies to the scientific and industrial communities. For regional programs,

such technologies will be appropriate for the needs and resources of particular regions of the United States.
- Qualified applicants: State and local governments; colleges and universities; profit organizations; nonprofit organizations
- Qualified technologies: Biomass
- For more information: See program number 81.079 at the CFDA website

Geothermal

3. Geothermal Technologies Program (GTP)

- Administered by: EERE
- Authorization: Geothermal Energy Research, Development, and Demonstration Act (P.L. 93-410); Department of Energy Organization Act (P.L. 95-91); Energy Tax Act of 1978 (P.L. 95-618); Energy Security Act of 1980 (P.L. 96-294); Renewable Energy and Energy Efficiency Technology Competitiveness Act of 1989 (P.L. 101-218); Solar, Wind, Waste, and Geothermal Power Production Incentives Act of 1990 (P.L. 101-575); Energy Policy Act of 1992 (EPACT; P.L. 102-486); Energy Policy Act of 2005 (EPACT 2005; P.L. 109-58); Energy Independence and Security Act of 2007 (EISA; P.L. 110-140); American Recovery and Reinvestment Act of 2009 (ARRA; P.L. 111-5)
- Annual funding: $68.2 million for FY2006; $5 million for FY2007; $19.3 million for FY2008; $43.3 million for FY2009; an additional $393 million appropriated in FY2009 from ARRA; $44 million for FY2010; $37 million for FY2011; $37 million for FY2012; $38.1 million for FY2013; and $60 million requested for FY2014
- Scheduled termination: None
- Description: This program partners the federal government with industry, academia, and research facilities to further the development of geothermal energy technologies. Competitive solicitations issued as Funding Opportunity Announcements (FOAs) are the principal mechanism used by the GTP to contract for cost-shared research, development, and demonstration projects.
- Qualified applicants: Profit organizations; colleges and universities
- Qualified technology: Geothermal
- For more information: See EERE's Geothermal Technologies Program website; and program number 81.087 at the CFDA website

Hydrogen and Fuel Cells

4. Hydrogen & Fuel Cell Technologies Program

- Administered by: EERE
- Authorization: Federal Energy Administration Act of 1974 (P.L. 93-275); Federal Nonnuclear Energy Research and Development Act of 1974 (P.L. 93-577); Energy Policy and Conservation Act of 1975 (EPCA; P.L. 94-163); Electric and Hybrid Vehicle Research, Development and Demonstration Act (P.L. 94-413); Department of Energy Organization Act of 1977 (P.L. 95-91); Automotive Propulsion Research and Development Act of 1978 (Title III of Department of Energy Act of 1978-Civilian Applications; P.L. 95-238); Methane Transportation Research, Development

and Demonstration Act of 1980 (P.L. 96-512); Energy Security Act of 1980 (P.L. 96-294); Alternative Motor Fuels Act of 1988 (P.L. 100-494); Spark M. Matsunaga Hydrogen Research, Development, and Demonstration Act of 1990 (P.L. 101-566); Energy Policy Act of 1992 (EPACT; P.L. 102-486); Hydrogen Future Act of 1996 (P.L. 104-271); Energy Policy Act of 2005 (EPACT 2005; P.L. 109-58); Energy Independence and Security Act of 2007 (EISA; P.L. 110-140); American Recovery and Reinvestment Act of 2009 (ARRA; P.L. 111-5)

- Annual funding: $153.4 million for FY2006; $190 million for FY2007; $206.2 million for FY2008; $164.6 million for FY2009; an additional $43 million appropriated in FY2009 from ARRA; $174 million for FY2010; $95.8 million for FY2011; $101.3 million for FY2012; $104.3 million for FY2013; and $100 million requested for FY2014
- Scheduled termination: None
- Description: This program partners with industry, academia, and national laboratories and works in close coordination with Vehicle Technologies and other programs at DOE to: overcome technical barriers through R&D of hydrogen production, delivery, and storage technologies; overcome technical barriers to fuel cell technologies for transportation, distributed stationary power, and portable power applications; address safety issues and facilitate the development of model codes and standards; validate and demonstrate hydrogen and fuel cells in real-world conditions; and educate key stakeholders whose acceptance of these technologies will determine their success in the marketplace.
- Qualified applicants: Federal government; national laboratories; colleges and universities; and profit organizations
- Qualified technologies: Hydrogen and fuel cells
- For more information: See EERE's Hydrogen and Fuel Cell Technologies website; and program number 81.087 at the CFDA website

Solar

5. Solar Energy Technologies Program (SETP)

- Administered by: EERE
- Authorization: Energy Policy and Conservation Act of 1975 (EPCA; P.L. 94- 163); Energy Conservation and Production Act of 1976 (ECPA; P.L. 94-385); Department of Energy Organization Act of 1977 (P.L. 95-91); Solar Photovoltaic Energy Research, Development and Demonstration Act of 1984 (P.L. 95-590); National Energy Conservation Policy Act of 1978 (NECPA; P.L. 95-619); Energy Security Act of 1980 (P.L. 96-294); Renewable Energy and Energy Efficiency Technology Competitiveness Act of 1989 (P.L. 101-218); Solar, Wind, Waste, and Geothermal Power Production Incentives Act of 1990 (P.L. 101-575); Solar, Wind, Waste, and Geothermal Power Production Incentives Technical Amendments Act of 1991 (P.L. 102-46); Energy Policy Act of 1992 (EPACT; P.L. 102-486); Energy Policy Act of 2005 (EPACT 2005; P.L. 109-58); Energy Independence and Security Act of 2007 (EISA; P.L. 110-140); American Recovery and Reinvestment Act of 2009 (ARRA; P.L. 111-5)

- Annual funding: $81.8 million for FY2006; $157 million for FY2007; $166.3 million for FY2008; $172.4 million for FY2009; an additional $116 million appropriated in FY2009 from ARRA; $247 million for FY2010; $259.6 million for FY2011; $284.7 million for FY2012; $290.7 million for FY2013; and $356 million requested for FY2014
- Scheduled termination: None
- Description: SETP partners with industry, national laboratories, and universities to develop and bring reliable and affordable solar energy technologies to the marketplace. This program finances R&D in four major subprograms: Photovoltaics (PV); Concentrating Solar Power (CSP); Systems Integration for Solar Technologies; and Market Transformation for Solar Technologies.
- Qualified applicants: Industry; national laboratories; colleges and universities
- Qualified technology: Solar
- For more information: See EERE's Solar Energy Technologies Program website; and program number 81.087 at the CFDA website

Water Power

6. Water Power Program (formerly Wind and Hydropower Technologies Program)

- Administered by: EERE
- Authorization: Energy Policy and Conservation Act of 1975 (EPCA; P.L. 94- 163); Renewable Energy and Energy Efficiency Technology Competitiveness Act of 1989 (P.L. 101-218); Solar, Wind, Waste, and Geothermal Power Production Incentives Act of 1990 (P.L. 101-575); Energy Policy Act of 1992 (EPACT; P.L. 102-486); Energy Policy Act of 2005 (EPACT 2005; P.L. 109-58); Energy Independence and Security Act of 2007 (EISA; P.L. 110-140); American Recovery and Reinvestment Act of 2009 (ARRA; P.L. 111-5)
- Annual funding: $495,000 for FY2006; $0 for FY2007; $9.7 million for FY2008; $39.1 million for FY2009; an additional $31.7 million appropriated in FY2009 from ARRA; $50 million for FY2010; $29.2 million for FY2011; $58.1 million for FY2012; $59.1 million for FY2013; and $55 million requested for FY2014
- Scheduled termination: None
- Description: This program partners with the national laboratories, industry, universities, and other federal agencies to promote the development and deployment of technologies capable of generating environmentally sustainable and cost-effective electricity from the nation's water resources (both conventional and marine and hydrokinetic technologies).
- Qualified applicants: Federal, state, local, and tribal governments; national laboratories; industry; small businesses; colleges and universities
- Qualified technologies: Hydroelectric; hydrokinetic energy; wave energy; tidal energy; ocean thermal energy conversion
- For more information: See EERE's Water Power Program; or program number 81.087 at the CFDA website

Wind Energy Program

7. Wind Energy Program (formerly Wind and Hydropower Technologies Program) Administered by: EERE

- Authorization: Energy Policy and Conservation Act of 1975 (EPCA; P.L. 94- 163); Renewable Energy and Energy Efficiency Technology Competitiveness Act of 1989 (P.L. 101-218); Solar, Wind, Waste, and Geothermal Power Production Incentives Act of 1990 (P.L. 101-575); Energy Policy Act of 1992 (EPACT; P.L. 102-486); Energy Policy Act of 2005 (EPACT 2005; P.L. 109-58); Energy Independence and Security Act of 2007 (EISA; P.L. 110-140); American Recovery and Reinvestment Act of 2009 (ARRA; P.L. 111-5)
- Annual funding: $38.3 million for FY2006; $48.7 million for FY2007; $49 million for FY2008; $54.4 million for FY2009; an additional $106.9 million appropriated in FY2009 from ARRA; $80 million for FY2010; $78.8 million for FY2011; $91.8 million for FY2012; $93.8 million for FY2013; and $144 million requested for FY2014
- Scheduled termination: None
- Description: This program partners with federal, state, and other stakeholder groups to conduct research and development activities through competitively selected, cost-shared research and development projects with industry to improve the performance, lower the costs, and accelerate the deployment of wind energy technologies.
- Qualified applicants: Federal, state, local, and tribal governments; national laboratories; industry; small businesses; colleges and universities
- Qualified technologies: Wind
- For more information: See EERE's Wind Energy Program website; and program number 81.087 at the CFDA website

Energy Efficiency

Buildings

8. Building Technologies Program

- Administered by: EERE
- Authorization: Energy Policy and Conservation Act of 1975 (EPCA; P.L. 94- 163); Energy Conservation and Production Act of 1976 (ECPA; P.L. 94-385); Department of Energy Organization Act of 1977 (P.L. 95-91); Energy Tax Act of 1978 (P.L. 95-618); National Energy Conservation Policy Act of 1978 (NECPA; P.L. 95-619); Power Plant and Industrial Fuel Use Act of 1978 (P.L. 95-620); Energy Security Act (P.L. 96-294); National Appliance Energy Supply Act of 1987 (P.L. 100-12); National Appliance Energy Conservation Amendments of 1988 (P.L. 100-357); Federal Energy Management Improvement Act of 1988 (P.L. 100-615); Energy Policy Act of 1992 (EPACT; P.L. 102-486); Energy Policy Act of 2005 (EPACT 2005; P.L. 109-58); Energy Independence and Security Act of 2007 (EISA; P.L. 110-140); American Recovery and Reinvestment Act of 2009 (ARRA; P.L. 111-5)

- Annual funding: $68.2 million for FY2006; $103 million for FY2007; $107.4 million for FY2008; $138.1 million for FY2009; an additional $319.2 million appropriated in FY2009 from ARRA; $222 million for FY2010; $207.3 million for FY2011; $214.7 million for FY2012; $220.5 million for FY2013; and $300 million requested for FY2014
- Scheduled termination: None
- Description: In partnership with the private sector, state and local governments, national laboratories, and universities, the Building Technologies Program works to improve the efficiency of buildings and the equipment, components, and systems within them. The program supports research and development (R&D) activities and provides tools, guidelines, training, and access to technical and financial resources.
- Qualified applicants: State and local governments; universities; national laboratories
- Qualified technologies: Energy-efficient innovations for building envelopes, equipment, lighting, daylighting, and windows; passive solar; photovoltaics; fuel cells; advanced sensors and controls; and combined heating, cooling, and power systems
- For more information: See EERE's Building Technologies Program website

9. Weatherization Assistance Program (WAP)

- Administered by: EERE
- Authorization: Energy Supply and Production Act of 1976 (ECPA; P.L. 94-385); National Energy Conservation Policy Act of 1978 (NECPA; P.L. 95-619); Energy Security Act of 1980 (P.L. 96-294); Energy Policy Act of 1992 (EPACT; P.L. 102-486); Energy Policy Act of 2005 (EPACT 2005; P.L. 109-58); Energy Independence and Security Act of 2007 (EISA; P.L. 110-140); American Recovery and Reinvestment Act of 2009 (ARRA; P.L. 111-5)
- Annual funding: $227.2 million for FY2008; $450 million for FY2009; an additional $5 billion appropriated in FY2009 from ARRA; $270 million for FY2010; $171 million for FY2011; $68 million for FY2012; $68.4 million for FY2013; and $184 million requested for FY2014
- Scheduled termination: None
- Description: This program reduces energy costs for low-income households by increasing the energy efficiency of their homes while ensuring their health and safety. DOE provides funding and technical guidance to states, which manage the day-to-day details of the program. Low-income families receive services from a network of more than 900 local weatherization service providers who install energy efficiency measures in the homes of qualifying homeowners free of charge.
- Qualified applicants: State and tribal governments, including U.S. territories
- Qualified technologies: Weatherization technologies include a wide range of energy efficiency measures for retrofitting homes and apartment buildings. Weatherization service providers choose the best package of efficiency measures for each home based on an energy audit of the home. Typical measures may include installing insulation, sealing ducts, tuning and repairing heating and cooling systems and if indicated, replacement of the same; mitigating air infiltration; and reducing electric base load consumption.

- For more information: See EERE's Weatherization Assistance Program website; and program number 81.042 at the CFDA website

Industrial

10. Advanced Manufacturing Office (AMO, formerly the Industrial Technologies Program - ITP)

- Administered by: EERE
- Authorization: Energy Policy and Conservation Act of 1975 (EPCA; P.L. 94- 163); Energy Conservation and Production Act of 1976 (ECPA; P.L. 94-385); Department of Energy Organization Act of 1977 (P.L. 95-91); National Energy Conservation Policy Act of 1978 (NECPA; P.L. 95-619); Powerplant and Industrial Fuel Use Act of 1978 (P.L. 95-620); Energy Security Act of 1980 (P.L. 96-294); Renewable Energy and Energy Efficiency Technology Competitiveness Act of 1989 (P.L. 101-218); Energy Policy Act of 1992 (EPACT; P.L. 102-486); Energy Policy Act of 2005 (EPACT 2005; P.L. 109-58); Energy Independence and Security Act of 2007 (EISA; P.L. 110-140); American Recovery and Reinvestment Act of 2009 (ARRA; P.L. 111-5)
- Annual funding: $55.9 million for FY2006; $55.8 million for FY2007; $63.2 million for FY2008; $88.2 million for FY2009; an additional $261.5 million appropriated in FY2009 from ARRA; $96 million for FY2010; $105.9 million for FY2011; $112.7 for FY2012; $116.3 million for FY2013; and $365 million requested for FY2014
- Scheduled termination: None
- Description: AMO works with industry to improve industrial energy efficiency and environmental performance while increasing productivity by: conducting R&D on new energy efficient technologies; supporting commercialization of emerging technologies; providing plants with access to proven technologies, energy assessments, software tools, and other resources; and promoting energy and carbon management in industry.
- Qualified applicants: Industrial organizations
- Qualified technologies: Crosscutting technologies that improve the efficiency of technologies that are common to many industrial processes and can benefit multiple industries. Crosscutting technology R&D areas include combustion; distributed energy; energy intensity processes; fuel and feedstock liability; industrial materials for the future; nanomanufacturing; and sensors and automation.
- For more information: See EERE's Advanced Manufacturing Office website

11. Inventions and Innovations Program

- Administered by: EERE
- Authorization: Federal Nonnuclear Energy Research and Development Policy Act (P.L. 93-577), Section 14; 42 USC 5913
- Annual funding: $2.8 million for FY2007; $145,000 for FY2008; $1.8 million for FY2009; $3 million for FY2010; $0 for FY2011; $940,000 for FY2012; an estimated $914,000 for FY2013; the FY2014 budget request does not provide details on this program
- Scheduled termination: None

- Description: This program provides financial and technical assistance for research and development of innovative, energy-saving ideas and inventions with future commercial market potential. Inventions and Innovations support energy efficiency and renewable energy technology development in focus areas that align with Office of Energy Efficiency and Renewable Energy programs.
- Qualified applicants: Individuals; small businesses
- Qualified technologies: Specific energy efficiency and renewable energy technologies not listed
- For more information: See program number 81.036 at the CFDA website. The U.S. Department of Energy's Inventions & Innovations website has been retired. To access information on financial opportunities and current solicitations, visit the Advanced Manufacturing Office's (formerly the Industrial Technologies Program's) financial opportunities website.

Vehicles

12. Vehicle Technologies Program

- Administered by: EERE
- Authorization: Department of Energy Organization Act of 1977 (P.L. 95-91); Energy Policy Act of 1992 (EPACT; P.L. 102-486); Energy Policy Act of 2005 (EPACT 2005; P.L. 109-58); Energy Independence and Security Act of 2007 (EISA; P.L. 110-140); American Recovery and Reinvestment Act of 2009 (ARRA; P.L. 111-5)
- Annual funding: \$178.4 million for FY2006; \$183.6 million for FY2007; \$208.4 million for FY2008; \$267.1 million for FY2009; an additional \$2.8 billion appropriated in FY2009 from ARRA; \$311.4 million for FY2010; \$293.2 million for FY2011; \$321 million for FY2012; \$330.8 million for FY2013; and \$575 million requested for FY2014
- Scheduled termination: None
- Description: The Vehicle Technologies Program works with industry leaders to develop and deploy advanced transportation technologies that could achieve significant improvements in vehicle fuel efficiency and displace oil with other fuels that ultimately can be domestically produced in a clean and cost-competitive manner. Program activities include research, development, demonstration, testing, technology validation, technology transfer, and education.
- Qualified applicants: Industry; colleges and universities; federal, state and local governments; national laboratories
- Qualified technologies: Hybrid electric systems; biofuels or fuels technology; advanced internal combustion engines; advanced propulsion materials
- For more information: See EERE's Vehicle Technology Program website; and EERE's Vehicle Technologies Program Factsheet

Other Energy Efficiency and Renewable Energy Programs

13. Conservation Research and Development Grants

- Administered by: EERE

- Authorization: Federal Nonnuclear Energy Research and Development Act of 1974 (P.L. 93-577); Department of Energy Organization Act of 1977 (P.L. 95- 91); Continuing Appropriations Act for FY1983 (P.L. 97-377); American Recovery and Reinvestment Act of 2009 (ARRA; P.L. 111-5)
- Annual funding: $87.5 million for FY2008; $203.7 million for FY2009; $1.96 billion for FY2010; $300 million for FY2011; $188.3 million for FY2012; an estimated $75.1 million for FY2013; the FY2014 budget request does not provide details on this program
- Scheduled termination: None
- Description: This program provides project grants to conduct balanced, long-term research efforts in buildings technologies, industrial technologies, vehicle technologies, and hydrogen and fuel cell technologies.
- Qualified applicant: State, local, and tribal governments; universities; profit organizations; and private nonprofit institutions/organizations
- Qualified technologies: Hydrogen and fuel cells; energy efficient technologies; advanced battery manufacturing
- For more information: See program number 81.086 at the CFDA website

14. Energy Efficiency and Renewable Energy Information Dissemination, Outreach, Training, and Technical Analysis/Assistance Grant Program

- Administered by: EERE
- Authorization: Energy Reorganization Act of 1974 (P.L. 93-438); Department of Energy Organization Act of 1977 (P.L. 95-91); Energy Policy Act of 1992 (EPACT; P.L. 102-486)
- Annual funding: $30 million for FY2007; $39.7 million for FY2008; $38 million for FY2009; $80.4 million for FY2010; an estimated $15 million for FY2011; $32.2 million for FY2012; an estimated $8.6 million for FY2013; the FY2014 budget request does not provide details on this program
- Scheduled termination: None
- Description: This program provides financial assistance for information dissemination, outreach, training and related technical analysis/assistance that will (1) stimulate increased energy efficiency in transportation, buildings, industry and the Federal sector and encourage increased use of renewable and alternative energy; and (2) accelerate the adoption of new technologies to increase energy efficiency and the use of renewable and alternative energy through the competitive solicitation of applications.
- Qualified applicants: State and local governments; Native American organizations; individuals; universities; profit organizations; private nonprofit organizations; public nonprofit organizations; and Alaskan Native corporations
- Qualified technologies: Specific energy efficiency and renewable energy technologies not listed
- For more information: See program number 81.117 at the CFDA website

15. Energy Efficiency and Renewable Energy Technology Deployment, Demonstration, and Commercialization Grant Program

- Administered by: EERE

- Authorization: Energy Policy Act of 1992 (EPACT; P.L. 102-486); Energy Policy Act of 2005 (EPACT 2005; P.L. 109-58); Energy Independence and Security Act of 2007 (EISA; P.L. 110-140); American Recovery and Reinvestment Act of 2009 (ARRA; P.L. 111-5)
- Annual Funding: $0 for FY2008; $21.8 million for FY2009; an estimated $7.2 million for FY2010. It is anticipated that all funds obligated under this program in FY2010 will be Recovery Act funds. $1 million for FY2011; $0 for FY2012; $0 for FY2013; the FY2014 budget request does not provide details on this program
- Scheduled termination: None
- Description: This program provides financial assistance for the technology deployment, demonstration, and commercialization of energy efficiency and renewable energy technologies. This includes biomass, building technologies, federal energy management, geothermal technologies, projects involving hydrogen, fuel cells and infrastructure technologies, industrial technologies, solar energy technologies, vehicle technologies, weatherization and intergovernmental technologies, and wind and hydropower technologies.
- Qualified applicants: State governments; profit organizations
- Qualified technologies: Biomass; geothermal; hydrogen and fuel cell technologies; solar; hydropower
- For more information: See program number 81.129 at the CFDA website

16. Renewable Energy Production Incentive (REPI)

- Administered by: EERE
- Authorization: Energy Policy Act of 1992 (EPACT; P.L. 102-486), Title XII, Section 1212; Energy Policy Act of 2005 (EPACT 2005; P.L. 109-58), Title II, Subtitle A, Section 202); 42 USC 13317
- Annual funding: $4.95 million for FY2006; $4.95 million for FY2007; $4.95 million for FY2008; $5 million for FY2009; $0 for FY2010; $0 for FY2011; $0 for FY2012; $0 for FY2013; and $0 requested for FY2014
- Scheduled termination: End of FY2026
- Description: This program provides incentive payments for electricity generated and sold by new qualifying renewable energy facilities. Qualifying systems are eligible for annual incentive payments of 1.5¢ per kilowatt-hour in 1993 dollars (indexed for inflation) for the first 10-year period of their operation, subject to the availability of annual appropriations in each federal fiscal year of operation.
- Qualified applicants: State, local, and tribal governments; public utilities; not-for-profit electrical cooperatives; Native American corporations
- Qualified technologies: Solar thermal electric; photovoltaics; landfill gas; wind; biomass; geothermal electric; anaerobic digestion; tidal energy; wave energy; ocean thermal
- For more information: See EERE's Renewable Energy Production Incentive Program website

17. Renewable Energy Research and Development Program

- Administered by: EERE

- Authorization: Department of Energy Organization Act of 1977 (P.L. 95-91); Department of Energy Act of 1978 - Civilian Applications (P.L. 95-238), Section 207; Renewable Energy and Energy Efficiency Technology Competitiveness Act of 1989 (P.L. 101-218); Energy Policy Act of 2005 (EPACT 2005; P.L. 109-58); Energy Independence and Security Act of 2007 (EISA; P.L. 111-140); American Recovery and Reinvestment Act of 2009 (ARRA; P.L. 111-5)
- Annual funding: $520 million for FY2008; $472.8 million for FY2009; $2.3 billion for FY2010 from ARRA funds; $114.7 million for FY2011; $233.2 million for FY2012; an estimated $141.5 million for FY2013; the FY2014 budget request does not provide details on this program. Breakdown of additional funds appropriated from ARRA:
  - Biomass—$800 million
  - Geothermal—$400 million
  - Hydrogen/Fuel Cell—$43.4 million
  - Solar—$117.6 million
  - Wind and Hydropower—$118 million
- Scheduled termination: None
- Description: This program provides financial assistance to conduct balanced research and development efforts in the following energy technologies: solar, biomass, hydrogen, fuel cells and infrastructure, wind and hydropower, hydrogen, and geothermal. Assistance may be used to develop and transfer renewable energy technologies to the scientific and industrial communities, states, and local governments.
- Qualified applicants: State, local, and tribal governments; colleges and universities; profit organizations; private nonprofit organizations
- Qualified technologies: Solar; biomass; hydrogen; fuel cells; wind; hydropower; geothermal
- For more information: See program number 81.087 at the CFDA website

18. State Energy Program (SEP)

- Administered by: EERE
- Authorization: Energy Policy and Conservation Act of 1975 (EPCA; P.L. 94- 163); Energy Supply and Production Act of 1976 (ECPA; P.L. 94-385); National Energy Conservation Policy Act of 1978 (NECPA; P.L. 95-619); State Energy Efficiency Programs Improvement Act of 1990 (P.L. 101-440); Energy Policy Act of 1992 (EPACT; P.L. 102-486); Energy Conservation Reauthorization Act of 1998 (P.L. 105-388); Energy Policy Act of 2005 (EPACT 2005; P.L. 109-58); Energy Independence and Security Act of 2007 (EISA; P.L. 110-140); American Recovery and Reinvestment Act of 2009 (ARRA; P.L. 111-5)
- Annual funding: $44.1 million for FY2008; $50 million for FY2009; an additional $3.1 billion appropriated in FY2009 from ARRA; $50 million for FY2010; $50 million for FY2011; $50 million for FY2012; $50.3 million for FY2013; and $57 million requested for FY2014
- Scheduled termination: None
- Description: SEP provides grants to states to design and carry out their own renewable energy and energy efficiency programs.

- Qualified applicants: State and tribal governments, including U.S. territories
- Qualified technologies: Emerging renewable energy and energy efficiency technologies
- For more information: See EERE's State Energy Program website; and program number 81.041 at the CFDA website

19. Tribal Energy Program

- Administered by: EERE
- Authorization: Energy Policy and Conservation Act of 1975 (EPCA; P.L. 94- 163); Energy Conservation and Production Act of 1976 (ECPA; P.L. 94-385); Department of Energy Organization Act of 1977 (P.L. 95-91); Energy Tax Act of 1978 (P.L. 95-618); National Energy Conservation Policy Act of 1978 (NECPA; P.L. 95-619); Power Plant and Industrial Fuel Use Act of 1978 (P.L. 95-620); Energy Security Act (P.L. 96-294); National Appliance Energy Supply Act of 1987 (P.L. 100-12); Federal Energy Management Improvement Act of 1988 (P.L. 100-615); Energy Policy Act of 1992 (EPACT; P.L. 102-486); Energy Policy Act of 2005 (EPACT 2005; P.L. 109-58); Energy Independence and Security Act of 2007 (EISA; P.L. 110-140); American Recovery and Reinvestment Act of 2009 (ARRA; P.L. 111-5)
- Annual funding: $3.96 million for FY2006; $3.96 million for FY2007; $5.95 million for FY2008; $6 million for FY2009; $10 million for FY2010; $7 million for FY2011; $10 million for FY2012; $10.1 for FY2013; and $7 million requested for FY2014
- Scheduled termination: None
- Description: This program promotes tribal energy sufficiency, economic growth, and employment on tribal lands through the development of renewable energy and energy efficiency technologies. The program provides financial assistance, technical assistance, education and training to tribes for the evaluation and development of renewable energy resources and energy efficiency measures.
- Qualified applicant: Tribal governments
- Qualified technologies: Energy efficient technologies: clothes washers; refrigerators/freezers; water heaters; lighting; lighting controls/sensors; chillers; furnaces; boilers; air conditioners; programmable thermostats; energy management; systems/building controls; caulking/weather-stripping; duct/air sealing; building insulation; windows; doors; siding; roofs; comprehensive measures/whole building; and other energy efficiency improvements may be eligible. Renewable energy technologies: passive solar space heat; solar water heat; solar space heat; photovoltaics; wind; biomass; hydroelectric; geothermal electric; geothermal heat pumps
- For more information: See EERE's Tribal Energy Program websitehttp://apps1.eere.energy and DSIRE's program summary for the Tribal Energy Program

Other DOE Offices/Cross-Cutting Programs

20. Advanced Research Projects Energy Financial Assistance Program (ARPA-E)

- Administered by: Advanced Research Projects Agency-Energy (ARPA-E)
- Authorization: America COMPETES Act (P.L. 110-69), Section 5012
- Annual funding: $15 million for FY2009; an additional $388.9 million in FY2009 from ARRA; $0 for FY2010; $165.6 million for FY2011; $275 million for FY2012; $276.7 million for FY2013; and $379 million requested for FY2014
- Scheduled termination: After ARPA-E has been in operation for six years, the Secretary of Energy shall offer to enter into a contract with the National Academy of Sciences under which the National Academy shall conduct an evaluation of how well ARPA-E is achieving the goals and mission of ARPA-E. The evaluation shall include the recommendation of the National Academy of Sciences on whether ARPA-E should be continued or terminated.
- Description: This program will fund organizations that have proposed sophisticated energy technology R&D projects that (1) translate scientific discoveries and cutting-edge inventions into technological innovations and (2) accelerate transformational technological advances in areas that industry by itself is not likely to undertake because of high technical or financial risk. Transformational energy technologies are those that have the potential to create new paradigms in how energy is produced, transmitted, used, or stored.
- Qualified applicants: ARPA-E welcomes submissions from any type of capable technology research and development entity. This includes, but is not limited to for-profit entities, academic institutions, research foundations, not-for-profit entities, collaborations, and consortia. The lead organization that will enter into the agreement with ARPA-E must be a U.S. entity.
- Qualified technologies: Transformational energy technologies
- For more information: See ARPA-E's Frequently Asked Questions (FAQ) website; and program number 81.135 at the CFDA website

21. Electricity Delivery and Energy Reliability, Research, Development and Analysis Grant Program (Office of Electricity Delivery and Energy Reliability - OE)

- Administered by: Office of Electricity Delivery and Energy Reliability (OE)
- Authorization: Department of Energy Organization Act of 1977 (P.L. 95-91); Energy Security Act of 1980 (P.L. 96-294); National Superconductivity and Competitiveness Act of 1988 (P.L. 100-697); Energy Policy Act of 1992 (EPACT; P.L. 102-486); Energy Policy Act of 2005 (EPACT 2005; P.L. 109-58); Energy Independence and Security Act of 2007 (EISA; P.L. 110-140); American Recovery and Reinvestment Act of 2009 (ARRA; P.L. 111-5)
- Annual funding: $82.8 million for FY2008; $83.1 million for FY2009; an additional $4.5 billion was appropriated to the Office of Electricity Delivery and Energy Reliability in FY2009 from ARRA. Approximately $4 billion of that total was used to implement smart grid programs authorized by EISA and accelerate the deployment of smart grid technologies across the transmission and distributions[4]; $121.4 million for FY2010; $102 million for FY2011; $96.2 million for FY2012; $99.8 million for FY2013; and $119.4 million requested for FY2014

- Scheduled termination: None
- Description: This grant program aims to develop cost-effective technology that enhances the reliability, efficiency, and resiliency of the electric grid.
- Qualified applicants: State, local, and tribal governments; universities; profit organizations; private nonprofit organizations; research organizations
- Qualified technologies: Specific technologies not listed
- For more information: See program number 81.122 at the CFDA website

22. *Federal Energy Management Program (FEMP)*

- Administered by: EERE
- Authorization: Energy Policy and Conservation Act of 1975 (EPCA; P.L. 94- 163); Energy Conservation and Production Act of 1976 (ECPA; P.L. 94-385); Department of Energy Organization Act (P.L. 95-91); National Energy Conservation Policy Act of 1978 (NECPA; P.L. 95-619); Federal Energy Management Improvement Act of 1988 (P.L. 100-615); Energy Policy Act of 1992 (EPACT; P.L. 102-486); Energy Policy Act of 2005 (EPACT 2005; P.L. 109-58); Energy Independence and Security Act of 2007 (EISA; P.L. 110-140)
- Annual funding: $19 million for FY2006; $19.5 million for FY2007; $19.8 million for FY2008; $22 million for FY2009; an additional $22.4 million in FY2009 from ARRA; $32 million for FY2010; $30.4 million for FY2011; $29.9 million for FY2012; $30.1 million for FY2013; and $36 million requested for FY2014
- Scheduled termination: None
- Description: FEMP assists federal agencies in developing and implementing energy efficient and renewable energy resources to meet energy management regulations and goals.
- Qualified applicants: Federal agencies
- Qualified technologies: Energy efficient technologies; solar; wind; incremental hydro; ocean; biomass; geothermal
- For more information: See EERE's Federal Energy Management Program website

23. *Financial Assistance Program (Office of Science)*

- Administered by: Office of Science
- Authorization: Atomic Energy Act of 1954 (P.L. 83-703), Section 31; Energy Reorganization Act of 1974 (P.L. 93-438), Title I, Section 107; Federal Nonnuclear Energy Research and Development Act of 1974 (P.L. 93-577); American Recovery and Reinvestment Act of 2009 (ARRA; P.L. 111-5)
- Annual funding: $974 million for FY2008; $1.4 billion for FY2009; $1.3 billion for FY2010; $1.3 billion for FY2011; $1 billion for FY2012; an estimated $962.6 million for FY2013; and an estimated $962.6 million for FY2014
- Scheduled termination: None
- Description: This program provides financial support for fundamental research in the basic sciences and advanced technology concepts and assessments in fields related to energy.
- Qualified applicants: State, local, and tribal governments; colleges and universities; profit commercial organizations; private nonprofit organizations; public nonprofit organizations; small businesses

- Qualified technologies: Specific advanced technologies not listed
- For more information: See program number 81.049 at the CFDA website; and the Office of Science's Funding Opportunities website

24. Loan Guarantee Program (Office of the Chief Financial Officer)

- Administered by: Office of the Chief Financial Officer
- Authorization: Energy Policy Act of 2005 (EPACT 2005; P.L. 109-58), Title XVII; American Recovery and Reinvestment Act of 2009 (ARRA; P.L. 111-5); 42 USC 16511 et seq.
- Annual funding:
  - Section 1703 Innovative Technology Loan Guarantee Program (permanent): $4.5 million for FY2008; $0 for FY2009; $0 for FY2010; $169.6 million for FY2011; $0 for FY2012; $0 for FY2013; and $0 requested for FY2014
  - Section 1705 Temporary Loan Guarantee Program: $0 for FY2008; $6 billion was appropriated for FY2009. However, $2 billion of that funding was transferred to the "cash for clunkers" automobile trade-in program by P.L. 111-47.[5] An additional $1.5 billion was rescinded for the Education Jobs and Medicaid Assistance Act, P.L. 111-226 (Section 308), leaving a total of $2.5 billion remaining from the FY2009 appropriations; $0 for FY2010; $0 for FY2011; $0 for FY2012; and $0 for FY2013
- Scheduled termination: None for the permanent (Section 1703) loan guarantee program. Projects authorized by the temporary loan guarantee (Section 1705) had to begin construction no later than September 30, 2011.
- Description: This program provides federal loan guarantees to encourage early commercial use in the United States of new or significantly improved technologies in energy projects that (1) avoid, reduce, or sequester air pollutants or anthropogenic emissions of greenhouse gases; and (2) employ new or significantly improved technologies as compared to commercial technologies in service in the United States at the time the guarantee is issued. Temporary loan guarantees can also be made under Section 1705 for rapid deployment of certain renewable and electric transmission projects.
- Qualified applicants: State, local, and tribal governments; universities; profit organizations; and public nonprofit organizations. No federal entity may apply
- Qualified technologies: Solar thermal electric; solar thermal process heat; photovoltaics; wind; hydroelectric; renewable transportation fuels; geothermal electric; fuel cells; manufacturing facilities; daylighting; tidal energy; wave energy; ocean thermal; biodiesel
- For more information: See program number 81.126 at the CFDA website; DSIRE's program summary for the Loan Guarantee Program; and DOE's Loan Guarantee Program website

25. Small Business Innovation Research Program (SBIR)/Small Business Technology Transfer Program (STTR)

- Administered by: EERE
- Authorization: Small Business Innovation Development Act of 1982 (P.L. 97- 219); Small Business Research and Development Act of 1992 (P.L. 102-564);

Consolidated Appropriations Act, 2001 (P.L. 106-554), Title I, Small Business Innovation Research Program Reauthorization Act of 2000; Small Business Technology Transfer Program Reauthorization Act of 2001 (P.L. 107-50); SBIR/STTR Reauthorization Act of 2011 (P.L. 112-81, Div. E, Title L)

- Annual funding: $24.2 million for FY2011; $29.1 million for FY2012; $33.3 million for FY2013; and $52.1 million requested for FY2014
- Scheduled termination: The SBIR/STTR Reauthorization Act of 2011 (P.L. 112- 81, Div. E, Title L) reauthorized the program through FY2017.
- Description: Small Business Innovation Research (SBIR) and Small Business Technology Transfers (STTR) are U.S. government programs in which federal agencies with large research and development (R&D) budgets set aside a small fraction of their funding for competitions among small businesses only. DOE's SBIR-STTR program is designed to stimulate technological innovation by small advanced technology firms and provide new, cost-effective scientific and engineering solutions to challenging problems. EERE funds appropriated for SBIR/STTR are allocated to larger EERE technology programs, detailed earlier in this report—including Biomass, Geothermal, Hydrogen & Fuel Cell, Solar Energy, Water Power; Wind Energy, Advanced Manufacturing, Building Technologies, and Vehicle Technologies.
- Qualified applicants: Small businesses
- Qualified technologies: Research areas include energy production (fossil, nuclear, renewable, and fusion energy); energy use (in buildings, vehicles, and industry); fundamental energy sciences (materials, life, environmental, and computational sciences, and nuclear and high energy physics); environmental management; and nuclear nonproliferation
- For more information: See EERE's Small Business Innovation Research/Small Business Technology Transfers (SBIR/STTR) website; and program number 10.212 (SBIR) at the CFDA website

II. U.S Department of the Treasury

Homeowner

1. Residential Energy Efficiency Tax Credit

- Administered by: Internal Revenue Service
- Authority: Energy Policy Act of 2005 (EPACT 2005; P.L. 109-58); Energy Improvement and Extension Act of 2008 (P.L. 110-343), Division B, Section 302; American Recovery and Reinvestment Act of 2009 (ARRA; P.L. 111-5), Division B, Section 1121; American Taxpayer Relief Act of 2012 (P.L. 112-240); IRS Form 5695 & Instructions: Residential Energy Credits; 26 USC 25C
- Scheduled Termination: December 31, 2013
- Description: The credit applied to energy efficiency improvements in the building envelope of existing homes and for the purchase of high-efficiency heating, cooling and water-heating equipment. Efficiency improvements or equipment must have

served a dwelling in the United States that is owned and used by the taxpayer as a primary residence. The maximum lifetime amount of homeowner credit for all improvements combined for 2011, 2012 and 2013 is $500 total.

- Qualified applicant: Residential
- Qualifying technologies: Water heaters; furnace; boilers; heat pumps; air conditioners; building insulation; windows; doors; roofs; circulating fans used in a qualifying furnace; biomass and stoves that use qualified biomass fuel
- For more information: See the Internal Revenue Service website

2. Residential Renewable Energy Tax Credit

- Administered by: Internal Revenue Service
- Authority: Energy Policy Act of 2005 (EPACT 2005; P.L. 109-58); Energy Improvement and Extension Act of 2008 (P.L. 110-343); American Recovery and Reinvestment Act of 2009 (ARRA; P.L. 111-5); IRS Form 5695 & Instructions: Residential Energy Credits; 26 USC 25D (amended)
- Scheduled Termination: December 31, 2016
- Description: A taxpayer may claim a credit of 30% of qualified expenditures for a system that serves a dwelling unit located in the United States and used as a residence by the taxpayer.
- Qualified applicant: Residential
- Qualifying technologies: Solar water heat; photovoltaics; wind; fuel cells; geothermal heat pumps; other solar electric technologies
- For more information: See IRS Form 5695 & Instructions: Residential Energy Credits

Business

3. Business Energy Investment Tax Credit

- Administered by: Internal Revenue Service
- Authority: Energy Improvement and Extension Act of 2008 (EISA; P.L. 110- 343), Division B; American Recovery and Reinvestment Act of 2009 (ARRA; P.L. 111-5), Division B, Section 1103
- Scheduled termination: December 31, 2016. Geothermal property, with the exception of geothermal heat pumps, has no stated expiration date. The credit for solar energy property returns to 10% after December 31, 2016.
- Description: Credit is 30% for solar, fuel cells and small wind & federal renewable electricity production tax credit-eligible technologies; 10% for geothermal, microturbines and CHP (Combined Heat and Power).
- Qualified Applicants: Commercial; industrial; utilities; agricultural
- Qualified Technologies: Solar water heat; solar space heat; solar thermal electric; solar thermal process heat; photovoltaics; wind; biomass; geothermal electric; fuel cells; geothermal heat pumps; CHP/Cogeneration; solar hybrid lighting; direct-use geothermal; microturbines
- For more information: See the DSIRE website.

4. Energy Efficient Commercial Buildings Tax Deduction

- Administered by: Internal Revenue Service
- Authority: Energy Policy Act of 2005 (EPACT 2005; P.L. 109-58), Title XIII, Subtitle C, Section 1331(a); Tax Relief and Health Care Act of 2006 (P.L. 109- 432), Division A, Title II, Section 204; Energy Improvement and Extension Act of 2008 (P.L. 110-343), Division B, Title III, Section 303; 26 USC § 179D
- Scheduled termination: December 31, 2013
- Description: A tax deduction of $1.80 per square foot is available to owners of new or existing buildings who install (1) interior lighting; (2) building envelope, or (3) heating, cooling, ventilation, or hot water systems that reduce the building's total energy and power cost by 50% or more in comparison to a building meeting minimum requirements set by ASHRAE Standard 90.1-2001. Energy savings must be calculated using qualified computer software approved by the IRS.
- Qualified applicants: Commercial; builder/developer; state government; federal government (deductions associated with government buildings are transferred to the designer)
- Qualified technologies: Equipment insulation; water heaters; lighting; lighting controls/sensors; chillers; furnaces; boilers; heat pumps; air conditioners; caulking/weather-stripping; duct/air sealing; building insulation; windows; doors; siding; roofs; comprehensive measures/whole building
- For more information: See the Energy Star website

5. Energy-Efficient New Homes Tax Credit for Home Builders

- Administered by: Internal Revenue Service
- Authority: Tax Technical Corrections Act of 2007 (P.L. 110-172), Section 11(a)(7); Energy Improvement and Extension Act (P.L. 110-343), Division B, Title III, Section 304; 26 USC § 45L 8/8/2005 (amended 2008); P.L. 111-312 (12/31/2011 extension)); American Taxpayer Relief Act of 2012 (P.L. 112-240).
- Scheduled termination: December 31, 2013
- Description: This program provided tax credits of up to $2,000 for builders of all new energy-efficient homes, including manufactured homes constructed in accordance with the Federal Manufactured Homes Construction and Safety Standards. This credit was created by the Energy Policy Act of 2005 for homes constructed in 2006 and 2007. It was renewed for homes constructed in 2008 and 2009, but then it expired and was not active in 2010. The Tax Relief, Unemployment Insurance Reauthorization, and Job Creation Act of 2010 (P.L. 111-312) retroactively reinstated this credit for homes acquired after December 31, 2009, and before January 1, 2012.
- Qualified applicant: Builder/developer
- Qualified technologies: Comprehensive measures/whole building
- For more information: See the IRS website; IRS Certification of Energy Efficient Home Credit Notice; and IRS Energy Efficient Home Credit; Manufactured Homes Notice

Industry

6. Energy Efficient Appliance Tax Credit for Manufacturers

- Administered by: Internal Revenue Service
- Authority: Energy Policy Act of 2005 (EPACT 2005; P.L. 109-58), Title XIII, Subtitle C, Section 1334(a); Energy Improvement and Extension Act of 2008 (P.L. 110-343), Division B, Section 305; 26 USC § 45M subsequently amended; H.R. 4853, Tax Relief, Unemployment Insurance Reauthorization, and Job Creation Act of 2010; American Taxpayer Relief Act of 2012, (P.L. 112-240)
- Scheduled termination: December 31, 2013
- Description: A tax credit for each manufacturer is limited to a total of $25 million for 2011, 2012 and 2013 combined.
- Qualified applicants: Industrial; appliance manufacturers
- Qualified technologies: Clothes washers; dishwashers; refrigerators
- For more information: See the IRS website; IRS form 8909

7. Qualifying Advanced Energy Manufacturing Investment Tax Credit (48C)

- Administered by: Internal Revenue Service
- Authority: American Recovery and Reinvestment Act of 2009 (ARRA; P.L. 111- 5), Division B, Section 1302; 26 USC 48C; IRS Notice 2013-12 Qualifying Advanced Energy Project Credit Phase II.
- Scheduled termination: Applications are no longer being accepted. Phase II concept papers were due to the U.S. Department of Energy (DOE) by April 9, 2013. DOE will review concept papers and select which companies will be allowed to submit a full application. Applications were due July 23, 2013.
- Description: The U.S. Treasury Department, in consultation with DOE, is no longer accepting applications for this tax credit. The applications were due to DOE by September 16, 2009, with final applications due to DOE October 16, 2009. Only applicants accepted and ranked by the DOE were allowed to submit final applications to the Internal Revenue Service (IRS) by December 16, 2009. Approved projects were announced in January 2010. This tax credit was designed to encourage a U.S.-based renewable energy manufacturing sector. Projects receiving awards are eligible for a tax credit of 30% of the qualified investment required for an advanced energy project.
- Qualified applicants: Commercial, industrial, manufacturing
- Qualifying technologies: Lighting; lighting controls/sensors; energy conservation technologies: smart grid; solar water heat; solar thermal electric; photovoltaics; wind; geothermal electric; fuel cells; geothermal heat pumps; batteries and energy storage; advanced transmission technologies that support renewable energy generation; renewable fuels; fuel cells using renewable fuels; microturbines
- For more information: See DOE's webpage for the 48C tax credit; the IRS' 48C webpage; and DSIRE's webpage for the tax credit

8. Renewable Electricity Production Tax Credit (PTC)

- Administered by: Internal Revenue Service

- Authority: Energy Policy Act of 1992 (EPACT; P.L. 102-486); Energy Policy Act of 2005 (EPACT 2005; P.L. 109-58), Title XIII, Section 1301; Tax Relief and Health Care Act of 2006 (P.L. 109-432), Division A, Section 201; Energy Improvement and Extension Act of 2008 (P.L. 110-343); American Recovery and Reinvestment Act of 2009 (ARRA; P.L. 111-5), Division B, Section 1101 and 1102; American Taxpayer Relief Act of 2012, (P.L. 112-240); IRS Notice 2013- 29; 26 USC 45 (amended)
- Scheduled termination: Projects must begin construction by December 31, 2013
- Description: The federal renewable electricity Production Tax Credit (PTC) is a per-kilowatt-hour tax credit for electricity generated by qualified energy resources and sold by the taxpayer to an unrelated person during the taxable year. P.L. 112-240 (American Taxpayer Relief Act of 2012) extended the PTC through the end of 2013 and allowed projects that begin construction by the end of 2013 to qualify for the PTC. Previously, the law required that qualifying projects be placed in service before the PTC expiration date.
- Qualified applicants: Commercial; industrial
- Qualifying technologies: Landfill gas; wind; biomass; hydroelectric; geothermal electric; municipal solid waste; hydrokinetic power (i.e., flowing water); anaerobic digestion; small hydroelectric; tidal energy; wave energy; ocean thermal
- For more information: See the IRS website and DSIRE website

9. Residential Energy Conservation Subsidy Exclusion (Corporate)

- Administered by: Internal Revenue Service
- Authority: Energy Policy Act of 1992 (EPACT; P.L. 102-486), Section 1912; Small Job Protection Act of 1996 (P.L. 104-188), Section 1617; 26 USC 136 (amended)
- Scheduled termination: None
- Description: Energy conservation subsidies provided by public utilities, either directly or indirectly, are nontaxable: "Gross income shall not include the value of any subsidy provided (directly or indirectly) by a public utility to a customer for the purchase or installation of any energy conservation measure."
- Qualified applicants: Residential; multi-family residential
- Qualifying technologies: Technologies installed to reduce electricity or natural gas consumption or improve the management of energy demand in a dwelling unit, including, but not limited to, solar water heat; solar space heat; photovoltaics; and other energy efficiency technologies not identified.
- For more information: See the IRS Publication 525 (2012), Taxable and Nontaxable Income website

10. Residential Energy Conservation Subsidy Exclusion (Personal)

- Administered by: Internal Revenue Service
- Authority: Energy Policy Act of 1992 (EPACT; P.L. 102-486); Small Job Protection Act of 1996 (P.L. 104-188), Section 1617; 26 USC 136 (amended)
- Scheduled termination: None
- Description: Energy conservation subsidies provided by public utilities, either directly or indirectly, are nontaxable: "Gross income shall not include the value of any subsidy provided (directly or indirectly) by a public utility to a customer for the purchase or installation of any energy conservation measure."

- Qualified applicant: Residential; multi-family residential
- Qualifying technologies: Technologies installed to reduce electricity or natural gas consumption or improve the management of energy demand in a dwelling unit, including, but not limited to, solar water heat; solar space heat; photovoltaics; and other energy efficiency technologies not identified.
- For more information: See the IRS Publication 525 (2012), Taxable and Nontaxable Income website

State, Local and Tribal Governments

11. Qualified Energy Conservation Bonds (QECBs)

- Administered by: Internal Revenue Service
- Authority: Energy Improvement and Extension Act of 2008 (P.L. 110-343), Division B, Section 301; American Recovery and Reinvestment Act of 2009, (ARRA; P.L. 111-5), Division B, Title I, Subtitle F, Part III, Section 1521(b)(1), (2), and Part IV, Section 1531(c)(2), and Part V, Section 1541(b)(2); 26 USC 54A ; 26 USC 54D; IRS Notice 2009-29; IRS Notice 2010-35; IRS Announcement 2010-54
- Scheduled termination: All funds have been allocated to the states. No new federal funding is available. There may be funding available in certain states.
- Description: QECBs may be used by state, local and tribal governments to finance certain types of energy projects. QECBs, as tax credit bonds, provide federally subsidized financing to all issuers. The original limit on the volume of energy conservation tax credit bonds to be issued by state and local governments was $800 million. The American Recovery and Reinvestment Act of 2009 expanded the allowable bond volume to $3.2 billion.
- Qualified applicants: State, local, and tribal governments
- Qualified technologies: Solar thermal electric; photovoltaics; landfill gas; wind; biomass; hydroelectric; geothermal electric; municipal solid waste; hydrokinetic power; anaerobic digestion; tidal energy; wave energy; ocean thermal
- For more information: See the DSIRE website

Cross-Cutting

12. Modified Accelerated Cost-Recovery System (MACRS)

- Administered by: Internal Revenue Service
- Authority: Economic Recovery Tax Act of 1981 (P.L. 97-34); Economic Stimulus Act of 2008 (P.L. 110-185); American Recovery and Reinvestment Act of 2009 (ARRA; P.L. 111-5), Division B, Title I, Subtitle C, Part I, Section 1201(a)(1)-(2)(D), (3)(A), (b)(1); Tax Relief, Unemployment Insurance Reauthorization, and Job Creation Act of 2010 (P.L. 111-312); 26 USC 168; IRS Rev. Proc. 2011-26; American Taxpayer Relief Act of 2012 (P.L. 112-240)
- Scheduled termination: None. The five-year schedule for most types of solar, geothermal, and wind property has been in place since 1986.

- Description: Under MACRS, businesses may recover investments in certain property through depreciation deductions. The MACRS establishes a set of class lives for various types of property, ranging from three to 50 years, over which the property may be depreciated. A number of renewable energy technologies are classified as five-year property (26 USC 168(e)(3)(B)(vi)) under MACRS.
- Qualified applicants: Commercial; industrial
- Qualified technologies: Solar water heat; solar space heat; solar thermal electric; solar thermal process heat; photovoltaics; landfill gas; wind; biomass; renewable transportation fuels; geothermal electric; fuel cells; geothermal heat pumps; municipal solid waste; CHP/cogeneration; solar hybrid lighting; direct use geothermal; anaerobic digestion; microturbines
- For more information: See IRS Publication 946, IRS Form 4562: Depreciation and Amortization, and Instructions for Form 4562

13. Alternative Motor Vehicle Credit

- Administered by: Internal Revenue Service
- Authority: Energy Policy Act of 2005 (EPACT 2005; P.L. 109-58); Energy Improvement and Extension Act of 2008 (P.L. 110-343), Division B, Section 205; American Recovery and Reinvestment Act of 2009 (ARRA; P.L. 111-5), Division B, Sections 1141-1144
- Scheduled Termination: December 31, 2014 for fuel cell vehicles and qualified plug-in electric drive motor vehicles; expired December 31, 2011 or earlier for all other vehicles.
- Description: Enacted in the Energy Policy Act of 2005, the provision includes separate credits for four distinct types of vehicles: using fuel cells, advanced lean burn technologies, qualified hybrid technology or qualified alternative fuels technologies.
- Qualified applicant: Taxpayers
- Qualifying technologies: Hybrid gasoline-electric; diesel; battery-electric; alternative fuel and fuel cell vehicles; advanced lean-burn technology vehicles; plug-in hybrid electric vehicles
- For more information: See the IRS website for the Alternative Motor Vehicle Credit

III. DEPARTMENT OF AGRICULTURE

1. Assistance to High Energy Cost Rural Communities Program

- Administered by: Rural Development (RD)
- Authorization: Rural Electrification Act of 1936 (P.L. 74-605); Grain Standards and Warehouse Improvement Act of 2000 (P.L. 106-472)
- Annual funding: $9.5 million for FY2012; and $7.6 million for FY2013
- Scheduled termination: None
- Description: This program provides financial assistance to rural communities with extremely high energy costs (exceeding 275% of the national average).

- Qualified applicants: State, local, and tribal governments (including U.S. territories); for-profit businesses; non-profit businesses; cooperatives; individuals
- Qualified technologies: Not specifically identified
- For more information: See CFDA program number 10.859 and the USDA program website

2. Bioenergy Program for Advanced Biofuels

- Administered by: Rural Development
- Authorization: P.L. 110-234, Food, Conservation, and Energy Act of 2008, Title IX, section 9001, subsection 9005
- Annual Funding:
  - Mandatory Farm Bill authorization: $55 million for FY2009; $55 million for FY2010; $85 million for FY2011; $105 million for FY2012 was authorized to remain available until expended. P.L. 112-55 limits mandatory spending to $65 million
  - Discretionary: $25 million authorized for FY2009-FY2013; No discretionary funding has been appropriated through FY2013
- Scheduled Termination: Mandatory funding authorized through FY2013.
- Description: To support and ensure an expanding production of advanced biofuels by providing payments to eligible advanced biofuel producers.
- Qualified applicants: Eligible advanced biofuels producers
- Qualified technologies: Payments will be made to eligible advanced biofuel producers for the production of fuel derived from renewable biomass, other than corn kernel starch, to include biofuel derived from cellulose, hemicellulose, or lignin; biofuel derived from sugar and starch (other than ethanol derived from corn kernel starch); biofuel derived from waste material, including crop residue, other vegetative waste material, animal waste, food waste and yard waste; diesel-equivalent fuel derived from renewable biomass, including vegetable oil and animal fat; biogas (including landfill gas and sewage waste treatment gas) produced through the conversion of organic matter from renewable biomass; butanol or other alcohols produced through the conversion of organic matter from renewable biomass; and other fuel derived from cellulosic biomass
- For more Information: See program number 10.867 on the CFDA website; USDA program website

3. Biomass Crop Assistance Program (BCAP; Sec. 9011)

- Administered by: Farm Services Agency (FSA)
- Authorization: Title IX of the Farm Security and Rural Investment Act of 2002 (FSRIA; P.L. 107-171) is amended by Title IX, Section 9001 of the Food, Conservation, and Energy Act of 2008 (P.L. 110-246), creating new Section 9011 under FSIRA

- Annual funding: Original mandatory funding authorization for FY2009-FY2012 authorizes "such sums as necessary." The Supplemental Appropriations Act of 2010 (P.L. 111-212) limits mandatory spending on BCAP by allowing no more than $552 million in FY2010 and $432 million in FY2011. The Department of Defense and Full-Year Continuing Appropriations Act, 2011 (P.L. 112-10), further reduced BCAP funding for FY2011 to $112 million. With respect to FY2012 funding, the President's FY2012 budget proposed to limit funding for CHST to $70 million. The remaining annual and establishment payment portion of BCAP would remain at such sums as necessary (SSAN). On June 16, 2011, the House passed an FY2012 appropriations bill (H.R. 2112) that would have eliminated funding for BCAP for FY2012. In contrast, the Senate FY2012 spending bill left BCAP mandatory spending untouched. In the final FY2012 Agriculture appropriations act (P.L. 112-55), BCAP mandatory spending was limited to $17 million. Under ATRA, no new mandatory funding was included for BCAP; however, discretionary funding of $20 million was authorized to be appropriated for FY2013.
- Scheduled termination: Funding authorized through FY2013
- Description: BCAP provides assistance to support the production of eligible biomass crops on land within approved BCAP project areas. In exchange for growing eligible crops, the FSA will provide annual payments through 10- to 15- year contracts. Under these contracts up to 75% of establishment costs may also be provided. FSA will also provide matching payments to eligible material owners at a rate of $1 for each $1 per dry ton paid by a qualified biomass conversion facility. Payments may not exceed $45 per ton for a two-year period and matching payments are available for no more than two years per participant.
- Qualified applicants: Eligible biomass material owners and eligible biomass producers
- Qualified technologies: Eligible material for a matching payment is renewable biomass, as defined by the 2008 farm bill, with several important exclusions including harvested grains, fiber or other commodities eligible to receive payments under the Commodity Title (Title I) of the 2008 farm bill (the residues of these commodities, however, are eligible and may qualify for payment); animal waste and animal waste by-products including fats, oils, greases and manure; food waste and yard waste; and algae. Eligible crops include renewable biomass, with the exception of crops eligible to receive a payment under Title I of the 2008 farm bill and plants that are invasive or noxious, or have the potential to become invasive or noxious.
- For more Information: See the USDA BCAP website

4. Biorefinery Assistance Program (Sec. 9003)

- Administered by: Rural Development
- Authorization: Title IX of the Farm Security and Rural Investment Act of 2002 (FSRIA, P.L. 107-171) is amended by Title IX, Section 9001 of the Food, Conservation and Energy Axt of 2008 (P.L. 110-246) creating new Section 9003 under FSIRA
- Annual Funding:

    - Mandatory authorization: $75 million for FY2009; $245 million for FY2010. $0 for FY2011; $0 for FY2012; any mandatory funding unspent from the FY2010 allocation ($245 million) remains available in FY2013.
    - Discretionary authorization: $150 million authorized annually for FY2009-FY2013. No discretionary funding has been appropriated for BAP through FY2013.
- Scheduled Termination: Funding authorized through FY2013
- Description: The purpose is to assist in the development of new and emerging technologies for the development of advanced biofuels, so as to increase the energy independence of the United States; promote resource conservation, public health, and the environment; diversify markets for agricultural and forestry products and agriculture waste material; and create jobs and enhance the economic development of the rural economy. Loan guarantees are made to fund the development, construction, and retrofitting of commercial-scale biorefineries using eligible technology. The maximum loan guarantee is $250 million.
- Qualified applicants: Individuals, tribal entities, state government entities, local government entities, corporations, farm cooperatives, farmer cooperative organizations, associations of agricultural producers, national laboratories, institutions of higher education, rural electric cooperatives, public power entities, and consortia of any of the previous entities
- Qualified technologies: Technologies being adopted in a viable commercial-scale operation of a biorefinery that produces an advanced biofuel; and technologies that have been demonstrated to have technical and economic potential for commercial application in a biorefinery that produces an advanced biofuel
- For more Information: See the USDA program website at; CFDA program number 10.865

5. Community Wood Energy Program

- Administered by: Forest Service
- Authorization: Title IX of the Farm Security and Rural Investment Act of 2002 (FSRIA, P.L. 107-171 is amended by Title IX Section 9001 of the Food, Conservation, and Energy Act of 2008 (P.L. 110-246), creating new Section 9013 under FSIRA
- Annual funding: Discretionary funding of $5 million annually was authorized to be appropriated for FY2009-FY2012. The Forest Service awarded $49 million in funding from the American Recovery and Reinvestment Act of 2009 (ARRA, P.L. 111-5) for wood-to-energy projects, and the appropriations committee reports in FY2010 and FY2011 have directed the use of $5 million in hazardous fuels funds for biomass energy projects. Under the American Taxpayer Relief Act of 2012 (ATRA; P.L. 112-240), discretionary funding of $15 million was authorized to be appropriated for FY2013.
- Scheduled Termination: Funding authorized through FY2013
- Description: Grants awarded for systems smaller than 5 million Btu per hour for heating (or 2 megawatts) for electric power production as directed by statute. At least

a 50% match is required from Non-Federal funds for grants. Technical assistance will be based on previous work and commitment to future work demonstrated by the applicant. The program is authorized $5 million annually when funded. Grant awards are limited to $50,000 by statute.

- Qualified applicants: State and local governments
- Qualified technology: Biomass

6. New Era Rural Technology Competitive Grants Program

- Administered by: National Institute of Food and Agriculture (NIFA)
- Authorization: National Agricultural Research, Extension, and Teaching Policy Act of 1977 (P.L. 95-113); Food, Conservation, and Energy Act of 2008 (P.L. 110-246)
- Annual funding: $875,000 for FY2010; an estimated $875,000 for FY2011; the Consolidated and Further Continuing Appropriations Act, P.L. 112-55, did not provide funding for the New Era Rural Technology Competitive Grants Program (RTP) in FY2012. As a result, NIFA will not offer the RTP funding opportunity in FY2012.
- Scheduled termination: Authorized through FY2013
- Description: This program provides grant funding for approved technology development, applied research, and training to develop an agriculture-based renewable energy workforce. The initiative shall support the following fields: (A) bioenergy; (B) pulp and paper manufacturing; and (C) agriculture-based renewable energy resources.
- Qualified applicants: Public or private nonprofit community colleges; advanced technology centers
- Qualified technologies: Biomass; bioenergy
- For more information: See the CFDA website, program number 10.314 and the USDA website

7. Repowering Assistance Program (RAP)

- Administered by: Rural Development
- Authorization: Title IX, Section 9003 of the Farm Security and Rural Investment Act of 2002 (FSIRA, P.L. 107-171 is amended by Title IX, Section 9001 of the Food, Conservation and Energy Act of 2008 (P.L. 110-246), creating new Section 9004 under FSIRA
- Annual funding:
    - Mandatory authorization: $35 million for FY2009, to remain available until expended. $15 million in FY2010 was appropriated through FY2012. Any mandatory funding unspent from the FY2009 allocation of $35 million is available in FY2013
    - Discretionary authorization: Discretionary funding of $15 million annually for FY2009-FY2013 was authorized to be appropriated under the 2008 farm bill and the ATRA extension; however, only $15 million in FY2010 has been

appropriated through FY2013. No new mandatory funding was included for RAP under the ATRA farm bill extension; however, any mandatory funding unspent from the FY2009 allocation of $35 million remains available through FY2013.

- Scheduled termination: Authorized through FY2013
- Description: The Repowering Assistance Program (RAP) makes payments to eligible biorefineries (those in existence on the date of enactment of the 2008 farm bill, June 18, 2008) to encourage the use of renewable biomass as a replacement for fossil fuels used to provide heat for processing or power in the operation of these eligible biorefineries. Not more than 5% of the funds shall be made available to eligible producers with a refining capacity exceeding 150 million gallons of advanced biofuel per year.
- Qualified applicants: Eligible biorefinery. The biorefinery must have been in existence on or before June 18, 2008
- Qualified technologies: Renewable biomass
- For more Information: See program number 10.866 on the CFDA website and the USDA program website

8. Rural Energy for America Program (REAP) Grants and Loans

- Administered by: Rural Development
- Authority: Title IX, Section 9006 of the Farm Security and Rural Investment Act of 2002 (FSIRA, P.L. 107-171) is amended by Title IX, Section 9001 of the Food Conservation, and Energy Act of 2008 (P.L. 110-246), creating new Section 9007 under FSIRA. The new Section 9007 converted the federal Renewable Energy Systems and Energy Efficiency Improvements Program into the Rural Energy for America Program (REAP)
- Annual funding:
  - Mandatory authorization: The FY2011 appropriations act (Department of Defense and Full-Year Continuing Appropriations Act, 2011; P.L. 112- 10) reduced REAP discretionary funds from $25 million to $5 million, but left REAP's mandatory funding of $70 million intact. The FY2012 Agriculture Appropriations Act (P.L. 112-55) limited REAP mandatory spending to $22 million while discretionary funding was authorized at $3.4 million, split evenly between grants and loan guarantees.
  - Discretionary authorization: $25 million authorized annually for FY2009-FY2013. Actual discretionary appropriations have been $5 million in FY2009, $40 million in FY2010, $5 million in FY2011, and $3.4 million in FY2012
- Scheduled termination: Authorized through FY2013
- Description: REAP promotes energy efficiency and renewable energy for agricultural producers and rural small businesses through the use of (1) grants and loan guarantees for energy efficiency improvements and renewable energy systems, and (2) grants for energy audits and renewable energy development assistance.
- Qualified applicants: Commercial; schools; state, local, and tribal governments; rural electric cooperatives; agricultural; public power entities

- Qualified technologies: Solar water heat; solar space heat; solar thermal electric; photovoltaics; wind; biomass; hydroelectric; renewable transportation fuels; geothermal electric; geothermal heat pumps; CHP/cogeneration; hydrogen; direct-use geothermal; anaerobic digestion; small hydroelectric; tidal energy; wave energy; ocean thermal; renewable fuels; fuel cells using renewable fuels; microturbines. Specific energy efficiency technologies not identified

9. Sustainable Agriculture Research and Education Program (SARE)

- Administered by: National Institute of Food and Agriculture; Agricultural Research Service; and other appropriate agencies
- Authorization: Food, Agriculture, Conservation and Trade Act of 1990 (P.L. 101-624); Food, Agriculture, Conservation and Trade Act Amendments of 1991 (P.L. 102-237); Federal Agriculture Improvement and Reform Act of 1996 (P.L. 104-127); Food, Conservation, and Energy Act of 2008 (P.L. 110-246)
- Annual funding: $12.5 million for FY2006; $12.4 million for FY2007; $9.1 million for FY2008; $14.5 million for FY2009; $14.5 million for FY2010, $19.2 million for FY2011; $13.5 for FY2012; and an estimated $12.5 million for FY2013.
- Scheduled termination: None
- Description: The purpose of the Sustainable Agriculture Research and Education Program (SARE) is, in part, to encourage research designed to increase our knowledge concerning agricultural production systems that conserve soil, water, energy, natural resources, and fish and wildlife habitat. SARE provides grants through the agricultural bioenergy feedstock and energy efficiency research and extension initiative for projects with the purpose of enhancing the production of biomass energy crops and the energy efficiency of agricultural operations.
- Qualified applicants: Federal and state governments; colleges and universities; state agricultural experiment stations; state cooperative extension services; nonprofit organizations; individuals with demonstrable expertise
- Qualified technologies: Biomass; biofuels; other technologies not identified.
- For more information: See CFDA program website, program number 10.215

IV. Department of the Interior

1. Energy and Mineral Development Program: Minerals and Mining on Indian Lands

- Administered by: Bureau of Indian Affairs. Energy and Mineral Development Program
- Authority: Indian Self-Determination and Education Assistance Act (P.L. 93- 638), 25 USC 450; Snyder Act of 1921 (P.L. 67-85), 25 USC 13; Indian Minerals Development Act (P.L. 97-382), 25 USC 2101 et seq.; Umatilla Basin Project Act (P.L. 101-557), 16 USC 1271 et seq.

- Annual funding: $12.972 million for FY2010; $12.87 million for FY2011; $12.7 million for FY2012; $12 million for FY2013
- Scheduled termination: None
- Description: Funding may be used to facilitate the inventory, assessment, promotion and marketing of both renewable and nonrenewable energy and mineral resources on Indian lands. Funds are awarded competitively to support assessment and inventory programs or to develop baseline data, but cannot be used for development purposes.
- Qualified applicants: Federally recognized Indian tribes; individual American Indian mineral owners
- Qualified technologies: Renewable energy technologies
- For more information: See program number 15.038 at the CFDA website

2. Tribal Energy Development Capacity Grant Program

- Administered by: Bureau of Indian Affairs
- Authorization: Energy Policy Act of 1992 (EPACT; P.L. 102-486); Tribal Energy Resource Development and Self-Determination Act of 2005 (Title V of Energy Policy Act of 2005; P.L. 109-58)
- Annual funding: $375,000 for FY2007; $1 million for FY2008; no estimate available for FY2009; $138,839 for FY2010; $250,000 for FY2011; $0 for FY2012; an estimated $400,000 for FY2013
- Scheduled termination: None
- Description: This program provides grants to Indian tribes to (1) develop and sustain the managerial and technical capacity needed to develop their energy resources; and (2) properly account for resulting energy production and revenues.
- Qualified applicant: Tribal governments
- Qualified technologies: Renewable energy technologies
- For more information: See program number 15.148 at the CFDA website; or contact IEED, the Division of Indian Energy at (202) 219-0740

V. Small Business Administration

1. 7(a) Loan Guarantees

- Administered by: Small Business Administration (SBA)
- Authority: Small Business Act of 1953 (Public Law 83-163)
- Scheduled termination: None
- Annual Funding: $168.0 million for FY2011 ($80.0 million for 7(a) loan guaranty credit subsidies and $88.0 million for administration); $233.0 million for FY2012 ($139.4 million for 7(a) loan guaranty credit subsidies and $93.6 million for administration); $303.1 million for FY2013 (approximately $214.2 million for 7(a) loan guaranty credit subsidies and $88.9 million for administration) in FY2013; and

$107.4 million budget requested for FY2014 (no funding is requested for 7(a) loan guaranty credit subsidies and $107.4 million for administration).

- Description: To guarantee loans from lenders to small businesses which are unable to obtain financing on reasonable terms and conditions in the private credit marketplace, but can demonstrate an ability to repay loans if granted, in a timely manner. Guaranteed loans are made available to for-profit small businesses. The SBA's 7(a) lending authority includes (1) regular 7(a); (2) SBAExpress Program; (3) Patriot Express Program; (4) the CapLines Program (5) Small/Rural Lender Advantage initiative; (6) Export Express Program; (7) Export Working Capital Program; (8) Preferred Lenders Program; (9) International Trade; and (10) Small Loan Advantage and Community Advantage initiatives.
- Qualified applicant: Small businesses (meeting the size and eligibility standards)
- Qualified technologies: Not specifically listed

2. 504 Loan Guarantees

- Administered by: Small Business Administration (SBA)
- Authority: Small Business Investment Act of 1958 (Public Law 85-699)
- Scheduled termination: None
- Annual Funding: $38.9 million for administration in FY2011; $107.3 million for FY2012 ($67.7 million for 504/CDC loan guaranty credit subsidies and $39.6 million for administration); $140.2 million for FY2013 (approximately $102.7 million for 504/CDC loan guaranty credit subsidies and $37.5 million for administration; and $146.5 million budget request for FY2014 ($107.0 million for 504/CDC loan guaranty credit subsidies and $39.5 million for administration.
- Description: provides long-term fixed rate financing for major fixed assets, such as land, buildings, equipment, and machinery. Of the total project costs, a third-party lender must provide at least 50% of the financing, the Certified Development Company provides up to 40% of the financing through a 100% SBA-guaranteed debenture, and the applicant provides at least 10% of the financing. Qualified projects are required to modernize or upgrade facilities by: (1) reducing energy use by at least 10 percent; or (2) employing sustainable design, or low-impact design, that reduces fossil fuel use; or (3) planning, equipping, and/or installing process upgrades or renewable energy sources—such as the small-scale (micropower) production of energy for individual buildings or communities consumption; or (4) supporting renewable fuels production by biodiesel and ethanol producers.
- Qualified applicant: Small businesses (meeting the size and eligibility standards)
- Qualified technologies: fossil fuels; energy efficiency equipment; renewable energy sources (unspecified); renewable fuels, including biodiesel and ethanol

VI. U.S. Department of Housing and Urban Development

1. Energy Efficient Mortgages (EEMs)

- Administered by: Federal Housing Administration (FHA) and Department of Veterans Affairs (VA). Conventional mortgages: Private lenders that sell mortgage loans to Fannie Mae or Freddie Mac may offer Energy Efficient Mortgages (EEMs)
- Authority: EEMs were initially introduced by lenders in the 1980s. In 1992, three pieces of legislation passed by Congress worked towards standardizing and expanding the use of EEMs. In 1992, Congress established an FHA Energy Efficient Mortgage Pilot Program (P.L. 102-550). The program was later expanded beyond five states to become a national program. The Housing Economic Recovery Act of 2008 (HERA; P.L. 110-289) increased the maximum amount that can be added to an FHA mortgage for energy efficient improvements. The 111th Congress also passed some incentives to encourage green home improvements in the American Recovery and Reinvestment Act of 2009 (ARRA; P.L. 111-5).
- Scheduled termination: None
- Description: Homeowners can take advantage of EEMs to finance a variety of energy efficiency measures, including renewable energy technologies, in a new or existing home. The U.S. federal government directly provides these loans through the FHA and VA lending programs. Fannie Mae and Freddie Mac will also purchase EEMs from primary lenders. Primary lenders may issue EEMs that do not conform to underwriting standards.
- Qualified applicants: The loan is available to anyone who meets the income requirements for FHA's Section 203 (b), provided the applicant can meet the monthly mortgage payments. New and existing owner-occupied homes of up to two units qualify for this loan. Cooperative units are not eligible. VA: available to qualified military personnel, reservists and veterans; Conventional: Applicants qualifying for a conventional mortgage are also eligible for an energy efficient mortgage
- Qualifying technologies: Passive solar space heat; solar water heat; solar space heat; photovoltaics; daylighting; and other technologies not specifically identified
- For more information: See the HUD, Energy Star and DSIRE websites

VII. Department of Labor

1. Program of Competitive Grants for Worker Training and Placement in High Growth and Emerging Industry Sectors

- Administered by: Employment Training Administration
- Authority: American Recovery and Reinvestment Act of 2009 (ARRA; P.L. 111- 5), Title VIII
- Annual funding: (Project Grants) $0 for FY2008; $750 million for FY2009 (ARRA) which remained available through June 30, 2010; $0 for FY2010- FY2013

- Scheduled termination: None
- Description: This program provides competitive grants for worker training and placement in high growth and emerging industry sectors.
- Qualified applicants: State, local, and tribal governments; colleges and universities; private nonprofit institutions/organizations
- For more information: See the U.S. Department of Labor's (DOL's) *Training and Employment Notice* for this program; and program number 17.275 at the CFDA website

VIII. Department of Veterans Affairs

1. Energy Efficient Mortgages (EEMs)

- Administered by: FHA and VA. Conventional mortgages: Private lenders that sell mortgage loans to Fannie Mae or Freddie Mac may offer EEMs
- Authority: EEMs were initially introduced by lenders in the 1980s. In 1992, three pieces of legislation passed by Congress worked towards standardizing and expanding the use of EEMs. In 1992, Congress established an FHA Energy Efficient Mortgage Pilot Program (P.L. 102-550). The program was later expanded beyond five states to become a national program. The Housing Economic Recovery Act of 2008 (HERA; P.L. 110-289) increased the maximum amount that can be added to an FHA mortgage for energy efficient improvements. The 111th Congress also passed some incentives to encourage green home improvements in the American Recovery and Reinvestment Act of 2009 (ARRA; P.L. 111-5)
- Scheduled termination: None
- Description: Homeowners can take advantage of EEMs to finance a variety of energy efficiency measures, including renewable energy technologies, in a new or existing home. The U.S. federal government directly provides these loans through the FHA and VA lending programs. Fannie Mae and Freddie Mac will also purchase EEMs from primary lenders. Primary lenders may issue EEMs that do not conform to underwriting standards.
- Qualified applicants: The loan is available to anyone who meets the income requirements for FHA's Section 203 (b), provided the applicant can meet the monthly mortgage payments. New and existing owner-occupied homes of up to two units qualify for this loan. Cooperative units are not eligible. VA: available to qualified military personnel, reservists and veterans; Conventional: applicants qualifying for a conventional mortgage are also eligible for an energy efficient mortgage
- Qualifying technologies: Passive solar space heat; solar water heat; solar space heat; photovoltaics; daylighting; and other technologies not specifically identified
- For more information: See the HUD, Energy Star and DSIRE websites

APPENDIX A. SUMMARY OF FEDERAL RENEWABLE ENERGY AND ENERGY EFFICIENCY INCENTIVES/INDEX OF PROGRAMS

Table A-1. Federal Incentives by Agency

Administering Agency	Program	Description	U.S. Code Citation	FY2013 Appropriations[a]	Expiration Date
Department of Energy	Advanced Manufacturing Office (formerly Industrial Technologies Program)	Develops and supports the commercialization of new energy efficient technologies to improve industrial efficiency while increasing productivity	42 USC §17111 et seq.	$116.3 million	None
	Advanced Research Projects Energy Financial Assistance Program (ARPA-E)	Grants to finance sophisticated energy technology R&D projects to accelerate transformation technology advances.	42 USC §16538	$276.7 million	Program evaluation after FY2012
	Bioenergy Technologies Program (formerly Biomass and Biorefinery Systems R&D Program)	Grants to develop cost-effective technologies and systems to transform domestic biomass resources into biofuels, bioproducts, and biopower.	42 USC §16232	$200.5 million	None
	Building Technologies Program	Provides financial and technical assistance to improve efficiency of buildings and the equipment, components and systems within them	42 USC §17061-17124	$220.5 million	None
	Conservation Research and Development Grant Program	Grants to finance long-terms R&D efforts in buildings technologies, Industrial technologies, vehicle technologies, and hydrogen/fuel cell technologies.	42 USC §5901 et seq.	Estimated $75.1 million	None
	Electricity Delivery and Energy Reliability, Research, Development and Analysis Grant Program	Grants to develop cost-effective technology to enhance the reliability, efficiency, and resiliency of the electric grid	42 USC §17381 et seq.	$99.8 million	None

Administering Agency	Program	Description	U.S. Code Citation	FY2013 Appropriationsa	Expiration Date
	Energy Efficiency and Renewable Energy Information Dissemination, Outreach, Training, and Technical Analysis/Assistance Program	Provides financial assistance to stimulate increased usage of energy efficiency/ renewable energy technologies and accelerate the adoption of these technologies	See Notes field[b]	Estimated $8.6 million	None
	Energy Efficiency and Renewable Energy Technology Deployment, Demonstration, and Commercialization Grant Program	Provides financial assistance for deployment, demonstration, and commercialization of energy efficiency and renewable energy technologies	42 USC §16191 et seq. and 42 USC §16231 et seq.	$0	None
	Federal Energy Management Program	Provides assistance to federal agencies in developing and implementing energy efficiency and renewable energy technologies to meet energy management goals	42 USC §17131 et seq.	$30.1 million	None
	Financial Assistance Program (Office of Science)	Grants support research in the basic sciences and advanced technology concepts and assessments in fields related to energy	42 USC §13503	Estimated $962.6 million	None
	Geothermal Technologies Program	Partners DOE with industry, academia, and research facilities to develop geothermal energy technologies	42 USC §16231 et seq. and 42 USC §17191 et seq.	$38.1 million	None
	Hydrogen & Fuel Cell Technologies Program	Partners DOE with industry, academia, and national laboratories to develop hydrogen and fuel cell technologies for the marketplace	42 USC §16151 et seq.	$104.3 million	None
	Inventions and Innovations Program	Provides financial and technical assistance to develop innovative cost-effective ideas and inventions with future commercial value. Focus on energy efficiency and renewable energy technologies.	42 USC § 5913	Estimated $914,000	None

Table A-1. (Continued)

Administering Agency	Program	Description	U.S. Code Citation	FY2013 Appropriations[a]	Expiration Date
	Loan Guarantee Program	Loan guarantees to encourage commercial use of new or significantly improved technologies that avoid, reduce or sequester air pollutants or greenhouse gas emissions	42 USC §16511 et seq.	$0 for the Innovative Technology Loan Guarantee Program (Section 1703) $0 for the Temporary Loan Guarantee Program (Section 1705)	None
	Regional Biomass Energy Programs	Provides financial assistance to increase America's use of fuels, chemicals, materials, and power made from domestic biomass	See Notes fieldb	$0	None
	Renewable Energy Production Incentive	Provides incentive payments for electricity generated and sold by new qualifying renewable energy facilities	42 USC §13317	$0	End of FY2026
	Renewable Energy Research and Development Program	Provides financial assistance to conduct R&D efforts in renewable energy technologies	42 USC §16231 et. seq.	$141.5 million	None
	Small Business Innovation Research/Small Business Technology Transfer Programs	Grants for small businesses to develop and commercialize energy technologies, including energy efficiency and renewable energy technologies	15 USC §638	$33.3 million	None
	Solar Energy Technologies Program	Program partners with industry, universities, and national laboratories to finance R&D and bring reliable and affordable solar energy technologies to the marketplace	42 USC §16231 et seq. and 42 USC §17171 et seq.	$290.7 million	None
	State Energy Program	Provides grants to states to design and implement their own renewable energy and energy efficiency programs	42 USC §6321 et seq.	$50.3 million	None

Administering Agency	Program	Description	U.S. Code Citation	FY2013 Appropriations[a]	Expiration Date
	Tribal Energy Program	Provides financial and technical assistance, education, and training to tribes to evaluate and develop renewable energy sources and energy efficiency measures	25 USC §3501 et seq.	$10.1 million	None
	Vehicle Technologies Program	Program partners with industry leaders to develop and deploy advanced transportation technologies to improve vehicle fuel efficiency and domestically produce clean and affordable alternative fuels	42 USC §17011 et seq.	$330.8 million	None
	Water Power Program (formerly Wind and Hydropower Technologies Program)	Program partners with industry, states, federal entities, and other stakeholders on R&D projects to improve the performance, lower costs, and accelerate the deployment of water power technologies	42 USC §16231 et. seq and 42 USC §17211 et seq.	$59.1 million	None
	Weatherization Assistance Program	Provides financial and technical assistance to states to increase the energy efficiency of low-income households	42 USC §6861 et seq.	$68.4 million	None
	Wind Energy Program (formerly Wind and Hydropower Technologies Program)	Program partners with industry, states, federal entities, and other stakeholders on R&D projects to improve the performance, lower costs, and accelerate the deployment of wind energy technologies	42 USC §16231 et. seq	$93.8 million	None
Internal Revenue Service	Business Energy Investment Tax Credit	Provides a tax credit for 30% of total expenditures on eligible systems placed in service, except geothermal systems, microturbines, and combined heat and power systems (10%)	26 USC §48	N/A	12/31/2016 for most eligible systems (except geothermal and solar thermal)

Table A-1. (Continued)

Administering Agency	Program	Description	U.S. Code Citation	FY2013 Appropriations[a]	Expiration Date
	Energy-Efficient Appliance Tax Credit for Manufacturers	Provides a tax credit to manufacturers for appliances that meet Energy Star 2007 requirements	26 USC §45M (amended)	N/A	12/31/2011
	Energy Efficient Commercial Buildings Tax Deduction	Tax deduction for certain qualifying systems and buildings	26 USC §179D (amended)	N/A	12/31/2013
	Energy-Efficient New Homes Tax Credit for Home Builders	Provides tax credits of up to $2,000 for builders of new, energy-efficient homes	26 USC §45L (amended)	N/A	12/31/2011
	Qualified Energy Conservation Bonds (QECBs)	Bond authority is allocated to state, local, and tribal governments to finance a broad range of energy efficiency and renewable energy projects	26 USC § 54A; 26 USC §54D; 26 USC § 6431	N/A	N/A
	Qualifying Advanced Energy Manufacturing Investment Tax Credit	Provides tax credits to encourage a U.S. based renewable energy manufacturing sector	26 USC §48C	N/A	Applications no longer accepted; Phase concept papers were due to DOE by 4/9/2013; final applications were due to DOE on 7/23/2013.
	Renewable Energy Production Tax Credit (PTC)	Provides a per-kilowatt-hour tax credit for electricity generated by qualified renewable energy technologies and sold during the tax year	26 USC §45 (amended)	N/A	Generally, 10 years after the date the facility placed in service (with exceptions for some technology types)
	Residential Energy Conservation Subsidy Exclusion (Corporate)	Corporate tax exemption for energy-conservation subsidies are provided by public utilities, either directly or indirectly	26 USC §136 (amended)	N/A	None

Administering Agency	Program	Description	U.S. Code Citation	FY2013 Appropriations[a]	Expiration Date
Residential Energy Conservation Subsidy Exclusion (Personal)	Personal tax exemption for energy-conservation subsidies provided by public utilities, either directly or indirectly	26 USC §136 (amended)	N/A	None	Residential Energy Conservation Subsidy Exclusion (Personal)
Residential Energy Efficiency Tax Credit	Provides tax credit to residents/individuals for the installation of qualified energy efficient equipment to existing homes (primary residence)	26 USC §25C	N/A	12/31/2011	Residential Energy Efficiency Tax Credit
	Residential Renewable Energy Tax Credit	Provides a tax credit to residents/ individuals for the installation of qualified renewable energy systems to existing homes. Home must serve as owner's primary residence.	26 USC §25D (amended)	N/A	12/31/2016
	Alternative Motor Vehicle Credit	Provides tax credit for hybrid and lean-burn vehicles.	26 USC §30B	N/A	Varies by technology type: See Table A-2 below
Department of Agriculture	Assistance to High Energy Cost Rural Communities Program	Provides financial assistance to rural communities with high energy costs	7 USC. §918a	$7.6 million	None
	Bioenergy Program for Advanced Biofuels	Supports and ensures an expanding production of advanced biofuels by providing payments to advanced biofuels producers	7 USC §8105	$105 million in mandatory funding to remain available until expended; P.L. 112-55 limits mandatory spending to $65 million; $0 in discretionary spending for FY2013	Authorized through FY2012

Table A-1. (Continued)

Administering Agency	Program	Description	U.S. Code Citation	FY2013 Appropriations[a]	Expiration Date
	Biomass Crop Assistance Program (BCAP)	Provides assistance to support the production of eligible biomass crops on land within approved project areas	7 USC §8111	No new mandatory funding for FY2013; discretionary funding authorized $20 million for FY2013	Authorized through FY2012
	Biorefinery Assistance Program	Assists in the development of new technologies for development of biofuels	7 USC §8103	$0 for FY2013; mandatory funding unspent from FY2010 allocation of $245 million remains available in FY2013	Authorized through FY2012
	Community Wood Energy Program	Provides grants to states and local governments to develop community wood energy plans or acquire or upgrade community wood energy systems	7 USC §8113	$0 mandatory funding; $15 million in discretionary funding authorized to be appropriated for FY2013	Authorized through FY2012
	New Era Rural Technology Competitive Grants Program	Provides grant funding for approved technology development, applied research, and training to develop bioenergy and agriculture-based renewable energy resources	7 USC §3319e	$0	Authorized through FY2012
	Repowering Assistance Program	Provides financial incentives to biorefineries in existence on June 18, 2008, to replace the use of fossil fuels used to produce heat or power by installing new systems that use renewable biomass or to produce new energy from renewable biomass	7 USC §8104	$15 million in FY2012 was appropriated through FY2012; Any mandatory funding unspent from the FY2009 allocation of $35 million remains	Authorized through FY2012

Administering Agency	Program	Description	U.S. Code Citation	FY2013 Appropriations[a]	Expiration Date
				available through FY2013; $0 in discretionary funding for FY2013.	
	Rural Energy for America Program	Provides grants and loan guarantees to promote energy efficiency and renewable energy to agricultural producers and rural small businesses	7 USC §8107	$22 million in mandatory funding; and $3.4 million in discretionary funding	Authorized through 2012
	Sustainable Agriculture Research and Education	Provides grants for research projects with the purpose of enhancing biomass energy crop production and increasing the energy efficiency of agricultural operations	7 USC §5801 et seq.	Estimated $12.5 million	None
Department of Housing and Urban Development	Energy Efficient Mortgages	Provides backing of loans for energy efficient mortgages to finance the installation of energy efficiency or renewable energy technologies in new or existing homes	12 USC §1701z-16	N/A	None
Department of the Interior	Energy and Mineral Development Program: Minerals and Mining on Indian Lands	Facilitate the inventory, assessment, promotion and marketing of both renewable and nonrenewable energy and mineral resources on Indian lands	25 USC §450; 25 USC §13; 25 USC §2101 et seq; 16 USC. §1271 et seq.	$12 million	None
	Tribal Energy Development Capacity Grant	Grants to Indian tribes to develop and sustain the managerial and technical capacity needed to develop their energy resources and properly account for resulting energy production and revenues	25 USC §3502	Estimated $400,000	None
Department of Labor	Program of Competitive Grants for Worker Training and Placement in High Growth and Emerging Industry Sectors	Intended to preserve and create jobs; promote economic recovery; assist those most impacted by the recession; provide investments and invest in infrastructure	See Notes fieldb	$0	None

Table A-1. (Continued)

Administering Agency	Program	Description	U.S. Code Citation	FY2013 Appropriations[a]	Expiration Date
Department of Veterans Affairs	Energy Efficient Mortgages	Provides backing of loans for energy efficient mortgages to finance the installation of energy efficiency or renewable energy technologies in new or existing homes	12 USC §1701z-16	N/A	None
Small Business Administration	7(a) Loan Guarantees	Provides guaranteed loans from lenders to small businesses	15 USC § 636(a)	$214.2 million	None
	504 Loan Guarantees	Provides long-term fixed rate financing for major fixed assets, such as land, buildings, equipment, and machinery.	16 USC §685	$107 million	None

Source: CRS.

[a] FY2013 Appropriations data compiled by CRS using executive agency budget justifications, congressional committee reports, and program descriptions from the online edition of the *Catalog of Federal Domestic Assistance*. Reflects the FY2013 Continuing Resolution but does not incorporate the reductions from the Budget Control Act sequester.

[b] Some programs are not specifically identified or codified in the United States Code.

Table A-2. Alternative Motor Vehicle Credit (26 USC §30B)

Type of Credit	Expiration Date
Fuel Cell Motor Vehicle Credit	December 31, 2014
Qualified Plug-In Electric Drive Motor Vehicle Credit	December 31, 2014
Qualified Plug-In Electric Motor Vehicle Conversion Credit	December 31, 2011
Advanced Lean Burn Technology Motor Vehicle Credit	December 31, 2010
Qualified Alternative Fuel Motor Vehicle Credit	December 31, 2010
Qualified Hybrid Motor Vehicle Credit	December 31, 2010

Source: U.S. Code and IRS.

APPENDIX B. INDEX OF PROGRAMS BY APPLICANT ELIGIBILITY AND TECHNOLOGY TYPE

Table B-1. Index of Programs by Applicant Eligibility

Applicant Eligibility	Program Numbers
Advanced Technology Centers	III-6
Agricultural/Extension/Biofuel Producers	II-3, III-2, III-3, III-4, III-7, III-8, III-9
Alaska Native Corporations	I-14
Builder/Developer	II-4, II-5
Commercial/Industrial/For-Profit	I-1, I-2, I-3, I-4, I-5, I-6, I-7, I-10, I-12, I-13, I-14, I-15, I-17, I-20, I-21, I-23, I-24, II-3, II-4, II-6, II-8, II-12, III-1, III-2, III-3, III-4, III-7, III-8
Cooperative/Collaborative/Consortia	I-16, I-20, III-1, III-4, III-8
Federal Government	I-4, I-6, I-7, I-12, I-22, II-4, III-9
Higher Education (Colleges and Universities)	I-1, I-2, I-3, I-4, I-5, I-6, I-7, I-8, I-12, I-13, I-14, I-17, I-20, I-21, I-23, I-24, III-4, III-6, III-9, VII-1
Local Government	I-2, I-6, I-7, I-8, I-12, I-13, I-14, I-16, I-17, I-21, I-23, I-24, III-1, III-4, III-5, III-8, VII-1
National Laboratories	I-4, I-5, I-6, I-7, I-8, I-12, III-4
Nonprofit	I-2, I-13, I-14, I-16, I-17, I-20, I-21, I-23, I-24, III-1, III-9, VII-1
Other/Cross-Cutting	I-20, II-12, II-13
Research Organization	I-20, I-21
Residential/Individual	I-11, I-14, II-1, II-2, II-9, II-10, II-13, III-1, III-4, III-9, IV-1, V-1
Schools	III-8
Small Businesses	I-6, I-7, I-11, I-23, I-25, V-1, V-2
State Government	I-2, I-6, I-7, I-8, I-9, I-12, I-13, I-14, I-15, I-16, I-17, I-18, I-21, I-23 , I-24, II-4, II-11, III-1, III-4, III-5, III-8, III-9, VII-1
Tribal Government	I-6, I-9, I-13, I-14, I-16, I-17, I-18, I-19, I-21, I-23, I-24, II-11, III-1, III-4, III-8, IV-1, IV-2, VII-1
U.S. Territories	I-9, I-18
Utilities	I-16, II-3, III-4, III-8
Veterans	VI-1, VIII-1

Source: CRS.

a. Program numbers correspond to agency (roman numeral) and (arabic) number assigned to each program as displayed in the Table of Contents.

Table B-2. Index of Programs by Technology Type

Qualified Technologies	Program Numbers[a]
Advanced Batteries	I-13
Air Conditioners	1-9, I-19, II-1, II-4
Alternative Vehicles/Vehicle Technologies	I-4, I-12, II-7, II-12, II-13
Anaerobic Digestion	I-16, II-7, II-11, II-12, III-8
Batteries (Energy Storage)	II-7
Biodiesel / Biofuels	I-1, I-12, I-24, III-2, III-4, III-9
Boilers	1-9, I-19, II-1, II-4
Biomass	I-1, I-2, I-15, I-16, I-17, I-19, I-22, II-1, II-3, II-8, II-11, II-12,
	III-2, III-3, III-4, III-5, III-6, III-7, III-8, III-9
Caulking/Weather Stripping	I-9, I-19, II-4
Chillers	I-19, II-4
Clothes Washers	I-19, II-6
Combined Systems/CHP/Energy Management Systems	I-8, I-19, II-3, II-12, III-8
Comprehensive/Whole Building	I-19, II-4, II-5
Dishwashers	II-6
Doors	I-19, II-1, II-4
Duct/Air Sealing	I-9, I-19, II-4
Equipment (Energy Efficient)	I-8
Fuel Cells	I-4, I-8, I-13, I-15, I-17, I-24, II-2, II-3, II-7, II-12, II-13, III-8
Furnaces	1-9, I-19, II-1, II-4
Geothermal (All)	I-3, I-15, I-17, I-22, II-3, II-12, III-8
—Geothermal (Direct Use)	II-3, II-12, III-8
—Geothermal (Electric)	I-16, I-19, I-24, II-3, II-7, II-8, II-11, 11-12, III-8
—Geothermal (Heat Pumps)	I-19, II-2, II-3, II-7, II-12, III-8
Heat Pumps	II-1, II-4
Hybrid Electric	I-12, II-13
Hydrogen	I-4, I-13, I-15, I-17, II-8, III-8
Hydropower (All)	I-6, I-15, I-17, I-22, II-8, II-11
—Hydroelectric	I-6, I-19, I-24, II-8, II-11, III-8
—Hydrokinetic	I-6, II-8, II-11
—Ocean	I-6, I-16, I-22, I-24, II-8, II-11, III-8
—Tidal	I-6, I-16, I-24, II-8, II-11, III-8
—Wave	I-6, I-16, I-24, II-8, II-11, III-8
Insulation	I-9, I-19, II-1, II-4,
Landfill Gas	I-16, II-8, II-11, II-12
Lighting/Lighting Sensors	I-8, I-19, I-24, II-3, II-4, II-7, II-12, VI-1, VIII-1
Manufacturing Facilities	I-24
Microturbines	II-3, II-7, III-8
Municipal Solid Waste	II-8, II-11, II-12
Other Technologiesb	I-9, I-11,1-13, I-14, I-18, I-19, I-20, I-21, I-23, I-25, II-7, II-9, II-10, III-1, III-8, III-9, IV-1, IV-2, V-1, V-2, VI-1, VIII-1
Programmable Thermostats	1-9, I-19

Qualified Technologies	Program Numbers[a]
Refrigerators/Freezers	I-19, II-6
Renewable Transportation Fuels	I-24, II-12, III-8
Roofs	I-19, II-1, II-4
Siding	I-19, II-4
Smart Grid	I-21, II-7
Solar (All)	I-5, I-8, I-15, I-17, I-22, II-2, II-3, II-12, III-8
—Photovoltaics	1-5, I-8, I-16, I-19, I-24, II-2, II-3, II-7, II-9, II-10, II-11,II-12, III-8, VI-1, VIII-1
—Solar Space Heat	I-19, II-2, II-3, II-9, II-10, II-12, III-8, VI-1, VIII-1
—Solar Thermal Electric/Process	I-16, I-24, II-2, II-3, II-7, II-11, II-12, III-8
—Solar Water Heat	II-2, II-3, II-7, II-9, II-10, II-12, III-8, VI-1, VIII-1
Water Heaters	I-19, II-1, II-4
Wind	I-7, I-16, I-17, I-19, I-22, I-24, II-2, II-3, II-7, II-8, II-11,II-12, III-8
Windows	I-8, I-9, I-19, II-1, II-4

Source: CRS.

[a] Program numbers correspond to agency (roman numeral) and (arabic) number assigned to each program as displayed in the Table of Contents.

[b] Other technologies include cross-cutting and advanced technologies, other unspecified technologies, all energy efficiency and/or renewable energy technologies, or not specifically identified.

APPENDIX C. EXPIRED FEDERAL ENERGY EFFICIENCY AND RENEWABLE ENERGY INCENTIVE PROGRAMS

1. Assisted Housing Stability and Energy and Green Retrofit Investments Program (Recovery Act Funded)

- Administered by: Department of Housing and Urban Development (HUD)
- Authority: American Recovery and Reinvestment Act of 2009 (ARRA; P.L. 111- 5), Title XII
- Annual funding: (Project Grants) $0 for FY2009; $235 million for FY2010 ; $0 for FY2011—All obligations were to be made by September 30, 2010
- Scheduled termination: All obligations were to be made by September 30, 2010. Receiving property owners were required to spend the funds on the specific improvements within two years of receipt.
- Description: This program will provide funding for energy and green retrofit investments to certain eligible assisted, affordable multifamily properties. Funding includes incentives for participating property owners, a set-aside for administrative functions, and a set-aside for due diligence and underwriting support. Assistance will be for specific retrofit purposes.
- Qualified applicant: Residential
- Qualifying technologies: Specific technologies not identified
- For more information: See program number 14.318 at the CFDA website

2. Clean Renewable Energy Bonds (CREBs)

- Administered by: Internal Revenue Service
- Authority: Established by the Energy Policy Act of 2005 (EPACT 2005; P.L. 109-58); Tax Relief and Health Care Act of 2006 (P.L. 109-432); Energy Improvement and Extension Act of 2008 (P.L. 110-343), Division B; American Recovery and Reinvestment Act of 2009 (ARRA; P.L. 111-5); 26 USC 54 (old CREBs); 26 USC 54A (new CREBs); 26 USC 54C (new CREBs); IRS Notice 2009-33; IRS Announcement 2010-54
- Annual funding: EPACT originally allocated $800 million of tax credit bonds to be issued between January 1, 2006, and December 31, 2007. Following the enactment of the federal Tax Relief and Health Care Act of 2006, the IRS made an additional $400 million in CREBs financing available for 2008 through Notice 2007-26. In November 2006, the IRS announced that the original $800 million allocation had been reserved for a total of 610 projects. The additional $400 million (plus surrendered volume from the previous allocation) was allocated to 312 projects in February 2008. Of the $1.2 billion total of tax-credit bond volume cap allocated to fund renewable-energy projects, state and local government borrowers were limited to $750 million of the volume cap, with the rest reserved for qualified municipal or cooperative electric companies. The Energy Improvement and Extension Act of 2008 (Div. A, Section107) allocated $800 million for new CREBs. In February 2009, the American Recovery and Reinvestment Act of 2009 (Div. B, Section 1111) allocated an additional $1.6 billion to expand the total new CREBs allocation to $2.4 billion
- Scheduled termination: The IRS is no longer accepting applications for new CREB bonds. The deadline for new CREB applications from electric cooperatives expired November 1, 2010 and bonds for government entities and public power providers was fully allocated in October 2009.
- Description: The IRS is not currently accepting applications for new CREB bond volume. CREBs are used to finance renewable energy projects. CREBs are issued, theoretically, with a 0% interest rate. The borrower pays back only the principal of the bond and the bondholder receives federal tax credits in lieu of the traditional bond interest.
- Qualified applicants: State, local, and tribal governments; municipal utility; rural electric cooperative
- Qualified technologies: Solar thermal electric; photovoltaics; landfill gas; wind; biomass; hydroelectric; geothermal electric; municipal solid waste; hydrokinetic power; anaerobic digestion; tidal energy; wave energy; ocean thermal
- For more information: See the DSIRE website ; Internal Revenue Service Bulletin 2007-14; and Website at; and Internal Revenue Service Notice 2009-33

3. Energy Efficient Appliance Rebate Program (EEARP)

- Administered by: EERE
- Authorization: Energy Policy Act of 2005 (EPACT 2005; P.L. 109-58), Title I, Part B; American Recovery and Reinvestment Act of 2009 (ARRA; P.L. 111-5)

- Annual funding: $0 for FY2008; $298.5 million in FY2009 from ARRA; $0 for FY2010; $0 for FY2011; $0 for FY2012; $0 requested for FY2013
- Scheduled termination: This program was authorized through FY2010. An act of Congress is required to reauthorize this program.
- Description: The program provided financial and technical assistance to states to establish residential Energy Star rated appliance rebate programs. The program's objectives were: to reduce fossil fuel emissions created as a result of activities within the jurisdictions of eligible entities; and to improve energy efficiency in the residential sector.
- Qualified applicants: State governments, including U.S territories and possessions
- Qualified technologies: Energy efficient appliances
- For more information: See program number 81.127 at the CFDA website

4. Energy Efficiency and Conservation Block Grants Program (EECBG)

- Administered by: EERE
- Authorization: Energy Independence and Security Act of 2007 (EISA; P.L. 110-140), Title V, Subtitle E; American Recovery and Reinvestment Act of 2009 (ARRA; P.L. 111-5)
- Annual funding: $0 for FY2008; $3.2 billion for FY2009 from ARRA; $0 for FY2010; $0 for FY2011; $0 for FY2012
- Scheduled termination: This program was authorized through FY2010. An act of Congress is required to reauthorize this program
- Description: This program is part of DOE's Weather and Intergovernmental Program. The EECBG Program provides formula and competitive grants to empower local communities to make strategic investments to meet the nation's long-term goals for energy independence and leadership on climate change. Grants can be used for energy efficiency and conservation programs and projects community-wide, as well as renewable energy installations on government buildings.
- Qualified applicants: State, local, and tribal governments, including U.S. territories
- Qualified technologies: Energy efficient equipment and lighting; combined heating and cooling systems; combined heat and power systems; solar; wind; fuel cells; biomass
- For more information: See EERE's Energy Efficiency and Conservation Block Grants Program website; and program number 81.128 at the CFDA website

5. Renewable Energy Grants (1603 Program)

- Administered by U.S. Department of the Treasury
- Authority: Section 707 of H.R. 4853, Tax Relief, Unemployment Insurance Reauthorization, and Job Creation Act of 2010; H.R. 1, American Recovery and Reinvestment Act of 2010 (ARRA; P.L. 111-5) Division B, Sec 1104 & 1603, U.S. Department of Treasury: Grant Program Guidance (amended)

- Scheduled Termination: Construction must have begun by December 31, 2011. Applications must have been submitted before October 1, 2012. A project may be eligible for a 1603 award if the developer began construction by the December 31, 2011 date or if the developer satisfied a 5% safe harbor by incurring 5% of the total eligible project costs before the December 31 deadline.
- Description: The purpose of the 1603 payment is to reimburse eligible applicants for a portion of the cost of installing specified energy property used in a trade or business or for the production of income.
- Qualified applicants: Commercial, Industrial, Agricultural
- Qualified technologies: Solar water heat; solar space heat; solar thermal electric; solar thermal process heat; photovoltaics; landfill gas; wind; biomass; hydroelectric; geothermal electric; fuel cells; geothermal heat pumps; municipal solid waste; CHP/cogeneration; solar hybrid lighting; hydrokinetic; anaerobic digestion; tidal energy; wave energy; ocean thermal; microturbines
- For more information: See the Treasury's 1603 website; 1603 program guidance

APPENDIX D. SUMMARY OF EXPIRED FEDERAL RENEWABLE ENERGY AND ENERGY EFFICIENCY INCENTIVES/INDEX OF PROGRAMS

Table D-1. Expired Federal Incentives by Agency

Administering Agency	Program	Description	U.S. Code Citation	Expiration Date
Department of Energy	Energy Efficiency and Conservation Block Grants Program	Grants to finance energy efficiency and conservation programs/projects in local communities and renewable energy installations on government buildings	42 USC §17151-17158	None
	Energy Efficient Appliance Rebate Program	Provided financial and technical assistance to states to establish residential Energy Star rated appliance rebate programs	42 USC §15821	End of FY2010
Internal Revenue Service	Clean Renewable Energy Bonds (CREBs)	Bonds finance renewable energy projects	26 USC §54 (Old CREBs);	12/31/2009 (old CREBs); 11/01/2010 (new CREBS
	Renewable Energy Grants (1603 Program)	Renewable energy grant program to reimburse eligible taxpayers for a portion of the expense of placing in service specified energy property	ARRA (P.L. 111-5) § 1603(a)	Construction must have begun by 12/31/2011 and Applications had to be submitted before 10/1/2012.
Department of Housing and Urban Development			26 USC §54A (New CREBs)	All obligations were to be made by 9/30/2010.

Source: CRS.

Note: Some programs are not specifically identified or codified in the United States Code.

End Notes

[1] For more details, see CRS Report R42498, *Energy and Water Development: FY2013 Appropriations*, coordinated by Carl E. Behrens.

[2] The American Taxpayer Relief Act (ATRA, P.L. 112-240), enacted on January 2, 2013, made a number of significant changes to the procedures in the BCA that apply during FY2013.

[3] H.Rept. 113-135, p. 87.

[4] For more information, see Department of Energy, *FY2011 Congressional Budget Request*, vol. 3, p. 500, at http://www.cfo.doe.gov/budget/11budget/Content/Volume 3.pdf.

[5] For more information, see CRS Report R40669, *Energy and Water Development: FY2010 Appropriations*, coordinated by Carl E. Behrens.

INDEX

A

B

C

D

E

F

G

H

I

J

K

L

M

N

O

P

Q

R

S

T

U

V

W

Y